STUDENT SOLUTIONS MANUAL

to accompany

CALCULUS

SINGLE AND MULTIVARIABLE　　　　**THIRD EDITION**

Deborah Hughes-Hallett
University of Arizona

Andrew M. Gleason
Harvard University

William G. McCallum
University of Arizona

et al.

JOHN WILEY & SONS, INC.

COVER PHOTO © Eddie Hironaka/The Image Bank.

To order books or for customer service call 1-800-CALL-WILEY (225-5945).

ISBN 0-471-44192-9

Printed in the United States of America

10 9 8 7 6 5 4 3 2

Printed and bound by Victor Graphics, Inc.

CONTENTS

CHAPTER ONE

Solutions for Section 1.1

Exercises

1. $f(35)$ means the value of P corresponding to $t = 35$. Since t represents the number of years since 1950, we see that $f(35)$ means the population of the city in 1985. So, in 1985, the city's population was 12 million.

5. Rewriting the equation as

$$y = -\frac{12}{7}x + \frac{2}{7}$$

shows that the line has slope $-12/7$ and vertical intercept $2/7$.

9. The slope is $(3 - 2)/(2 - 0) = 1/2$. So the equation of the line is $y = (1/2)x + 2$.

13. The line parallel to $y = mx + c$ also has slope m, so its equation is

$$y = m(x - a) + b.$$

The line perpendicular to $y = mx + c$ has slope $-1/m$, so its equation will be

$$y = -\frac{1}{m}(x - a) + b.$$

17. Since the function goes from $x = -2$ to $x = 2$ and from $y = -2$ to $y = 2$, the domain is $-2 \le x \le 2$ and the range is $-2 \le y \le 2$.

21. The value of $f(t)$ is real provided $t^2 - 16 \ge 0$ or $t^2 \ge 16$. This occurs when either $t \ge 4$, or $t \le -4$. Solving $f(t) = 3$, we have

$$\sqrt{t^2 - 16} = 3$$
$$t^2 - 16 = 9$$
$$t^2 = 25$$

so

$$t = \pm 5.$$

25. We know that N is proportional to $1/l^2$, so

$$N = \frac{k}{l^2}, \quad \text{for some constant } k.$$

Problems

29.

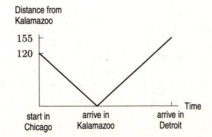

33. **(a)** This could be a linear function because w increases by 5 as h increases by 1.
  (b) We find the slope m and the intercept b in the linear equation $w = b + mh$. We first find the slope m using the first two points in the table. Since we want w to be a function of h, we take

$$m = \frac{\Delta w}{\Delta h} = \frac{171 - 166}{69 - 68} = 5.$$

  Substituting the first point and the slope $m = 5$ into the linear equation $w = b + mh$, we have $166 = b + (5)(68)$, so $b = -174$. The linear function is

$$w = 5h - 174.$$

  The slope, $m = 5$ is in units of pounds per inch.
  (c) We find the slope and intercept in the linear function $h = b + mw$ using $m = \Delta h / \Delta w$ to obtain the linear function

$$h = 0.2w + 34.8.$$

  Alternatively, we could solve the linear equation found in part (b) for h. The slope, $m = 0.2$, has units inches per pound.

37. **(a)** $R = k(350 - H)$, where k is a positive constant.
  If H is greater than $350°$, the rate is negative, indicating that a very hot yam will cool down toward the temperature of the oven.
  (b) Letting H_0 equal the initial temperature of the yam, the graph of R against H looks like:

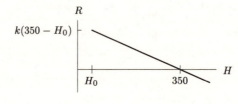

  Note that by the temperature of the yam, we mean the average temperature of the yam, since the yam's surface will be hotter than its center.

Solutions for Section 1.2

Exercises

1. The graph shows a concave up function.

5. Initial quantity = 5; growth rate = 0.07 = 7%.

9. **(a)** The function is linear with initial population of 1000 and slope of 50, so $P = 1000 + 50t$.
  (b) This function is exponential with initial population of 1000 and growth rate of 5%, so $P = 1000(1.05)^t$.

Problems

13. **(a)** Advertising is generally cheaper in bulk; spending more money will give better and better marginal results initially, (Spending $5,000 could give you a big newspaper ad reaching 200,000 people; spending $100,000 could give you a series of TV spots reaching 50,000,000 people.) A graph is shown below, left.
  (b) The temperature of a hot object decreases at a rate proportional to the difference between its temperature and the temperature of the air around it. Thus, the temperature of a very hot object decreases more quickly than a cooler object. The graph is decreasing and concave up. (We are assuming that the coffee is all at the same temperature.)

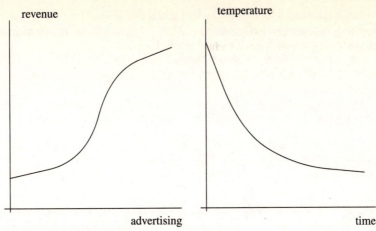

17. (a) Using $Q = Q_0(1 - r)^t$ for loss, we have

$$Q = 10{,}000(1 - 0.1)^{10} = 10{,}000(0.9)^{10} = 3486.78.$$

The investment was worth \$3486.78 after 10 years.

(b) Measuring time from the moment at which the stock begins to gain value and letting $Q_0 = 3486.78$, the value after t years is

$$Q = 3486.78(1 + 0.1)^t = 3486.78(1.1)^t.$$

We can estimate the value of t when $Q = 10{,}000$ by tracing along a graph of Q, giving $t \approx 11$. It will take about 11 years to get the investment back to \$10,000.

21. The difference, D, between the horizontal asymptote and the graph appears to decrease exponentially, so we look for an equation of the form

$$D = D_0 a^x$$

where $D_0 = 4 = $ difference when $x = 0$. Since $D = 4 - y$, we have

$$4 - y = 4a^x \quad \text{or} \quad y = 4 - 4a^x = 4(1 - a^x)$$

The point $(1, 2)$ is on the graph, so $2 = 4(1 - a^1)$, giving $a = \frac{1}{2}$.
Therefore $y = 4(1 - (\frac{1}{2})^x) = 4(1 - 2^{-x})$.

25. Since $e^{-0.5t} = (e^{-0.5})^t \approx (0.61)^t$, we have $P = 2(0.61)^t$. This is exponential decay since -0.5 is negative. We can also see that this is decay because $0.61 < 1$.

29. Direct calculation reveals that each 1000 foot increase in altitude results in a longer takeoff roll by a factor of about 1.096. Since the value of d when $h = 0$ (sea level) is $d = 670$, we are led to the formula

$$d = 670(1.096)^{h/1000},$$

where d is the takeoff roll, in feet, and h is the airport's elevation, in feet.

Alternatively, we can write

$$d = d_0 a^h,$$

where d_0 is the sea level value of d, $d_0 = 670$. In addition, when $h = 1000$, $d = 734$, so

$$734 = 670a^{1000}.$$

Solving for a gives

$$a = \left(\frac{734}{670}\right)^{1/1000} = 1.00009124,$$

so

$$d = 670(1.00009124)^h.$$

33. (a) This is the graph of a linear function, which increases at a constant rate, and thus corresponds to $k(t)$, which increases by 0.3 over each interval of 1.

(b) This graph is concave down, so it corresponds to a function whose increases are getting smaller, as is the case with $h(t)$, whose increases are 10, 9, 8, 7, and 6.

(c) This graph is concave up, so it corresponds to a function whose increases are getting bigger, as is the case with $g(t)$, whose increases are 1, 2, 3, 4, and 5.

37. Because the population is growing exponentially, the time it takes to double is the same, regardless of the population levels we are considering. For example, the population is 20,000 at time 3.7, and 40,000 at time 6.0. This represents a doubling of the population in a span of $6.0 - 3.7 = 2.3$ years.

How long does it take the population to double a second time, from 40,000 to 80,000? Looking at the graph once again, we see that the population reaches 80,000 at time $t = 8.3$. This second doubling has taken $8.3 - 6.0 = 2.3$ years, the same amount of time as the first doubling.

Further comparison of any two populations on this graph that differ by a factor of two will show that the time that separates them is 2.3 years. Similarly, during any 2.3 year period, the population will double. Thus, the doubling time is 2.3 years.

Suppose $P = P_0 a^t$ doubles from time t to time $t + d$. We now have $P_0 a^{t+d} = 2P_0 a^t$, so $P_0 a^t a^d = 2P_0 a^t$. Thus, canceling P_0 and a^t, d must be the number such that $a^d = 2$, no matter what t is.

Solutions for Section 1.3

Exercises

1. **(a)** $g(2 + h) = (2 + h)^2 + 2(2 + h) + 3 = 4 + 4h + h^2 + 4 + 2h + 3 = h^2 + 6h + 11$.
   (b) $g(2) = 2^2 + 2(2) + 3 = 4 + 4 + 3 = 11$, which agrees with what we get by substituting $h = 0$ into (a).
   (c) $g(2 + h) - g(2) = (h^2 + 6h + 11) - (11) = h^2 + 6h$.

5. $m(z + 1) - m(z) = (z + 1)^2 - z^2 = 2z + 1$.

9. **(a)** $f(25)$ is q corresponding to $p = 25$, or, in other words, the number of items sold when the price is 25.
   (b) $f^{-1}(30)$ is p corresponding to $q = 30$, or the price at which 30 units will be sold.

13. The function is not invertible since there are many horizontal lines which hit the function twice.

17. This looks like a shift of the graph $y = x^3$. The graph is shifted to the right 2 units and down 1 unit, so a possible formula is $y = (x - 2)^3 - 1$.

Problems

21. Not invertible. Given a certain number of customers, say $f(t) = 1500$, there could be many times, t, during the day at which that many people were in the store. So we don't know which time instant is the right one.

25. $f(g(1)) = f(2) \approx 0.4$.

29. Using the same way to compute $g(f(x))$ as in Problem 26, we get the following table. Then we can plot the graph of $g(f(x))$.

x	$f(x)$	$g(f(x))$
-3	3	-2.6
-2.5	0.1	0.8
-2	-1	-1.4
-1.5	-1.3	-1.8
-1	-1.2	-1.7
-0.5	-1	-1.4
0	-0.8	-1
0.5	-0.6	-0.6
1	-0.4	-0.3
1.5	-0.1	0.3
2	0.3	1.1
2.5	0.9	2
3	1.6	2.2

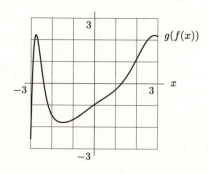

33. Since $B = y - 1$ and $n = 2B^2 - B$, substitution gives

$$n = 2B^2 - B = 2(y - 1)^2 - (y - 1) = 2y^2 - 5y + 3.$$

Solutions for Section 1.4

Exercises

1. The function e^x has a vertical intercept of 1, so must be A. The function $\ln x$ has an x-intercept of 1, so must be D. The graphs of x^2 and $x^{1/2}$ go through the origin. The graph of $x^{1/2}$ is concave down so it corresponds to graph C and the graph of x^2 is concave up so it corresponds to graph B.

5. Isolating the exponential term

$$20 = 50(1.04)^x$$
$$\frac{20}{50} = (1.04)^x$$

Taking logs of both sides

$$\log \frac{2}{5} = \log(1.04)^x$$
$$\log \frac{2}{5} = x \log(1.04)$$
$$x = \frac{\log(2/5)}{\log(1.04)} \approx -23.4.$$

9. $\ln(2^x) = \ln(e^{x+1})$
$$x \ln 2 = (x+1) \ln e$$
$$x \ln 2 = x + 1$$
$$0.693x = x + 1$$
$$x = \frac{1}{0.693 - 1} \approx -3.26$$

13. Using the rules for ln, we get

$$\ln 7^{x+2} = \ln e^{17x}$$
$$(x+2) \ln 7 = 17x$$
$$x(\ln 7 - 17) = -2 \ln 7$$
$$x = \frac{-2 \ln 7}{\ln 7 - 17} \approx 0.26.$$

17. $t = \frac{\log a}{\log b}.$

21. $t = \ln \frac{a}{b}.$

25. Using the identity $e^{\ln x} = x$, we have $5A^2$.

29. Since $(1.5)^t = \left(e^{\ln 1.5}\right)^t = e^{(\ln 1.5)t} = e^{0.41t}$, we have $P = 15e^{0.41t}$. Since 0.41 is positive, this is exponential growth.

33. If $p(t) = (1.04)^t$, then, for p^{-1} the inverse of p, we should have

$$(1.04)^{p^{-1}(t)} = t,$$
$$p^{-1}(t) \log(1.04) = \log t,$$
$$p^{-1}(t) = \frac{\log t}{\log(1.04)} \approx 58.708 \log t.$$

Problems

37. (a) The initial dose is 10 mg.
   (b) Since $0.82 = 1 - 0.18$, the decay rate is 0.18, so 18% leaves the body each hour.
   (c) When $t = 6$, we have $A = 10(0.82)^6 = 3.04$. The amount in the body after 6 hours is 3.04 mg.

(d) We want to find the value of t when $A = 1$. Using logarithms:

$$1 = 10(0.82)^t$$
$$0.1 = (0.82)^t$$
$$\ln(0.1) = t \ln(0.82)$$
$$t = 11.60 \text{ hours.}$$

After 11.60 hours, the amount is 1 mg.

41. In ten years, the substance has decayed to 40% of its original mass. In another ten years, it will decay by an additional factor of 40%, so the amount remaining after 20 years will be $100 \cdot 40\% \cdot 40\% = 16$ kg.

45. Let $t =$ number of years since 1980. Then the number of vehicles, V, in millions, at time t is given by

$$V = 170(1.04)^t$$

and the number of people, P, in millions, at time t is given by

$$P = 227(1.01)^t.$$

There is an average of one vehicle per person when $\dfrac{V}{P} = 1$, or $V = P$. Thus, we must solve for t the equation:

$$170(1.04)^t = 227(1.01)^t,$$

which implies

$$\left(\frac{1.04}{1.01}\right)^t = \frac{(1.04)^t}{(1.01)^t} = \frac{227}{170}$$

Taking logs on both sides,

$$t \log \frac{1.04}{1.01} = \log \frac{227}{170}.$$

Therefore,

$$t = \frac{\log\left(\frac{227}{170}\right)}{\log\left(\frac{1.04}{1.01}\right)} \approx 9.9 \text{ years.}$$

So there was, according to this model, about one vehicle per person in 1990.

49. Since the amount of strontium-90 remaining halves every 29 years, we can solve for the decay constant;

$$0.5P_0 = P_0 e^{-29k}$$
$$k = \frac{\ln(1/2)}{-29}.$$

Knowing this, we can look for the time t in which $P = 0.10P_0$, or

$$0.10P_0 = P_0 e^{\ln(0.5)t/29}$$
$$t = \frac{29 \ln(0.10)}{\ln(0.5)} = 96.34 \text{ years.}$$

Solutions for Section 1.5

Exercises

1.

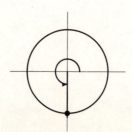

$$\sin\left(\frac{3\pi}{2}\right) = -1 \quad \text{is negative.}$$

$$\cos\left(\frac{3\pi}{2}\right) = 0$$

$$\tan\left(\frac{3\pi}{2}\right) \quad \text{is undefined.}$$

5.

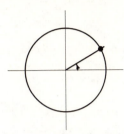

$$\sin\left(\frac{\pi}{6}\right) \quad \text{is positive.}$$

$$\cos\left(\frac{\pi}{6}\right) \quad \text{is positive.}$$

$$\tan\left(\frac{\pi}{6}\right) \quad \text{is positive.}$$

9. $-1 \text{ radian} \cdot \frac{180°}{\pi \text{ radians}} = -\left(\frac{180°}{\pi}\right) \approx -60°$

$$\sin(-1) \quad \text{is negative}$$

$$\cos(-1) \quad \text{is positive}$$

$$\tan(-1) \quad \text{is negative}$$

13. (a) We determine the amplitude of y by looking at the coefficient of the cosine term. Here, the coefficient is 1, so the amplitude of y is 1. Note that the constant term does not affect the amplitude.

(b) We know that the cosine function $\cos x$ repeats itself at $x = 2\pi$, so the function $\cos(3x)$ must repeat itself when $3x = 2\pi$, or at $x = 2\pi/3$. So the period of y is $2\pi/3$. Here as well the constant term has no effect.

(c) The graph of y is shown in the figure below.

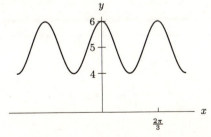

17. The graph is an inverted sine curve with amplitude 1 and period 2π, shifted up by 2, so it is given by $f(x) = 2 - \sin x$.

21. The graph is a sine curve which has been shifted up by 2, so $f(x) = (\sin x) + 2$.

25. The graph is a cosine curve with period $2\pi/5$ and amplitude 2, so it is given by $f(x) = 2\cos(5x)$.

Problems

29. Using the fact that 1 revolution $= 2\pi$ radians and 1 minute = 60 seconds, we have

$$200\frac{\text{rev}}{\text{min}} = (200) \cdot 2\pi\frac{\text{rad}}{\text{min}} = 200 \cdot 2\pi\frac{1}{60}\frac{\text{rad}}{\text{sec}}$$
$$\approx \frac{(200)(6.283)}{60}$$
$$\approx 20.94 \text{ radians per second.}$$

Similarly, 500 rpm is equivalent to 52.36 radians per second.

33. (a) Reading the graph of θ against t shows that $\theta \approx 5.2$ when $t = 1.5$. Since the coordinates of P are $x = 5\cos\theta$, $y = 5\sin\theta$, when $t = 1.5$ the coordinates are

$$(x, y) \approx (5\cos 5.2, 5\sin 5.2) = (2.3, -4.4).$$

(b) As t increases from 0 to 5, the angle θ increases from 0 to about 6.3 and then decreases to 0 again. Since $6.3 \approx 2\pi$, this means that P starts on the x-axis at the point $(5, 0)$, moves counterclockwise the whole way around the circle (at which time $\theta \approx 2\pi$), and then moves back clockwise to its starting point.

37. The US voltage has a maximum value of 156 volts and has a period of 1/60 of a second, so it executes 60 cycles a second. The European voltage has a higher maximum of 339 volts, and a slightly longer period of 1/50 seconds, so it oscillates at 50 cycles per second.

41.

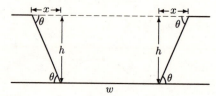

Figure 1.1

Figure 1.1 shows that the cross-sectional area is one rectangle of area hw and two triangles. Each triangle has height h and base x, where

$$\frac{h}{x} = \tan\theta \quad \text{so} \quad x = \frac{h}{\tan\theta}.$$
$$\text{Area of triangle} = \frac{1}{2}xh = \frac{h^2}{2\tan\theta}$$

$$\text{Total area} = \text{Area of rectangle} + 2(\text{Area of triangle})$$
$$= hw + 2 \cdot \frac{h^2}{2\tan\theta} = hw + \frac{h^2}{\tan\theta}.$$

Solutions for Section 1.6

Exercises

1. Exponential growth dominates power growth as $x \to \infty$, so $10 \cdot 2^x$ is larger.

5. (I) Degree ≥ 3, leading coefficient negative.
   (II) Degree ≥ 4, leading coefficient positive.
   (III) Degree ≥ 4, leading coefficient negative.
   (IV) Degree ≥ 5, leading coefficient negative.
   (V) Degree ≥ 5, leading coefficient positive.

9. **(a)** A polynomial has the same end behavior as its leading term, so this polynomial behaves as $-5x^4$ globally. Thus we have:
$$f(x) \to -\infty \text{ as } x \to -\infty, \quad \text{and} \quad f(x) \to -\infty \text{ as } x \to +\infty.$$

(b) Polynomials behave globally as their leading term, so this rational function behaves globally as $(3x^2)/(2x^2)$, or $3/2$. Thus we have:
$$f(x) \to 3/2 \text{ as } x \to -\infty, \quad \text{and} \quad f(x) \to 3/2 \text{ as } x \to +\infty.$$

(c) We see from a graph of $y = e^x$ that
$$f(x) \to 0 \text{ as } x \to -\infty, \quad \text{and} \quad f(x) \to +\infty \text{ as } x \to +\infty.$$

Problems

13. $f(x) = k(x+2)(x-2)^2(x-5) = k(x^4 - 7x^3 + 6x^2 + 28x - 40)$, where $k < 0$. ($k \approx -\frac{1}{15}$ if the scales are equal; otherwise one can't tell how large k is.)

17. **(a)** (i) The water that has flowed out of the pipe in 1 second is a cylinder of radius r and length 3 cm. Its volume is
$$V = \pi r^2(3) = 3\pi r^2.$$

(ii) If the rate of flow is k cm/sec instead of 3 cm/sec, the volume is given by
$$V = \pi r^2(k) = \pi r^2 k.$$

(b) (i) The graph of V as a function of r is a quadratic. See Figure 1.2.

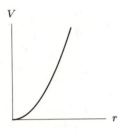

Figure 1.2

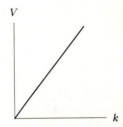

Figure 1.3

(ii) The graph of V as a function of k is a line. See Figure 1.3.

21. Let $D(v)$ be the stopping distance required by an Alpha Romeo as a function of its velocity. The assumption that stopping distance is proportional to the square of velocity is equivalent to the equation
$$D(v) = kv^2$$
where k is a constant of proportionality. To determine the value of k, we use the fact that $D(70) = 177$.
$$D(70) = k(70)^2 = 177.$$
Thus,
$$k = \frac{177}{70^2} \approx 0.0361.$$
It follows that
$$D(35) = \left(\frac{177}{70^2}\right)(35)^2 = \frac{177}{4} = 44.25 \text{ ft}$$
and
$$D(140) = \left(\frac{177}{70^2}\right)(140)^2 = 708 \text{ ft}.$$
Thus, at half the speed it requires one fourth the distance, whereas at twice the speed it requires four times the distance, as we would expect from the equation. (We could in fact have figured it out that way, without solving for k explicitly.)

25. **(a)** (i) If $(1, 1)$ is on the graph, we know that
$$1 = a(1)^2 + b(1) + c = a + b + c.$$

(ii) If $(1, 1)$ is the vertex, then the axis of symmetry is $x = 1$, so

$$-\frac{b}{2a} = 1,$$

and thus

$$a = -\frac{b}{2}, \text{ so } b = -2a.$$

But to be the vertex, $(1, 1)$ must also be on the graph, so we know that $a + b + c = 1$. Substituting $b = -2a$, we get $-a + c = 1$, which we can rewrite as $a = c - 1$, or $c = 1 + a$.

(iii) For $(0, 6)$ to be on the graph, we must have $f(0) = 6$. But $f(0) = a(0^2) + b(0) + c = c$, so $c = 6$.

(b) To satisfy all the conditions, we must first, from (a)(iii), have $c = 6$. From (a)(ii), $a = c - 1$ so $a = 5$. Also from (a)(ii), $b = -2a$, so $b = -10$. Thus the completed equation is

$$y = f(x) = 5x^2 - 10x + 6,$$

which satisfies all the given conditions.

29. Consider the end behavior of the graph; that is, as $x \to +\infty$ and $x \to -\infty$. The ends of a degree 5 polynomial are in Quadrants I and III if the leading coefficient is positive or in Quadrants II and IV if the leading coefficient is negative. Thus, there must be at least one root. Since the degree is 5, there can be no more than 5 roots. Thus, there may be 1, 2, 3, 4, or 5 roots. Graphs showing these five possibilities are shown in Figure 1.4.

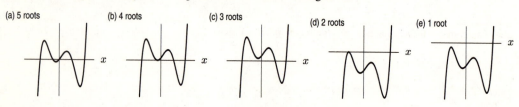

Figure 1.4

33. The graphs of both these functions will resemble that of x^3 on a large enough window. One way to tackle the problem is to graph them both (along with x^3 if you like) in successively larger windows until the graphs come together. In Figure 1.5, f, g and x^3 are graphed in four windows. In the largest of the four windows the graphs are indistinguishable, as required. Answers may vary.

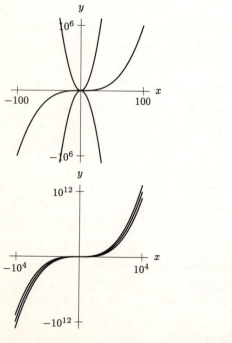

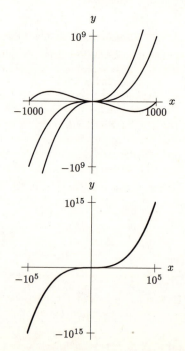

Figure 1.5

Solutions for Section 1.7

Exercises

1. Yes, because $2x + x^{3/2}$ is defined for all x.

5. Yes, because $2x - 5$ is positive for $3 \leq x \leq 4$.

9. No, because $e^x - 1 = 0$ at $x = 0$.

Problems

13. For any value of k, the function is continuous at every point except $x = 2$. We choose k to make the function continuous at $x = 2$.

   Since $3x^2$ takes the value $3(2^2) = 12$ at $x = 2$, we choose k so that kx goes through the point $(2, 12)$. Thus $k = 6$.

17.

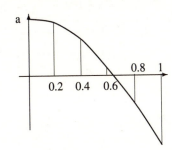
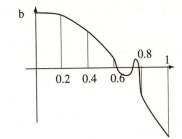
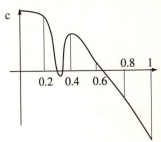

Solutions for Chapter 1 Review

Exercises

1. (a) The domain of f is the set of values of x for which the function is defined. Since the function is defined by the graph and the graph goes from $x = 0$ to $x = 7$, the domain of f is $[0, 7]$.
   (b) The range of f is the set of values of y attainable over the domain. Looking at the graph, we can see that y gets as high as 5 and as low as -2, so the range is $[-2, 5]$.
   (c) Only at $x = 5$ does $f(x) = 0$. So 5 is the only root of $f(x)$.
   (d) Looking at the graph, we can see that $f(x)$ is decreasing on $(1, 7)$.
   (e) The graph indicates that $f(x)$ is concave up at $x = 6$.
   (f) The value $f(4)$ is the y-value that corresponds to $x = 4$. From the graph, we can see that $f(4)$ is approximately 1.
   (g) This function is not invertible, since it fails the horizontal-line test. A horizontal line at $y = 3$ would cut the graph of $f(x)$ in two places, instead of the required one.

5. The amplitude is 2. The period is $2\pi/5$. See Figure 1.6.

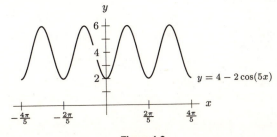

Figure 1.6

9. $y = -kx(x + 5) = -k(x^2 + 5x)$, where $k > 0$ is any constant.

13. $x = ky(y - 4) = k(y^2 - 4y)$, where $k > 0$ is any constant.

17. There are many solutions for a graph like this one. The simplest is $y = 1 - e^{-x}$, which gives the graph of $y = e^x$, flipped over the x-axis and moved up by 1. The resulting graph passes through the origin and approaches $y = 1$ as an upper bound, the two features of the given graph.

21. The graph appears to have a vertical asymptote at $t = 0$, so $f(t)$ is not continuous on $[-1, 1]$.

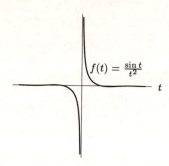

Problems

25. We will let

$$T = \text{amount of fuel for take-off,}$$
$$L = \text{amount of fuel for landing,}$$
$$P = \text{amount of fuel per mile in the air,}$$
$$m = \text{the length of the trip in miles.}$$

Then Q, the total amount of fuel needed, is given by

$$Q(m) = T + L + Pm.$$

29. Assuming the US population grows exponentially, we have

$$248.7 = 226.5e^{10k}$$
$$k = \frac{\ln(1.098)}{10} = 0.00935.$$

We want to find the time t in which

$$300 = 226.5e^{0.00935t}$$
$$t = \frac{\ln(1.324)}{0.00935} = 30 \text{ years.}$$

Thus, the population will go over 300 million around the year 2010.

33. (a) Let the height of the can be h. Then

$$V = \pi r^2 h.$$

The surface area consists of the area of the ends (each is πr^2) and the curved sides (area $2\pi rh$), so

$$S = 2\pi r^2 + 2\pi rh.$$

Solving for h from the formula for V, we have

$$h = \frac{V}{\pi r^2}.$$

Substituting into the formula for S, we get

$$S = 2\pi r^2 + 2\pi r \cdot \frac{V}{\pi r^2} = 2\pi r^2 + \frac{2V}{r}.$$

(b) For large r, the $2V/r$ term becomes negligible, meaning $S \approx 2\pi r^2$, and thus $S \to \infty$ as $r \to \infty$.

(c) The graph is in Figure 1.7.

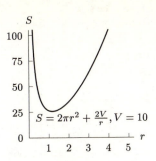

Figure 1.7

37. (a) Yes, f is invertible, since f is increasing everywhere.

(b) $f^{-1}(400)$ is the year in which 400 million motor vehicles were registered in the world. From the picture, we see that $f^{-1}(400)$ is around 1979.

(c) Since the graph of f^{-1} is the reflection of the graph of f over the line $y = x$, we get Figure 1.8.

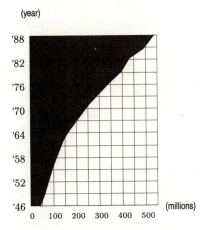

Figure 1.8: Graph of f^{-1}

41. (a) $r(p) = kp(A - p)$, where $k > 0$ is a constant.

(b) $p = A/2$.

CAS Challenge Problems

45. (a) As $x \to \infty$, the term e^{6x} dominates and tends to ∞. Thus, $f(x) \to \infty$ as $x \to \infty$.

As $x \to -\infty$, the terms of the form e^{kx}, where $k = 6, 5, 4, 3, 2, 1$, all tend to zero. Thus, $f(x) \to 16$ as $x \to -\infty$.

(b) A CAS gives

$$f(x) = (e^x + 1)(e^{2x} - 2)(e^x - 2)(e^{2x} + 2e^x + 4).$$

Since e^x is always positive, the factors $(e^x + 1)$ and $(e^{2x} + 2e^x + 4)$ are never zero. The other factors each lead to a zero, so there are two zeros.

(c) The zeros are given by

$$e^{2x} = 2 \quad \text{so} \quad x = \frac{\ln 2}{2}$$
$$e^x = 2 \quad \text{so} \quad x = \ln 2.$$

Thus, one zero is twice the size of the other.

49. Using the trigonometric expansion capabilities of your computer algebra system, you get something like

$$\cos(4x) = \cos^4(x) - 6\cos^2(x)\sin^2(x) + \sin^4(x).$$

Answers may vary.

(a) To get rid of the powers of cosine, use the identity $\cos^2(x) = 1 - \sin^2(x)$. This gives

$$\cos(4x) = \cos^4(x) - 6\cos^2(x)\left(1 - \cos^2(x)\right) + \left(1 - \cos^2(x)\right)^2.$$

Finally, using the CAS to simplify,

$$\cos(4x) = 1 - 8\cos^2(x) + 8\cos^4(x).$$

(b) This time we use $\sin^2(x) = 1 - \cos^2(x)$ to get rid of powers of sine. We get

$$\cos(4x) = \left(1 - \sin^2(x)\right)^2 - 6\sin^2(x)\left(1 - \sin^2(x)\right) + \sin^4(x) = 1 - 8\sin^2(x) + 8\sin^4(x).$$

CHECK YOUR UNDERSTANDING

1. False. A line can be put through any two points in the plane. However, if the line is vertical, it is not the graph of a function.

5. True. The highest degree term in a polynomial determines how the polynomial behaves when x is very large in the positive or negative direction. When n is odd, x^n is positive when x is large and positive but negative when x is large and negative. Thus if a polynomial $p(x)$ has odd degree, it will be positive for some values of x and negative for other values of x. Since every polynomial is continuous, the Intermediate Value Theorem then guarantees that $p(x) = 0$ for some value of x.

9. False. Suppose $y = 5^x$. Then increasing x by 1 increases y by a factor of 5. However increasing x by 2 increases y by a factor of 25, not 10, since

$$y = 5^{x+2} = 5^x \cdot 5^2 = 25 \cdot 5^x.$$

(Other examples are possible.)

13. True. The period is $2\pi/(200\pi) = 1/100$ seconds. Thus, the function executes 100 cycles in 1 second.

17. False. For $x < 0$, as x increases, x^2 decreases, so e^{-x^2} increases.

21. True. If $b > 1$, then $ab^x \to 0$ as $x \to -\infty$. If $0 < b < 1$, then $ab^x \to 0$ as $x \to \infty$. In either case, the function $y = a + ab^x$ has $y = a$ as the horizontal asymptote.

25. False. A counterexample is given by $f(x) = x^2$ and $g(x) = x + 1$. The function $f(g(x)) = (x+1)^2$ is not even because $f(g(1)) = 4$ and $f(g(-1)) = 0 \neq 4$.

29. Let $f(x) = \dfrac{1}{(x-1)(x-2)(x-3)\cdots(x-16)(x-17)}$. This function has an asymptote corresponding to every factor in the denominator. Other answers are possible.

33. This is impossible. If $a < b$, then $f(a) < f(b)$, since f is increasing, and $g(a) > g(b)$, since g is decreasing, so $-g(a) < -g(b)$. Therefore, if $a < b$, then $f(a) - g(a) < f(b) - g(b)$, which means that $f(x) + g(x)$ is increasing.

37. False. For example, let $f(x) = \log x$. Then $f(x)$ is increasing on $[1, 2]$, but $f(x)$ is concave down. (Other examples are possible.)

41. False. For example, let $f(x) = \begin{cases} 1 & x \leq 3 \\ 2 & x > 3 \end{cases}$, then $f(x)$ is defined at $x = 3$ but it is not continuous at $x = 3$. (Other examples are possible.)

CHAPTER TWO

Solutions for Section 2.1

Exercises

1. For t between 2 and 5, we have

$$\text{Average velocity} = \frac{\Delta s}{\Delta t} = \frac{400 - 135}{5 - 2} = \frac{265}{3} \text{ km/hr.}$$

The average velocity on this part of the trip was $265/3$ km/hr.

5. Using $h = 0.1, 0.01, 0.001$, we see

$$\frac{(3 + 0.1)^3 - 27}{0.1} = 27.91$$

$$\frac{(3 + 0.01)^3 - 27}{0.01} = 27.09$$

$$\frac{(3 + 0.001)^3 - 27}{0.001} = 27.009.$$

These calculations suggest that $\lim_{h \to 0} \dfrac{(3 + h)^3 - 27}{h} = 27$.

9. For $-0.5 \leq \theta \leq 0.5$, $0 \leq y \leq 3$, the graph of $y = \dfrac{\sin(2\theta)}{\theta}$ is shown in Figure 2.1. Therefore, $\lim_{\theta \to 0} \dfrac{\sin(2\theta)}{\theta} = 2$.

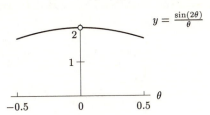

$$y = \frac{\sin(2\theta)}{\theta}$$

Figure 2.1

Problems

13.

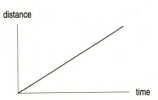

17. Between 1804 and 1927, the world's population increased 1 billion people in 123 years, for an average rate of change of $1/123$ billion people per year. We convert this to people per minute:

$$\frac{1,000,000,000}{123} \text{ people/year} \cdot \frac{1}{60 \cdot 24 \cdot 365} \text{ years/minute} = 15.47 \text{ people/minute.}$$

Between 1804 and 1927, the population of the world increased at an average rate of 15.47 people per minute. Similarly, we find the following:

Between 1927 and 1960, the increase was 57.65 people per minute.

Between 1960 and 1974, the increase was 135.90 people per minute.

Between 1974 and 1987, the increase was 146.35 people per minute.

Between 1987 and 1999, the increase was 158.55 people per minute.

Solutions for Section 2.2

Exercises

1. **(a)** As x approaches -2 from either side, the values of $f(x)$ get closer and closer to 3, so the limit appears to be about 3.

(b) As x approaches 0 from either side, the values of $f(x)$ get closer and closer to 7. (Recall that to find a limit, we are interested in what happens to the function near x but not at x.) The limit appears to be about 7.

(c) As x approaches 2 from either side, the values of $f(x)$ get closer and closer to 3 on one side of $x = 2$ and get closer and closer to 2 on the other side of $x = 2$. Thus the limit does not exist.

(d) As x approaches 4 from either side, the values of $f(x)$ get closer and closer to 8. (Again, recall that we don't care what happens right at $x = 4$.) The limit appears to be about 8.

5. From Table 2.1, it appears the limit is 0. This is confirmed by Figure 2.2. An appropriate window is $-0.005 < x < 0.005$, $-0.01 < y < 0.01$.

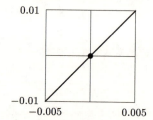

Table 2.1

x	$f(x)$
0.1	0.1987
0.01	0.0200
0.001	0.0020
0.0001	0.0002

x	$f(x)$
-0.0001	-0.0002
-0.001	-0.0020
-0.01	-0.0200
-0.1	-0.1987

Figure 2.2

9. From Table 2.2, it appears the limit is 1. This is confirmed by Figure 2.3. An appropriate window is $-0.0198 < x < 0.0198$, $0.99 < y < 1.01$.

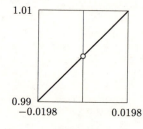

Table 2.2

x	$f(x)$
0.1	1.0517
0.01	1.0050
0.001	1.0005
0.0001	1.0001

x	$f(x)$
-0.0001	1.0000
-0.001	0.9995
-0.01	0.9950
-0.1	0.9516

Figure 2.3

13. From Table 2.3, it appears the limit is 0. Figure 2.4 confirms this. An appropriate window is $1.55 < x < 1.59$, $-0.01 < y < 0.01$.

Table 2.3

x	$f(x)$
1.6708	-0.0500
1.5808	-0.0050
1.5718	-0.0005
1.5709	-0.0001
1.5707	0.0001
1.5698	0.0005
1.5608	0.0050
1.4708	0.0500

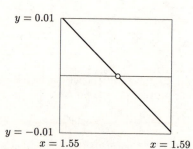

Figure 2.4

17. $\displaystyle \lim_{h \to 0} \frac{1}{h}\left(\frac{1}{(1+h)^2} - 1\right) = \lim_{h \to 0} \frac{1 - (1 + 2h + h^2)}{h(1+h)^2} = \lim_{h \to 0} \frac{-2 - h}{(1+h)^2} = -2$

21. $f(x) = \dfrac{|x-2|}{x} = \begin{cases} \dfrac{x-2}{x}, & x > 2 \\[2mm] -\dfrac{x-2}{x}, & x < 2 \end{cases}$

Figure 2.5 confirms that $\lim\limits_{x \to 2^+} f(x) = \lim\limits_{x \to 2^-} f(x) = \lim\limits_{x \to 2} f(x) = 0.$

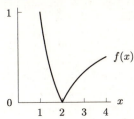

Figure 2.5

Problems

25. The only change is that, instead of considering all x near c, we only consider x near to and greater than c. Thus the phrase "$|x - c| < \delta$" must be replaced by "$c < x < c + \delta$." Thus, we define

$$\lim_{x \to c^+} f(x) = L$$

to mean that for any $\epsilon > 0$ (as small as we want), there is a $\delta > 0$ (sufficiently small) such that if $c < x < c + \delta$, then $|f(x) - L| < \epsilon$.

29. We use values of h approaching, but not equal to, zero. If we let $h = 0.01, 0.001, 0.0001, 0.00001$, we calculate the values 2.7048, 2.7169, 2.7181, and 2.7183. If we let $h = -0.01 -0.001, -0.0001, -0.00001$, we get values 2.7320, 2.7196, 2.7184, and 2.7183. These numbers suggest that the limit is the number $e = 2.71828\ldots$. However, these calculations cannot tell us that the limit is exactly e; for that a proof is needed.

33. Divide numerator and denominator by x^3, giving

$$f(x) = \frac{2x^3 - 16x^2}{4x^2 + 3x^3} = \frac{2 - 16/x}{4/x + 3},$$

so

$$\lim_{x \to \infty} f(x) = \lim_{x \to \infty} \frac{2 - 16/x}{4/x + 3} = \frac{\lim_{x \to \infty}(2 - 16/x)}{\lim_{x \to \infty}(4/x + 3)} = \frac{2}{3}.$$

37. Because the denominator equals 0 when $x = 4$, so must the numerator. This means $k^2 = 16$ and the choices for k are 4 or -4.

41. For the numerator, $\lim\limits_{x \to -\infty} \left(e^{2x} - 5\right) = -5$. If $k > 0$, $\lim\limits_{x \to -\infty} \left(e^{kx} + 3\right) = 3$, so the quotient has a limit of $-5/3$. If $k = 0$, $\lim\limits_{x \to -\infty} \left(e^{kx} + 3\right) = 4$, so the quotient has limit of $-5/4$. If $k < 0$, the limit of the quotient is given by $\lim\limits_{x \to -\infty} \left(e^{2x} - 5\right)/\left(e^{kx} + 3\right) = 0.$

45. (a) Since $\sin(n\pi) = 0$ for $n = 1, 2, 3, \ldots$ the sequence of x-values

$$\frac{1}{\pi}, \frac{1}{2\pi}, \frac{1}{3\pi}, \ldots$$

works. These x-values $\to 0$ and are zeroes of $f(x)$.

(b) Since $\sin(n\pi/2) = 1$ for $n = 1, 5, 9 \ldots$ the sequence of x-values

$$\frac{2}{\pi}, \frac{2}{5\pi}, \frac{2}{9\pi}, \ldots$$

works.

(c) Since $\sin(n\pi)/2 = -1$ for $n = 3, 7, 11, \ldots$ the sequence of x-values

$$\frac{2}{3\pi}, \frac{2}{7\pi}, \frac{2}{11\pi} \ldots$$

works.

(d) Any two of these sequences of x-values show that if the limit were to exist, then it would have to have two (different) values: 0 and 1, or 0 and -1, or 1 and -1. Hence, the limit can not exist.

Solutions for Section 2.3

Exercises

1. The derivative, $f'(2)$, is the rate of change of x^3 at $x = 2$. Notice that each time x changes by 0.001 in the table, the value of x^3 changes by 0.012. Therefore, we estimate

$$f'(2) = \frac{\text{Rate of change}}{\text{of } f \text{ at } x = 2} \approx \frac{0.012}{0.001} = 12.$$

The function values in the table look exactly linear because they have been rounded. For example, the exact value of x^3 when $x = 2.001$ is 8.012006001, not 8.012. Thus, the table can tell us only that the derivative is approximately 12. Example 5 on page 82 shows how to compute the derivative of $f(x)$ exactly.

5. $f'(1) = \lim\limits_{h \to 0} \dfrac{\log(1 + h) - \log 1}{h} = \lim\limits_{h \to 0} \dfrac{\log(1 + h)}{h}$

   Evaluating $\frac{\log(1+h)}{h}$ for $h = 0.01, 0.001$, and 0.0001, we get $0.43214, 0.43408, 0.43427$, so $f'(1) \approx 0.43427$. The corresponding secant lines are getting steeper, because the graph of $\log x$ is concave down. We thus expect the limit to be more than 0.43427. If we consider negative values of h, the estimates are too large. We can also see this from the graph below:

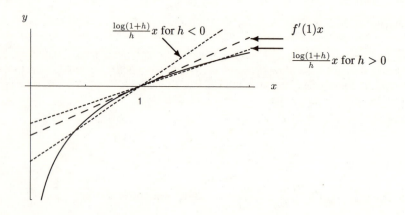

9. (a) The average rate of change from $x = a$ to $x = b$ is the slope of the line between the points on the curve with $x = a$ and $x = b$. Since the curve is concave down, the line from $x = 1$ to $x = 3$ has a greater slope than the line from $x = 3$ to $x = 5$, and so the average rate of change between $x = 1$ and $x = 3$ is greater than that between $x = 3$ and $x = 5$.

   (b) Since f is increasing, $f(5)$ is the greater.

   (c) As in part (a), f is concave down and f' is decreasing throughout so $f'(1)$ is the greater.

13.

$$f'(1) = \lim\limits_{h \to 0} \frac{f(1 + h) - f(1)}{h} = \lim\limits_{h \to 0} \frac{((1 + h)^3 + 5) - (1^3 + 5)}{h}$$

$$= \lim\limits_{h \to 0} \frac{1 + 3h + 3h^2 + h^3 + 5 - 1 - 5}{h} = \lim\limits_{h \to 0} \frac{3h + 3h^2 + h^3}{h}$$

$$= \lim\limits_{h \to 0} (3 + 3h + h^2) = 3.$$

17. As we saw in the answer to Problem 11, the slope of the tangent line to $f(x) = x^3$ at $x = -2$ is 12. When $x = -2$, $f(x) = -8$ so we know the point $(-2, -8)$ is on the tangent line. Thus the equation of the tangent line is $y = 12(x + 2) - 8 = 12x + 16$.

Problems

21. The coordinates of A are $(4, 25)$. See Figure 2.6. The coordinates of B and C are obtained using the slope of the tangent line. Since $f'(4) = 1.5$, the slope is 1.5

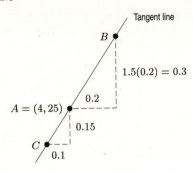

Figure 2.6

From A to B, $\Delta x = 0.2$, so $\Delta y = 1.5(0.2) = 0.3$. Thus, at C we have $y = 25 + 0.3 = 25.3$. The coordinates of B are $(4.2, 25.3)$.

From A to C, $\Delta x = -0.1$, so $\Delta y = 1.5(-0.1) = -0.15$. Thus, at C we have $y = 25 - 0.15 = 24.85$. The coordinates of C are $(3.9, 24.85)$.

25. Figure 2.7 shows the quantities in which we are interested.

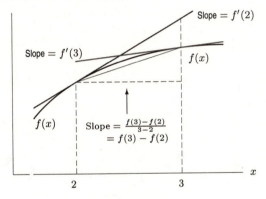

Figure 2.7

The quantities $f'(2)$, $f'(3)$ and $f(3) - f(2)$ have the following interpretations:

- $f'(2) = $ slope of the tangent line at $x = 2$
- $f'(3) = $ slope of the tangent line at $x = 3$
- $f(3) - f(2) = \frac{f(3) - f(2)}{3 - 2} = $ slope of the secant line from $f(2)$ to $f(3)$.

From Figure 2.7, it is clear that $0 < f(3) - f(2) < f'(2)$. By extending the secant line past the point $(3, f(3))$, we can see that it lies above the tangent line at $x = 3$.

Thus
$$0 < f'(3) < f(3) - f(2) < f'(2).$$

29. (a)

$$f'(0) = \lim_{h \to 0} \frac{\overbrace{\sin h}^{h \text{ in degrees}} - \overbrace{\sin 0}^{0}}{h} = \frac{\sin h}{h}.$$

To four decimal places,

$$\frac{\sin 0.2}{0.2} \approx \frac{\sin 0.1}{0.1} \approx \frac{\sin 0.01}{0.01} \approx \frac{\sin 0.001}{0.001} \approx 0.01745$$

so $f'(0) \approx 0.01745$.

(b) Consider the ratio $\frac{\sin h}{h}$. As we approach 0, the numerator, $\sin h$, will be much smaller in magnitude if h is in degrees than it would be if h were in radians. For example, if $h = 1°$ radian, $\sin h = 0.8415$, but if $h = 1$ degree, $\sin h = 0.01745$. Thus, since the numerator is smaller for h measured in degrees while the denominator is the same, we expect the ratio $\frac{\sin h}{h}$ to be smaller.

33. We want to approximate $P'(0)$ and $P'(2)$. Since for small h

$$P'(0) \approx \frac{P(h) - P(0)}{h},$$

if we take $h = 0.01$, we get

$$P'(0) \approx \frac{1.15(1.014)^{0.01} - 1.15}{0.01} = 0.01599 \text{ billion/year}$$
$$= 16.0 \text{ million people/year}$$
$$P'(2) \approx \frac{1.15(1.014)^{2.01} - 1.15(1.014)^2}{0.01} = 0.0164 \text{ billion/year}$$
$$= 16.4 \text{ million people/year}$$

Solutions for Section 2.4

Exercises

1. The graph is that of the line $y = -2x + 2$. The slope, and hence the derivative, is -2.

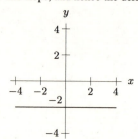

5.

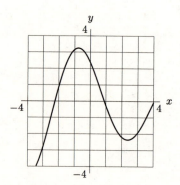

9.

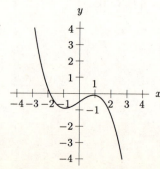

13. Since $1/x = x^{-1}$, using the power rule gives

$$\frac{d}{dx}(x^{-1}) = (-1)x^{-2} = -\frac{1}{x^2}.$$

Using the definition of the derivative, we have

$$k'(x) = \lim_{h \to 0} \frac{k(x+h) - k(x)}{h} = \lim_{h \to 0} \frac{\frac{1}{x+h} - \frac{1}{x}}{h} = \lim_{h \to 0} \frac{x - (x+h)}{h(x+h)x}$$

$$= \lim_{h \to 0} \frac{-h}{h(x+h)x} = \lim_{h \to 0} \frac{-1}{(x+h)x} = -\frac{1}{x^2}.$$

17.

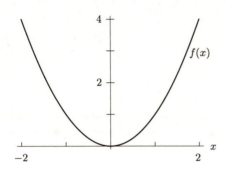
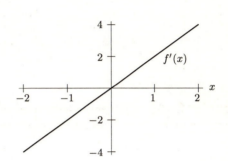

Problems

21. We know that $f'(x) \approx \dfrac{f(x+h) - f(x)}{h}$. For this problem, we'll take the average of the values obtained for $h = 1$

and $h = -1$; that's the average of $f(x+1) - f(x)$ and $f(x) - f(x-1)$ which equals $\dfrac{f(x+1) - f(x-1)}{2}$. Thus,

$f'(0) \approx f(1) - f(0) = 13 - 18 = -5.$
$f'(1) \approx [f(2) - f(0)]/2 = [10 - 18]/2 = -4.$
$f'(2) \approx [f(3) - f(1)]/2 = [9 - 13]/2 = -2.$
$f'(3) \approx [f(4) - f(2)]/2 = [9 - 10]/2 = -0.5.$
$f'(4) \approx [f(5) - f(3)]/2 = [11 - 9]/2 = 1.$
$f'(5) \approx [f(6) - f(4)]/2 = [15 - 9]/2 = 3.$
$f'(6) \approx [f(7) - f(5)]/2 = [21 - 11]/2 = 5.$
$f'(7) \approx [f(8) - f(6)]/2 = [30 - 15]/2 = 7.5.$
$f'(8) \approx f(8) - f(7) = 30 - 21 = 9.$

The rate of change of $f(x)$ is positive for $4 \le x \le 8$, negative for $0 \le x \le 3$. The rate of change is greatest at about $x = 8$.

25. This function is decreasing for $x < 2$ and increasing for $x > 2$ and so the derivative is negative for $x < 2$ and positive for $x > 2$. One possible graph is shown in Figure 2.8.

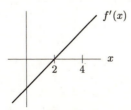

Figure 2.8

29.

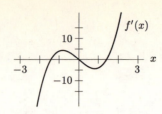

33. From the given information we know that f is increasing for values of x less than -2, is decreasing between $x = -2$ and $x = 2$, and is constant for $x > 2$. Figure 2.9 shows a possible graph—yours may be different.

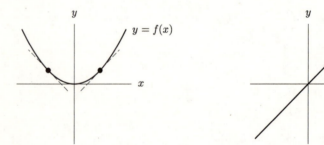

Figure 2.9

37. (a) Graph II
   (b) Graph I
   (c) Graph III

41. If $f(x)$ is even, its graph is symmetric about the y-axis. So the tangent line to f at $x = x_0$ is the same as that at $x = -x_0$ reflected about the y-axis.

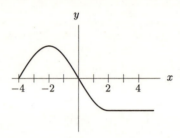

So the slopes of these two tangent lines are opposite in sign, so $f'(x_0) = -f'(-x_0)$, and f' is odd.

Solutions for Section 2.5

Exercises

1. (a) As the cup of coffee cools, the temperature decreases, so $f'(t)$ is negative.
   (b) Since $f'(t) = dH/dt$, the units are degrees Celsius per minute. The quantity $f'(20)$ represents the rate at which the coffee is cooling, in degrees per minute, 20 minutes after the cup is put on the counter.

5. Since B is measured in dollars and t is measured in years, dB/dt is measured in dollars per year. We can interpret dB as the extra money added to your balance in dt years. Therefore dB/dt represents how fast your balance is growing, in units of dollars/year.

9. The units of $f'(x)$ are feet/mile. The derivative, $f'(x)$, represents the rate of change of elevation with distance from the source, so if the river is flowing downhill everywhere, the elevation is always decreasing and $f'(x)$ is always negative. (In fact, there may be some stretches where the elevation is more or less constant, so $f'(x) = 0$.)

Problems

13. Since $f(t) = 1.15(1.014)^t$, we have
$$f(6) = 1.15(1.014)^6 = 1.25.$$
To estimate $f'(6)$, we use a small interval around 6:
$$f'(6) \approx \frac{f(6.001) - f(6)}{6.001 - 6} = \frac{1.15(1.014)^{6.001} - 1.15(1.014)^6}{0.001} = 0.0174.$$
We see that $f(6) = 1.25$ billion people and $f'(6) = 0.0174$ billion people per year. This model tells us that the population of China was about 1,250,000,000 people in 1999 and was growing at a rate of about 17,400,000 people per year at that time.

17. Units of $g'(55)$ are mpg/mph. The statement $g'(55) = -0.54$ means that at 55 miles per hour the fuel efficiency (in miles per gallon, or mpg) of the car decreases at a rate of approximately one half mpg as the velocity increases by one mph.

21. **(a)** The units of compliance are units of volume per units of pressure, or liters per centimeter of water.
(b) The increase in volume for a 5 cm reduction in pressure is largest between 10 and 15 cm. Thus, the compliance appears maximum between 10 and 15 cm of pressure reduction. The derivative is given by the slope, so
$$\text{Compliance} \approx \frac{0.70 - 0.49}{15 - 10} = 0.042 \text{ liters per centimeter.}$$
(c) When the lung is nearly full, it cannot expand much more to accommodate more air.

Solutions for Section 2.6

Exercises

1. **(a)** Since the graph is below the x-axis at $x = 2$, $f(2)$ is negative.
(b) Since $f(x)$ is decreasing at $x = 2$, $f'(2)$ is negative.
(c) Since $f(x)$ is concave up at $x = 2$, $f''(2)$ is positive.

5. The function is everywhere increasing and concave up. One possible graph is shown in Figure 2.10.

Figure 2.10

9. $f'(x) = 0$
$f''(x) = 0$

13. $f'(x) < 0$
$f''(x) < 0$

Problems

17. **(a)** $dP/dt > 0$ and $d^2P/dt^2 > 0$.

(b) $dP/dt < 0$ and $d^2P/dt^2 > 0$ (but dP/dt is close to zero).

21. **(a)** The EPA will say that the rate of discharge is still rising. The industry will say that the rate of discharge is increasing less quickly, and may soon level off or even start to fall.
(b) The EPA will say that the rate at which pollutants are being discharged is levelling off, but not to zero — so pollutants will continue to be dumped in the lake. The industry will say that the rate of discharge has decreased significantly.

Solutions for Section 2.7

Exercises

1. (a) Function f is not continuous at $x = 1$.
  (b) Function f appears not differentiable at $x = 1, 2, 3$.

5. Yes.

Problems

9. We can see from Figure 2.11 that the graph of f oscillates infinitely often between the curves $y = x^2$ and $y = -x^2$ near the origin. Thus the slope of the line from $(0, 0)$ to $(h, f(h))$ oscillates between h (when $f(h) = h^2$ and $\frac{f(h)-0}{h-0} = h$) and $-h$ (when $f(h) = -h^2$ and $\frac{f(h)-0}{h-0} = -h$) as h tends to zero. So, the limit of the slope as h tends to zero is 0, which is the derivative of f at the origin. Another way to see this is to observe that

$$\lim_{h\to 0} \frac{f(h) - f(0)}{h} = \lim_{h\to 0}\left(\frac{h^2 \sin(\frac{1}{h})}{h}\right)$$
$$= \lim_{h\to 0} h \sin(\frac{1}{h})$$
$$= 0,$$

since $\lim_{h\to 0} h = 0$ and $-1 \le \sin(\frac{1}{h}) \le 1$ for any h. Thus f is differentiable at $x = 0$, and $f'(0) = 0$.

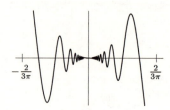

Figure 2.11

13. (a) Since

$$\lim_{r\to r_0^-} E = kr_0$$

and

$$\lim_{r\to r_0^+} E = \frac{kr_0^2}{r_0} = kr_0$$

and

$$E(r_0) = kr_0,$$

we see that E is continuous at r_0.

(b) The function E is not differentiable at $r = r_0$ because the graph has a corner there. The slope is positive for $r < r_0$ and the slope is negative for $r > r_0$.

(c)

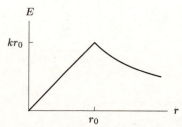

17. Since $f(x) = x$ is continuous, Theorem 2.2 on page 95 shows that products of the form $f(x) \cdot f(x) = x^2$ and $f(x) \cdot x^2 = x^3$, etc., are continuous. By a similar argument, x^n is continuous for any $n > 0$.

Solutions for Chapter 2 Review

Exercises

1.

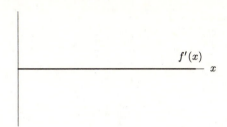

5.

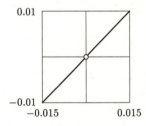

9. From Table 2.4, it appears the limit is 0. This is confirmed by Figure 2.12. An appropriate window is $-0.015 < x < 0.015$, $-0.01 < y < 0.01$.

Table 2.4

x	$f(x)$		x	$f(x)$
0.1	0.0666		-0.0001	-0.0001
0.01	0.0067		-0.001	-0.0007
0.001	0.0007		-0.01	-0.0067
0.0001	0		-0.1	-0.0666

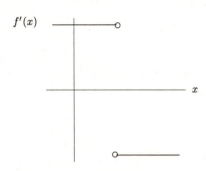

Figure 2.12

13. $\lim_{h \to 0} \dfrac{(a+h)^2 - a^2}{h} = \lim_{h \to 0} \dfrac{a^2 + 2ah + h^2 - a^2}{h} = \lim_{h \to 0} (2a + h) = 2a$

17. We combine terms in the numerator and multiply top and bottom by $\sqrt{a} + \sqrt{a+h}$.

$$\frac{1}{\sqrt{a+h}} - \frac{1}{\sqrt{a}} = \frac{\sqrt{a} - \sqrt{a+h}}{\sqrt{a+h}\sqrt{a}} = \frac{(\sqrt{a} - \sqrt{a+h})(\sqrt{a} + \sqrt{a+h})}{\sqrt{a+h}\sqrt{a}(\sqrt{a} + \sqrt{a+h})}$$

$$= \frac{a - (a+h)}{\sqrt{a+h}\sqrt{a}(\sqrt{a} + \sqrt{a+h})}$$

Therefore $\lim_{h \to 0} \dfrac{1}{h}\left(\dfrac{1}{\sqrt{a+h}} - \dfrac{1}{\sqrt{a}}\right) = \lim_{h \to 0} \dfrac{-1}{\sqrt{a+h}\sqrt{a}(\sqrt{a} + \sqrt{a+h})} = \dfrac{-1}{2(\sqrt{a})^3}$

Problems

21. Since $f(2) = 3$ and $f'(2) = 1$, near $x = 2$ the graph looks like the segment shown in Figure 2.13.

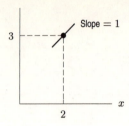

Figure 2.13

   (a) If $f(x)$ is even, then the graph of $f(x)$ near $x = 2$ and $x = -2$ looks like Figure 2.14. Thus $f(-2) = 3$ and $f'(-2) = -1$.

   (b) If $f(x)$ is odd, then the graph of $f(x)$ near $x = 2$ and $x = -2$ looks like Figure 2.15. Thus $f(-2) = -3$ and $f'(-2) = 1$.

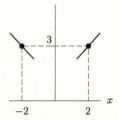

Figure 2.14: For f even

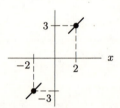

Figure 2.15: For f odd

25. (a) Since the point $A = (7, 3)$ is on the graph of f, we have $f(7) = 3$.

   (b) The slope of the tangent line touching the curve at $x = 7$ is given by

$$\text{Slope} = \frac{\text{Rise}}{\text{Run}} = \frac{3.8 - 3}{7.2 - 7} = \frac{0.8}{0.2} = 4.$$

Thus, $f'(7) = 4$.

29. $f(10) = 240{,}000$ means that if the commodity costs \$10, then 240,000 units of it will be sold. $f'(10) = -29{,}000$ means that if the commodity costs \$10 now, each \$1 increase in price will cause a decline in sales of 29,000 units.

33. By tracing on a calculator or solving equations, we find the following values of δ:

For $\epsilon = 0.1$, $\delta \leq 0.45$.
For $\epsilon = 0.001$, $\delta \leq 0.0447$.
For $\epsilon = 0.00001$, $\delta \leq 0.00447$.

37. (a) The population varies periodically with a period of 12 months (i.e. one year).

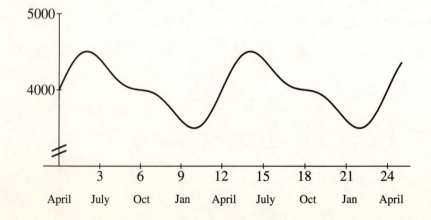

(b) The herd is largest about June 1st when there are about 4500 deer.

(c) The herd is smallest about February 1st when there are about 3500 deer.

(d) The herd grows the fastest about April 1st. The herd shrinks the fastest about July 15 and again about December 15.

(e) It grows the fastest about April 1st when the rate of growth is about 400 deer/month, i.e about 13 new fawns per day.

41. (a)

Table 2.5

x	$\frac{\sinh(x+0.001)-\sinh(x)}{0.001}$	$\frac{\sinh(x+0.0001)-\sinh(x)}{0.0001}$	so $f'(0) \approx$	$\cosh(x)$
0	1.00000	1.00000	1.00000	1.00000
0.3	1.04549	1.04535	1.04535	1.04534
0.7	1.25555	1.25521	1.25521	1.25517
1	1.54367	1.54314	1.54314	1.54308

(b) It seems that they are approximately the same, i.e. the derivative of $\sinh(x) = \cosh(x)$ for $x = 0, 0.3, 0.7$, and 1.

CAS Challenge Problems

45. (a) The CAS gives the same derivative, $1/x$, in all three cases.

(b) From the properties of logarithms, $g(x) = \ln(2x) = \ln 2 + \ln x = f(x) + \ln 2$. So the graph of g is the same shape as the graph of f, only shifted up by $\ln 2$. So the graphs have the same slope everywhere, and therefore the two functions have the same derivative. By the same reasoning, $h(x) = f(x) + \ln 3$, so h and f have the same derivative as well.

CHECK YOUR UNDERSTANDING

1. False. For example, the car could slow down or even stop at one minute after 2 pm, and then speed back up to 60 mph at one minute before 3 pm. In this case the car would travel only a few miles during the hour, much less than 50 miles.

5. True. By definition, Average velocity = Distance traveled/Time.

9. True. The derivative of a function is the limit of difference quotients. A few difference quotients can be computed from the table, but the limit can not be computed from the table.

13. True. Shifting a graph vertically does not change the shape of the graph and so it does not change the slopes of the tangent lines to the graph.

17. True. The second derivative $f''(x)$ is the derivative of $f'(x)$. Thus the derivative of $f'(x)$ is positive, and so $f'(x)$ is increasing.

21. True. Let $f(x) = |x - 3|$. Then $f(x)$ is continuous for all x but not differentiable at $x = 3$ because its graph has a corner there. Other answers are possible.

25. False. For example, $f(x) = |x|$ is not differentiable at $x = 0$, but it is continuous at $x = 0$.

29. True, by Property 2 of limits in Theorem 2.1.

33. True. Suppose instead that $\lim_{x \to 3} g(x)$ does not exist but $\lim_{x \to 3}(f(x)g(x))$ did exist. Since $\lim_{x \to 3} f(x)$ exists and is not zero, then $\lim_{x \to 3}((f(x)g(x))/f(x))$ exists, by Property 4 of limits in Theorem 2.1. Furthermore, $f(x) \neq 0$ for all x in some interval about 3, so $(f(x)g(x))/f(x) = g(x)$ for all x in that interval. Thus $\lim_{x \to 3} g(x)$ exists. This contradicts our assumption that $\lim_{x \to 3} g(x)$ does not exist.

37. False. Although x may be far from c, the value of $f(x)$ could be close to L. For example, suppose $f(x) = L$, the constant function.

CHAPTER THREE

Solutions for Section 3.1

Exercises

1. The derivative, $f'(x)$, is defined as

$$f'(x) = \lim_{h \to 0} \frac{f(x+h) - f(x)}{h}.$$

If $f(x) = 7$, then

$$f'(x) = \lim_{h \to 0} \frac{7 - 7}{h} = \lim_{h \to 0} \frac{0}{h} = 0.$$

5. $y' = 11x^{-12}$.

9. $y' = \frac{3}{4}x^{-1/4}$.

13. $f'(x) = ex^{e-1}$.

17. Dividing gives $g(t) = t^2 + k/t$ so $g'(t) = 2t - \dfrac{k}{t^2}$.

21. $y' = 15t^4 - \frac{5}{2}t^{-1/2} - \frac{7}{t^2}$.

25. $f(z) = \dfrac{z}{3} + \dfrac{1}{3}z^{-1} = \dfrac{1}{3}\left(z + z^{-1}\right)$, so $f'(z) = \dfrac{1}{3}\left(1 - z^{-2}\right) = \dfrac{1}{3}\left(\dfrac{z^2 - 1}{z^2}\right)$.

29. Since $4/3$, π, and b are all constants, we have

$$\frac{dV}{dr} = \frac{4}{3}\pi(2r)b = \frac{8}{3}\pi r b.$$

33. $g'(x) = -\dfrac{1}{2}(5x^4 + 2)$.

Problems

37. The x is in the exponent and we haven't learned how to handle that yet.

41. $y' = -2/3z^3$.   (power rule and sum rule)

45. The graph increases when $dy/dx > 0$:

$$\frac{dy}{dx} = 5x^4 - 5 > 0$$

$$5(x^4 - 1) > 0 \quad \text{so} \quad x^4 > 1 \quad \text{so} \quad x > 1 \text{ or } x < -1.$$

The graph is concave up when $d^2y/dx^2 > 0$:

$$\frac{d^2y}{dx^2} = 20x^3 > 0 \quad \text{so} \quad x > 0.$$

We need values of x where $\{x > 1 \text{ or } x < -1\}$ AND $\{x > 0\}$, which implies $x > 1$. Thus, both conditions hold for all values of x larger than 1.

49. Differentiating gives

$$f'(x) = 6x^2 - 4x \quad \text{so} \quad f'(1) = 6 - 4 = 2.$$

Thus the equation of the tangent line is $(y - 1) = 2(x - 1)$ or $y = 2x - 1$.

53. Since $f(x) = ax^n$, $f'(x) = anx^{n-1}$. We know that $f'(2) = (an)2^{n-1} = 3$, and $f'(4) = (an)4^{n-1} = 24$. Therefore,

$$\frac{f'(4)}{f'(2)} = \frac{24}{3}$$

$$\frac{(an)4^{n-1}}{(an)2^{n-1}} = \left(\frac{4}{2}\right)^{n-1} = 8$$

$$2^{n-1} = 8, \text{ and thus } n = 4.$$

Substituting $n = 4$ into the expression for $f'(2)$, we get $3 = a(4)(8)$, or $a = 3/32$.

57. (a) The average velocity between $t = 0$ and $t = 2$ is given by

$$\text{Average velocity} = \frac{f(2) - f(0)}{2 - 0} = \frac{-4.9(2^2) + 25(2) + 3 - 3}{2 - 0} = \frac{33.4 - 3}{2} = 15.2 \text{ m/sec.}$$

(b) Since $f'(t) = -9.8t + 25$, we have

$$\text{Instantaneous velocity} = f'(2) = -9.8(2) + 25 = 5.4 \text{ m/sec.}$$

(c) Acceleration is given $f''(t) = -9.8$. The acceleration at $t = 2$ (and all other times) is the acceleration due to gravity, which is -9.8 m/sec^2.

(d) We can use a graph of height against time to estimate the maximum height of the tomato. See Figure 3.1. Alternately, we can find the answer analytically. The maximum height occurs when the velocity is zero and $v(t) = -9.8t + 25 = 0$ when $t = 2.6$ sec. At this time the tomato is at a height of $f(2.6) = 34.9$. The maximum height is 34.9 meters.

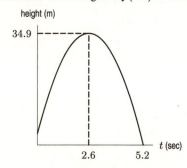

Figure 3.1

(e) We see in Figure 3.1 that the tomato hits ground at about $t = 5.2$ seconds. Alternately, we can find the answer analytically. The tomato hits the ground when

$$f(t) = -4.9t^2 + 25t + 3 = 0.$$

We solve for t using the quadratic formula:

$$t = \frac{-25 \pm \sqrt{(25)^2 - 4(-4.9)(3)}}{2(-4.9)}$$

$$t = \frac{-25 \pm \sqrt{683.8}}{-9.8}$$

$$t = -0.12 \quad \text{and} \quad t = 5.2.$$

We use the positive values, so the tomato hits the ground at $t = 5.2$ seconds.

61. $V = \frac{4}{3}\pi r^3$. Differentiating gives $\frac{dV}{dr} = 4\pi r^2 = $ surface area of a sphere.

The difference quotient $\frac{V(r+h) - V(r)}{h}$ is the volume between two spheres divided by the change in radius. Furthermore, when h is very small, the difference between volumes, $V(r + h) - V(r)$, is like a coating of paint of depth h applied to the surface of the sphere. The volume of the paint is about $h \cdot (\text{Surface Area})$ for small h: dividing by h gives back the surface area.

Thinking about the derivative as the rate of change of the function for a small change in the variable gives another way of seeing the result. If you increase the radius of a sphere a small amount, the volume increases by a very thin layer whose volume is the surface area at that radius multiplied by that small amount.

Solutions for Section 3.2

Exercises

1. $f'(x) = 2e^x + 2x$.

5. $y' = 10x + (\ln 2)2^x$.

9. $\dfrac{dy}{dx} = \dfrac{1}{3}(\ln 3)3^x - \dfrac{33}{2}(x^{-\frac{3}{2}})$.

13. $y = e^\theta e^{-1} \quad y' = \dfrac{d}{d\theta}(e^\theta e^{-1}) = e^{-1}\dfrac{d}{d\theta}e^\theta = e^\theta e^{-1} = e^{\theta-1}$.

17. $f'(x) = 3x^2 + 3^x \ln 3$

21. $f'(x) = (\ln \pi)\pi^x$.

25. $f'(z) = (2\ln 3)z + (\ln 4)e^z$.

29. We can take the derivative of the sum $x^2 + 2^x$, but not the product.

33. The exponent is x^2, and we haven't learned what to do about that yet.

Problems

37. Since $P = 1 \cdot (1.05)^t$, $\frac{dP}{dt} = \ln(1.05)1.05^t$. When $t = 10$,

$$\frac{dP}{dt} = (\ln 1.05)(1.05)^{10} \approx \$0.07947/\text{year} \approx 7.95¢/\text{year}.$$

41. (a) The rate of change of the population is $P'(t)$. If $P'(t)$ is proportional to $P(t)$, we have

$$P'(t) = kP(t).$$

   (b) If $P(t) = Ae^{kt}$, then $P'(t) = kAe^{kt} = kP(t)$.

45. The derivative of e^x is $\frac{d}{dx}(e^x) = e^x$. Thus the tangent line at $x = 0$, has slope $e^0 = 1$, and the tangent line is $y = x + 1$. A function which is always concave up will always stay above any of its tangent lines. Thus $e^x \geq x + 1$ for all x, as shown in the figure below.

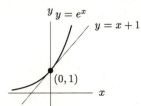

Solutions for Section 3.3

Exercises

1. By the product rule, $f'(x) = 2x(x^3 + 5) + x^2(3x^2) = 2x^4 + 3x^4 + 10x = 5x^4 + 10x$. Alternatively, $f'(x) = (x^5 + 5x^2)' = 5x^4 + 10x$. The two answers should, and do, match.

5. $y' = \frac{1}{2\sqrt{x}}2^x + \sqrt{x}(\ln 2)2^x$.

9. $y' = (3t^2 - 14t)e^t + (t^3 - 7t^2 + 1)e^t = (t^3 - 4t^2 - 14t + 1)e^t$.

13. $q'(r) = \dfrac{3(5r + 2) - 3r(5)}{(5r + 2)^2} = \dfrac{15r + 6 - 15r}{(5r + 2)^2} = \dfrac{6}{(5r + 2)^2}$

17. Using the quotient rule gives $\dfrac{dz}{dt} = \dfrac{(2t + 3)(t + 1) - (t^2 + 3t + 1)}{(t + 1)^2}$ or $\dfrac{dz}{dt} = \dfrac{t^2 + 2t + 2}{(t + 1)^2}$.

21. $\dfrac{d}{dz}\left(\dfrac{z^2 + 1}{\sqrt{z}}\right) = \dfrac{d}{dz}(z^{\frac{3}{2}} + z^{-\frac{1}{2}}) = \dfrac{3}{2}z^{\frac{1}{2}} - \dfrac{1}{2}z^{-\frac{3}{2}} = \dfrac{\sqrt{z}}{2}(3 - z^{-2})$.

25. $w'(x) = \dfrac{17e^x(2^x) - (\ln 2)(17e^x)2^x}{2^{2x}} = \dfrac{17e^x(2^x)(1 - \ln 2)}{2^{2x}} = \dfrac{17e^x(1 - \ln 2)}{2^x}$.

29. $w' = (3t^2 + 5)(t^2 - 7t + 2) + (t^3 + 5t)(2t - 7)$.

Problems

33. Since $f(0) = -5/1 = -5$, the tangent line passes through the point $(0, -5)$, so its vertical intercept is -5. To find the slope of the tangent line, we find the derivative of $f(x)$ using the quotient rule:

$$f'(x) = \frac{(x + 1) \cdot 2 - (2x - 5) \cdot 1}{(x + 1)^2} = \frac{7}{(x + 1)^2}.$$

At $x = 0$, the slope of the tangent line is $m = f'(0) = 7$. The equation of the tangent line is $y = 7x - 5$.

37. Since $\frac{d}{dx}e^{2x} = 2e^{2x}$ and $\frac{d}{dx}e^{3x} = 3e^{3x}$, we might guess that $\frac{d}{dx}e^{4x} = 4e^{4x}$.

41. (a) We have $h'(2) = f'(2) + g'(2) = 5 - 2 = 3$.
   (b) We have $h'(2) = f'(2)g(2) + f(2)g'(2) = 5(4) + 3(-2) = 14$.
   (c) We have $h'(2) = \dfrac{f'(2)g(2) - f(2)g'(2)}{(g(2))^2} = \dfrac{5(4) - 3(-2)}{4^2} = \dfrac{26}{16} = \dfrac{13}{8}$.

45. We want dR/dr_1. Solving for R:

$$\frac{1}{R} = \frac{1}{r_1} + \frac{1}{r_2} = \frac{r_2 + r_1}{r_1 r_2}, \quad \text{which gives } R = \frac{r_1 r_2}{r_2 + r_1}.$$

So, thinking of r_2 as a constant and using the quotient rule,

$$\frac{dR}{dr_1} = \frac{r_2(r_2 + r_1) - r_1 r_2(1)}{(r_2 + r_1)^2} = \frac{r_2^2}{(r_1 + r_2)^2}.$$

49. (a) $f'(x) = (x - 2) + (x - 1)$.
   (b) Think of f as the product of two factors, with the first as $(x - 1)(x - 2)$. (The reason for this is that we have already differentiated $(x - 1)(x - 2)$).

$$f(x) = [(x - 1)(x - 2)](x - 3).$$

Now $f'(x) = [(x - 1)(x - 2)]'(x - 3) + [(x - 1)(x - 2)](x - 3)'$
Using the result of a):

$$f'(x) = [(x - 2) + (x - 1)](x - 3) + [(x - 1)(x - 2)] \cdot 1$$
$$= (x - 2)(x - 3) + (x - 1)(x - 3) + (x - 1)(x - 2).$$

   (c) Because we have already differentiated $(x - 1)(x - 2)(x - 3)$, rewrite f as the product of two factors, the first being $(x - 1)(x - 2)(x - 3)$:

$$f(x) = [(x - 1)(x - 2)(x - 3)](x - 4)$$

Now $f'(x) = [(x - 1)(x - 2)(x - 3)]'(x - 4) + [(x - 1)(x - 2)(x - 3)](x - 4)'$.

$$f'(x) = [(x - 2)(x - 3) + (x - 1)(x - 3) + (x - 1)(x - 2)](x - 4)$$
$$+ [(x - 1)(x - 2)(x - 3)] \cdot 1$$
$$= (x - 2)(x - 3)(x - 4) + (x - 1)(x - 3)(x - 4)$$
$$+ (x - 1)(x - 2)(x - 4) + (x - 1)(x - 2)(x - 3).$$

From the solutions above, we can observe that when f is a product, its derivative is obtained by differentiating each factor in turn (leaving the other factors alone), and adding the results.

Solutions for Section 3.4

Exercises

1. $f'(x) = 99(x + 1)^{98} \cdot 1 = 99(x + 1)^{98}$.

5. $w' = 100(\sqrt{t} + 1)^{99} \left(\frac{1}{2\sqrt{t}}\right) = \frac{50}{\sqrt{t}}(\sqrt{t} + 1)^{99}$.

9. $g(x) = \pi e^{\pi x}$.

13. $k'(x) = 4(x^3 + e^x)^3(3x^2 + e^x)$.

17. $\dfrac{d}{dt} e^{(1+3t)^2} = e^{(1+3t)^2} \dfrac{d}{dt}(1 + 3t)^2 = e^{(1+3t)^2} \cdot 2(1 + 3t) \cdot 3 = 6(1 + 3t)e^{(1+3t)^2}$.

21. $y' = \frac{3}{2} e^{\frac{3}{2}w}$.

25. $y' = 1 \cdot e^{-t^2} + te^{-t^2}(-2t)$

29. $f'(t) = 1 \cdot e^{5-2t} + te^{5-2t}(-2) = e^{5-2t}(1 - 2t)$.

33. $y' = \dfrac{-(3e^{3x} + 2x)}{(e^{3x} + x^2)^2}$.

37. $f'(\theta) = -1(1 + e^{-\theta})^{-2}(e^{-\theta})(-1) = \dfrac{e^{-\theta}}{(1 + e^{-\theta})^2}$.

41. $f(y) = \left[10^{(5-y)}\right]^{\frac{1}{2}} = 10^{\frac{5}{2} - \frac{1}{2}y}$

$f'(y) = (\ln 10)\left(10^{\frac{5}{2} - \frac{1}{2}y}\right)\left(-\frac{1}{2}\right) = -\frac{1}{2}(\ln 10)(10^{\frac{5}{2} - \frac{1}{2}y})$.

45. Since a and b are constants, we have $f'(t) = ae^{bt}(b) = abe^{bt}$.

Problems

49. We have $f(2) = (2-1)^3 = 1$, so $(2, 1)$ is a point on the tangent line. Since $f'(x) = 3(x-1)^2$, the slope of the tangent line is
$$m = f'(2) = 3(2-1)^2 = 3.$$
The equation of the line is
$$y - 1 = 3(x - 2) \quad \text{or} \quad y = 3x - 5.$$

53. **(a)** $H(x) = F(G(x))$
   $H(4) = F(G(4)) = F(2) = 1$
   (b) $H(x) = F(G(x))$
   $H'(x) = F'(G(x)) \cdot G'(x)$
   $H'(4) = F'(G(4)) \cdot G'(4) = F'(2) \cdot 6 = 5 \cdot 6 = 30$
   (c) $H(x) = G(F(x))$
   $H(4) = G(F(4)) = G(3) = 4$
   (d) $H(x) = G(F(x))$
   $H'(x) = G'(F(x)) \cdot F'(x)$
   $H'(4) = G'(F(4)) \cdot F'(4) = G'(3) \cdot 7 = 8 \cdot 7 = 56$
   (e) $H(x) = \frac{F(x)}{G(x)}$
   $H'(x) = \frac{G(x) \cdot F'(x) - F(x) \cdot G'(x)}{[G(x)]^2}$
   $H'(4) = \frac{G(4) \cdot F'(4) - F(4) \cdot G'(4)}{[G(4)]^2} = \frac{2 \cdot 7 - 3 \cdot 6}{2^2} = \frac{14 - 18}{4} = \frac{-4}{4} = -1$

57. Yes. To see why, simply plug $x = \sqrt[3]{2t+5}$ into the expression $3x^2 \frac{dx}{dt}$ and evaluate it. To do this, first we calculate $\frac{dx}{dt}$. By the chain rule,
$$\frac{dx}{dt} = \frac{d}{dt}(2t+5)^{\frac{1}{3}} = \frac{2}{3}(2t+5)^{-\frac{2}{3}} = \frac{2}{3}[(2t+5)^{\frac{1}{3}}]^{-2}.$$
But since $x = (2t+5)^{\frac{1}{3}}$, we have (by substitution)
$$\frac{dx}{dt} = \frac{2}{3}x^{-2}.$$
It follows that $3x^2 \frac{dx}{dt} = 3x^2 \left(\frac{2}{3}x^{-2}\right) = 2.$

61. We have $f(0) = 6$ and $f(10) = 6e^{0.013(10)} = 6.833$. The derivative of $f(t)$ is
$$f'(t) = 6e^{0.013t} \cdot 0.013 = 0.078e^{0.013t},$$
and so $f'(0) = 0.078$ and $f'(10) = 0.089$.

These values tell us that in 1999 (at $t = 0$), the population of the world was 6 billion people and the population was growing at a rate of 0.078 billion people per year. In the year 2009 (at $t = 10$), this model predicts that the population of the world will be 6.833 billion people and growing at a rate of 0.089 billion people per year.

65. The ripple's area and radius are related by $A(t) = \pi[r(t)]^2$. Taking derivatives and using the chain rule gives
$$\frac{dA}{dt} = \pi \cdot 2r \frac{dr}{dt}.$$
We know that $dr/dt = 10$ cm/sec, so when $r = 20$ cm we have
$$\frac{dA}{dt} = \pi \cdot 2 \cdot 20 \cdot 10 \text{cm}^2/\text{sec} = 400\pi \text{cm}^2/\text{sec}.$$

69. Recall that $v = dx/dt$. We want to find the acceleration, dv/dt, when $x = 2$. Differentiating the expression for v with respect to t using the chain rule and substituting for v gives
$$\frac{dv}{dt} = \frac{d}{dx}(x^2 + 3x - 2) \cdot \frac{dx}{dt} = (2x + 3)v = (2x + 3)(x^2 + 3x - 2).$$
Substituting $x = 2$ gives
$$\text{Acceleration} = \frac{dv}{dt}\bigg|_{x=2} = (2(2) + 3)(2^2 + 3 \cdot 2 - 2) = 56 \text{ cm/sec}^2.$$

Solutions for Section 3.5

Exercises

1.

Table 3.1

x	$\cos x$	Difference Quotient	$-\sin x$
0	1.0	−0.0005	0.0
0.1	0.995	−0.10033	−0.099833
0.2	0.98007	−0.19916	−0.19867
0.3	0.95534	−0.296	−0.29552
0.4	0.92106	−0.38988	−0.38942
0.5	0.87758	−0.47986	−0.47943
0.6	0.82534	−0.56506	−0.56464

5. $f'(x) = \cos(3x) \cdot 3 = 3\cos(3x)$.

9. $f'(x) = (2x)(\cos x) + x^2(-\sin x) = 2x\cos x - x^2\sin x$.

13. $z' = e^{\cos\theta} - \theta(\sin\theta)e^{\cos\theta}$.

17.
$$f(x) = (1 - \cos x)^{\frac{1}{2}}$$
$$f'(x) = \frac{1}{2}(1 - \cos x)^{-\frac{1}{2}}(-(-\sin x))$$
$$= \frac{\sin x}{2\sqrt{1 - \cos x}}.$$

21. $f'(x) = 2 \cdot [\sin(3x)] + 2x[\cos(3x)] \cdot 3 = 2\sin(3x) + 6x\cos(3x)$

25. $y' = 5\sin^4\theta\cos\theta$.

29. $h'(t) = 1 \cdot (\cos t) + t(-\sin t) + \frac{1}{\cos^2 t} = \cos t - t\sin t + \frac{1}{\cos^2 t}$.

33. Using the power and quotient rules gives

$$f'(x) = \frac{1}{2}\left(\frac{1 - \sin x}{1 - \cos x}\right)^{-1/2}\left[\frac{-\cos x(1 - \cos x) - (1 - \sin x)\sin x}{(1 - \cos x)^2}\right]$$

$$= \frac{1}{2}\sqrt{\frac{1 - \cos x}{1 - \sin x}}\left[\frac{-\cos x(1 - \cos x) - (1 - \sin x)\sin x}{(1 - \cos x)^2}\right]$$

$$= \frac{1}{2}\sqrt{\frac{1 - \cos x}{1 - \sin x}}\left[\frac{1 - \cos x - \sin x}{(1 - \cos x)^2}\right].$$

37. $w' = (\ln 2)(2^{2\sin x + e^x})(2\cos x + e^x)$.

41. $f'(w) = -2\cos w\sin w - \sin(w^2)(2w) = -2(\cos w\sin w + w\sin(w^2))$.

Problems

45. We begin by taking the derivative of $y = \sin(x^4)$ and evaluating at $x = 10$:

$$\frac{dy}{dx} = \cos(x^4) \cdot 4x^3.$$

Evaluating $\cos(10{,}000)$ on a calculator (in radians) we see $\cos(10{,}000) < 0$, so we know that $dy/dx < 0$, and therefore the function is decreasing.

Next, we take the second derivative and evaluate it at $x = 10$;

$$\frac{d^2y}{dx^2} = \underbrace{\cos(x^4) \cdot (12x^2)}_{\text{negative}} + \underbrace{4x^3 \cdot (-\sin(x^4))(4x^3)}_{\substack{\text{positive, but much} \\ \text{larger in magnitude}}}.$$

From this we can see that $d^2y/dx^2 > 0$, thus the graph is concave up.

49. (a) When $\sqrt{\frac{k}{m}}t = \frac{\pi}{2}$ the spring is farthest from the equilibrium position. This occurs at time $t = \frac{\pi}{2}\sqrt{\frac{m}{k}}$

$v = A\sqrt{\frac{k}{m}}\cos\left(\sqrt{\frac{k}{m}}t\right)$, so the maximum velocity occurs when $t = 0$

$a = -A\frac{k}{m}\sin\left(\sqrt{\frac{k}{m}}t\right)$, so the maximum acceleration occurs when $\sqrt{\frac{k}{m}}t = \frac{3\pi}{2}$, which is at time $t = \frac{3\pi}{2}\sqrt{\frac{m}{k}}$

(b) $T = \frac{2\pi}{\sqrt{k/m}} = 2\pi\sqrt{\frac{m}{k}}$

(c) $\dfrac{dT}{dm} = \dfrac{2\pi}{\sqrt{k}} \cdot \dfrac{1}{2}m^{-\frac{1}{2}} = \dfrac{\pi}{\sqrt{km}}$

Since $\dfrac{dT}{dm} > 0$, an increase in the mass causes the period to increase.

53. (a) If $f(x) = \sin x$, then

$$f'(x) = \lim_{h \to 0} \frac{\sin(x + h) - \sin x}{h}$$
$$= \lim_{h \to 0} \frac{(\sin x \cos h + \sin h \cos x) - \sin x}{h}$$
$$= \lim_{h \to 0} \frac{\sin x(\cos h - 1) + \sin h \cos x}{h}$$
$$= \sin x \lim_{h \to 0} \frac{\cos h - 1}{h} + \cos x \lim_{h \to 0} \frac{\sin h}{h}.$$

(b) $\frac{\cos h - 1}{h} \to 0$ and $\frac{\sin h}{h} \to 1$, as $h \to 0$. Thus, $f'(x) = \sin x \cdot 0 + \cos x \cdot 1 = \cos x$.

(c) Similarly,

$$g'(x) = \lim_{h \to 0} \frac{\cos(x + h) - \cos x}{h}$$
$$= \lim_{h \to 0} \frac{(\cos x \cos h - \sin x \sin h) - \cos x}{h}$$
$$= \lim_{h \to 0} \frac{\cos x(\cos h - 1) - \sin x \sin h}{h}$$
$$= \cos x \lim_{h \to 0} \frac{\cos h - 1}{h} - \sin x \lim_{h \to 0} \frac{\sin h}{h}$$
$$= -\sin x.$$

Solutions for Section 3.6

Exercises

1. $f'(t) = \frac{2t}{t^2+1}$.

5. $f'(z) = -1(\ln z)^{-2} \cdot \frac{1}{z} = \frac{-1}{z(\ln z)^2}$.

9. $f'(x) = \frac{1}{e^x+1} \cdot e^x$.

13. $f'(x) = \frac{1}{e^{7x}} \cdot (e^{7x})7 = 7$.
   (Note also that $\ln(e^{7x}) = 7x$ implies $f'(x) = 7$.)

17. $f'(y) = \dfrac{2y}{\sqrt{1-y^4}}$.

21. $g'(t) = \dfrac{-\sin(\ln t)}{t}$.

25. Using the chain rule gives $r'(t) = \dfrac{2}{\sqrt{1-4t^2}}$.

29. Using the quotient rule gives

$$f'(x) = \frac{1 + \ln x - x(\frac{1}{x})}{(1 + \ln x)^2}$$
$$= \frac{\ln x}{(1 + \ln x)^2}.$$

33. Using the chain rule gives

$$T'(u) = \left[\frac{1}{1 + \left(\frac{u}{1+u}\right)^2}\right]\left[\frac{(1+u) - u}{(1+u)^2}\right]$$

$$= \frac{(1+u)^2}{(1+u)^2 + u^2}\left[\frac{1}{(1+u)^2}\right]$$

$$= \frac{1}{1 + 2u + 2u^2}.$$

Problems

37. Let

$$g(x) = \arcsin x$$

so

$$\sin[g(x)] = x.$$

Differentiating,

$$\cos[g(x)] \cdot g'(x) = 1$$

$$g'(x) = \frac{1}{\cos[g(x)]}$$

Using the fact that $\sin^2 \theta + \cos^2 \theta = 1$, and $\cos[g(x)] \geq 0$, since $-\frac{\pi}{2} \leq g(x) \leq \frac{\pi}{2}$, we get

$$\cos[g(x)] = \sqrt{1 - (\sin[g(x)])^2}.$$

Therefore,

$$g'(x) = \frac{1}{\sqrt{1 - (\sin[g(x)])^2}}$$

Since $\sin[g(x)] = x$, we have

$$g'(x) = \frac{1}{\sqrt{1 - x^2}}, -1 < x < 1.$$

41. pH $= 2 = -\log x$ means $\log x = -2$ so $x = 10^{-2}$. Rate of change of pH with hydrogen ion concentration is

$$\frac{d}{dx}\text{pH} = -\frac{d}{dx}(\log x) = \frac{-1}{x(\ln 10)} = -\frac{1}{(10^{-2})\ln 10} = -43.4$$

45. (a) Let $g(x) = ax^2 + bx + c$ be our quadratic and $f(x) = \ln x$. For the best approximation, we want to find a quadratic with the same value as $\ln x$ at $x = 1$ and the same first and second derivatives as $\ln x$ at $x = 1$. $g'(x) = 2ax + b, g''(x) = 2a, f'(x) = \frac{1}{x}, f''(x) = -\frac{1}{x^2}$.

$$g(1) = a(1)^2 + b(1) + c \quad f(1) = 0$$
$$g'(1) = 2a(1) + b \quad f'(1) = 1$$
$$g''(1) = 2a \quad f''(1) = -1$$

Thus, we obtain the equations

$$a + b + c = 0$$
$$2a + b = 1$$
$$2a = -1$$

We find $a = -\frac{1}{2}$, $b = 2$ and $c = -\frac{3}{2}$. Thus our approximation is:

$$g(x) = -\frac{1}{2}x^2 + 2x - \frac{3}{2}$$

(b) From the graph below, we notice that around $x = 1$, the value of $f(x) = \ln x$ and the value of $g(x) = -\frac{1}{2}x^2 + 2x - \frac{3}{2}$ are very close.

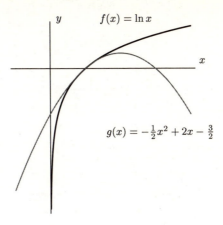

(c) $g(1.1) = 0.095 \quad g(2) = 0.5$
Compare with $f(1.1) = 0.0953$, $f(2) = 0.693$.

49. If V is the volume of the balloon and r is its radius, then

$$V = \frac{4}{3}\pi r^3.$$

We want to know the rate at which air is being blown into the balloon, which is the rate at which the volume is increasing, dV/dt. We are told that

$$\frac{dr}{dt} = 2 \text{ cm/sec} \quad \text{when} \quad r = 10 \text{ cm}.$$

Using the chain rule, we have

$$\frac{dV}{dt} = \frac{dV}{dr} \cdot \frac{dr}{dt} = 4\pi r^2 \frac{dr}{dt}.$$

Substituting gives

$$\frac{dV}{dt} = 4\pi(10)^2 2 = 800\pi = 2513.3 \text{ cm}^3/\text{sec}.$$

53. (a) Using Pythagoras' theorem, we see

$$z^2 = 0.5^2 + x^2$$

so

$$z = \sqrt{0.25 + x^2}.$$

(b) We want to calculate dz/dt. Using the chain rule, we have

$$\frac{dz}{dt} = \frac{dz}{dx} \cdot \frac{dx}{dt} = \frac{2x}{2\sqrt{0.25 + x^2}} \frac{dx}{dt}.$$

Because the train is moving at 0.8 km/hr, we know that

$$\frac{dx}{dt} = 0.8 \text{ km/hr}.$$

At the moment we are interested in $z = 1$ km so

$$1^2 = 0.25 + x^2$$

giving

$$x = \sqrt{0.75} = 0.866 \text{ km}.$$

Therefore

$$\frac{dz}{dt} = \frac{2(0.866)}{2\sqrt{0.25 + 0.75}} \cdot 0.8 = 0.866 \cdot 0.8 = 0.693 \text{ km/min}.$$

(c) We want to know $d\theta/dt$, where θ is as shown in Figure 3.2. Since

$$\frac{x}{0.5} = \tan\theta$$

we know

$$\theta = \arctan\left(\frac{x}{0.5}\right),$$

so

$$\frac{d\theta}{dt} = \frac{1}{1 + (x/0.5)^2} \cdot \frac{1}{0.5}\frac{dx}{dt}.$$

We know that $dx/dt = 0.8$ km/min and, at the moment we are interested in, $x = \sqrt{0.75}$. Substituting gives

$$\frac{d\theta}{dt} = \frac{1}{1 + 0.75/0.25} \cdot \frac{1}{0.5} \cdot 0.8 = 0.4 \text{ radians/min}.$$

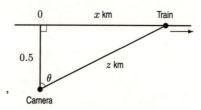

Figure 3.2

Solutions for Section 3.7

Exercises

1. We differentiate implicitly both sides of the equation with respect to x.

$$2x + 2y\frac{dy}{dx} = 0,$$

$$\frac{dy}{dx} = -\frac{2x}{2y} = -\frac{x}{y}.$$

5. We differentiate implicity with respect to x.

$$y + x\frac{dy}{dx} - 1 - \frac{3dy}{dx} = 0$$

$$(x-3)\frac{dy}{dx} = 1 - y$$

$$\frac{dy}{dx} = \frac{1-y}{x-3}$$

9. We differentiate implicitly both sides of the equation with respect to x.

$$e^{x^2} + \ln y = 0$$

$$2xe^{x^2} + \frac{1}{y}\frac{dy}{dx} = 0$$

$$\frac{dy}{dx} = -2xye^{x^2}.$$

13. $\dfrac{2}{3}x^{-1/3} + \dfrac{2}{3}y^{-1/3} \cdot \dfrac{dy}{dx} = 0$, $\dfrac{dy}{dx} = -\dfrac{x^{-1/3}}{y^{-1/3}} = -\dfrac{y^{1/3}}{x^{1/3}}$.

17. Differentiating $\sin(xy) = x$ with respect to x gives

$$(y + xy')\cos(xy) = 1$$

or

$$xy'\cos(xy) = 1 - y\cos(xy)$$

so that

$$y' = \frac{1 - y\cos(xy)}{x\cos(xy)}.$$

As we move along the curve to the point $(1, \frac{\pi}{2})$, the value of $dy/dx \to \infty$, which tells us the tangent to the curve at $(1, \frac{\pi}{2})$ has infinite slope; the tangent is the vertical line $x = 1$.

21. First we must find the slope of the tangent, $\dfrac{dy}{dx}$, at $(1, e^2)$. Differentiating implicitly, we have:

$$\frac{1}{xy}\left(x\frac{dy}{dx} + y\right) = 2$$

$$\frac{dy}{dx} = \frac{2xy - y}{x}.$$

Evaluating dy/dx at $(1, e^2)$ yields $(2(1)e^2 - e^2)/1 = e^2$. Using the point-slope formula for the equation of the line, we have:

$$y - e^2 = e^2(x - 1),$$

or

$$y = e^2 x.$$

Problems

25. **(a)** Taking derivatives implicitly, we get

$$\frac{2}{25}x + \frac{2}{9}y\frac{dy}{dx} = 0$$

$$\frac{dy}{dx} = \frac{-9x}{25y}$$

(b) The slope is not defined anywhere along the line $y = 0$. This ellipse intersects that line in two places, $(-5, 0)$ and $(5, 0)$. (These are, of course, the "ends" of the ellipse where the tangent is vertical.)

29. Let the point of intersection of the tangent line with the smaller circle be (x_1, y_1) and the point of intersection with the larger be (x_2, y_2). Let the tangent line be $y = mx + c$. Then at (x_1, y_1) and (x_2, y_2) the slopes of $x^2 + y^2 = 1$ and $y^2 + (x - 3)^2 = 4$ are also m. The slope of $x^2 + y^2 = 1$ is found by implicit differentiation: $2x + 2yy' = 0$ so $y' = -x/y$. Similarly, the slope of $y^2 + (x - 3)^2 = 4$ is $y' = -(x - 3)/y$. Thus,

$$m = \frac{y_2 - y_1}{x_2 - x_1} = -\frac{x_1}{y_1} = -\frac{(x_2 - 3)}{y_2},$$

where $y_1 = \sqrt{1 - x_1^2}$ and $y_2 = \sqrt{4 - (x_2 - 3)^2}$. The positive values for y_1 and y_2 follow from Figure 3.3 and from our choice of $m > 0$. We obtain

$$\frac{x_1}{\sqrt{1 - x_1^2}} = \frac{x_2 - 3}{\sqrt{4 - (x_2 - 3)^2}}$$

$$\frac{x_1^2}{1 - x_1^2} = \frac{(x_2 - 3)^2}{4 - (x_2 - 3)^2}$$

$$x_1^2[4 - (x_2 - 3)^2] = (1 - x_1^2)(x_2 - 3)^2$$

$$4x_1^2 - (x_1^2)(x_2 - 3)^2 = (x_2 - 3)^2 - x_1^2(x_2 - 3)^2$$

$$4x_1^2 = (x_2 - 3)^2$$

$$2|x_1| = |x_2 - 3|.$$

From the picture $x_1 < 0$ and $x_2 < 3$. This gives $x_2 = 2x_1 + 3$ and $y_2 = 2y_1$. From

$$\frac{y_2 - y_1}{x_2 - x_1} = -\frac{x_1}{y_1},$$

substituting $y_1 = \sqrt{1 - x_1^2}$, $y_2 = 2y_1$ and $x_2 = 2x_1 + 3$ gives

$$x_1 = -\frac{1}{3}.$$

From $x_2 = 2x_1 + 3$ we get $x_2 = 7/3$. In addition, $y_1 = \sqrt{1 - x_1^2}$ gives $y_1 = 2\sqrt{2}/3$, and finally $y_2 = 2y_1$ gives $y_2 = 4\sqrt{2}/3$.

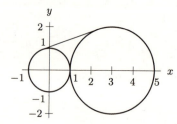

Figure 3.3

Solutions for Section 3.8

Exercises

1. Between times $t = 0$ and $t = 1$, x goes at a constant rate from 0 to 1 and y goes at a constant rate from 1 to 0. So the particle moves in a straight line from $(0, 1)$ to $(1, 0)$. Similarly, between times $t = 1$ and $t = 2$, it goes in a straight line to $(0, -1)$, then to $(-1, 0)$, then back to $(0, 1)$. So it traces out the diamond shown in Figure 3.4.

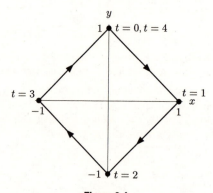

Figure 3.4

5. The particle moves clockwise: For $0 \leq t \leq \frac{\pi}{2}$, we have $x = \cos t$ decreasing and $y = -\sin t$ decreasing. Similarly, for the time intervals $\frac{\pi}{2} \leq t \leq \pi, \pi \leq t \leq \frac{3\pi}{2}$, and $\frac{3\pi}{2} \leq t \leq 2\pi$, we see that the particle moves clockwise.

9. Let $f(t) = \ln t$. Then $f'(t) = \frac{1}{t}$. The particle is moving counterclockwise when $f'(t) > 0$, that is, when $t > 0$. Any other time, when $t \leq 0$, the position is not defined.

13. One possible answer is $x = 2 + 5\cos t, y = 1 + 5\sin t, 0 \leq t \leq 2\pi$.

17. The parameterization $x = -3\cos t$, $y = 7\sin t$, $0 \leq t \leq 2\pi$, starts at the right point but sweeps out the ellipse in the wrong direction (the y-coordinate becomes positive as t increases). Thus, a possible parameterization is $x = -3\cos(-t) = -3\cos t, y = 7\sin(-t) = -7\sin t, 0 \leq t \leq 2\pi$.

21. We have $dx/dt = 2t$ and $dy/dt = 3t^2$. Therefore, the speed of the particle is

$$v = \sqrt{\left(\frac{dx}{dt}\right)^2 + \left(\frac{dy}{dt}\right)^2} = \sqrt{((2t)^2 + (3t^2)^2)} = |t| \cdot \sqrt{(4 + 9t^2)}.$$

The particle comes to a complete stop when its speed is 0, that is, if $t\sqrt{4 + 9t^2} = 0$, and so when $t = 0$.

25. At $t = 2$, the position is $(2^2, 2^3) = (4, 8)$, the velocity in the x-direction is $2 \cdot 2 = 4$, and the velocity in the y-direction is $3 \cdot 2^2 = 12$. So we want the line going through the point $(4, 8)$ at the time $t = 2$, with the given x- and y-velocities:

$$x = 4 + 4(t - 2), \quad y = 8 + 12(t - 2).$$

Problems

29. (a) The curve is a spiral as shown in Figure 3.5.

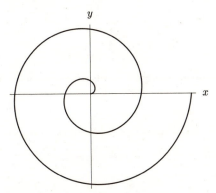

Figure 3.5: The spiral $x = t \cos t, y = t \sin t$
for $0 \leq t \leq 4\pi$

(b) At $t = 2$, the position is $(2 \cos 2, 2 \sin 2) = (-0.8323, 1.8186)$, and at $t = 2.01$ the position is $(2.01 \cos 2.01, 2.01 \sin 2.01) = (-0.8546, 1.8192)$. The distance between these points is

$$\sqrt{(-0.8546 - (-0.8323))^2 + (1.8192 - 1.8186)^2} \approx 0.022.$$

Thus the speed is approximately $0.022/0.01 \approx 2.2$.

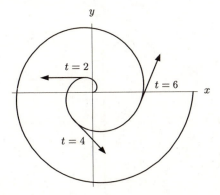

Figure 3.6: The spiral $x = t \cos t, y = t \sin t$ and three velocity vectors

(c) Evaluating the exact formula

$$v = \sqrt{(\cos t - t \sin t)^2 + (\sin t + t \cos t)^2}$$

gives :

$$v(2) = \sqrt{(-2.235)^2 + (0.077)^2} = 2.2363.$$

33. (a) C_1 has center at the origin and radius 5, so $a = b = 0, k = 5$ or -5.
    (b) C_2 has center at $(0, 5)$ and radius 5, so $a = 0, b = 5, k = 5$ or -5.
    (c) C_3 has center at $(10, -10)$, so $a = 10, b = -10$. The radius of C_3 is $\sqrt{10^2 + (-10)^2} = \sqrt{200}$, so $k = \sqrt{200}$ or $k = -\sqrt{200}$.

37. For $0 \le t \le 2\pi$

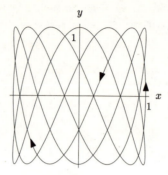

Solutions for Section 3.9

Exercises

1. With $f(x) = \sqrt{1 + x}$, the chain rule gives $f'(x) = 1/(2\sqrt{1 + x})$, so $f(0) = 1$ and $f'(0) = 1/2$. Therefore the tangent line approximation of f near $x = 0$,

$$f(x) \approx f(0) + f'(0)(x - 0),$$

becomes

$$\sqrt{1 + x} \approx 1 + \frac{x}{2}.$$

This means that, near $x = 0$, the function $\sqrt{1 + x}$ can be approximated by its tangent line $y = 1 + x/2$. (See Figure 3.7.)

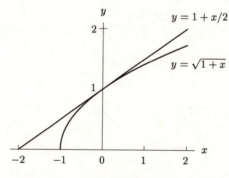

Figure 3.7

5. Let $f(x) = e^{-x}$. Then $f'(x) = -e^{-x}$. So $f(0) = 1, f'(0) = -e^0 = -1$. Therefore, $e^{-x} \approx f(0) + f'(0)x = 1 - x$.

Problems

9. (a) Let $f(x) = 1/(1 + x)$. Then $f'(x) = -1/(1 + x)^2$ by the chain rule. So $f(0) = 1$, and $f'(0) = -1$. Therefore, for x near 0, $1/(1 + x) \approx f(0) + f'(0)x = 1 - x$.
    (b) We know that for small y, $1/(1 + y) \approx 1 - y$. Let $y = x^2$; when x is small, so is $y = x^2$. Hence, for small x, $1/(1 + x^2) \approx 1 - x^2$.
    (c) Since the linearization of $1/(1 + x^2)$ is the line $y = 1$, and this line has a slope of 0, the derivative of $1/(1 + x^2)$ is zero at $x = 0$.

13. (a) Considering l as a constant, we have

$$T = f(g) = 2\pi\sqrt{\frac{l}{g}}.$$

Then,

$$f'(g) = 2\pi\sqrt{l}\left(-\frac{1}{2}g^{-3/2}\right) = -\pi\sqrt{\frac{l}{g^3}}.$$

Thus, local linearity gives

$$f(g + \Delta g) \approx f(g) - \pi\sqrt{\frac{l}{g^3}}(\Delta g).$$

Since $T = f(g)$ and $\Delta T = f(g + \Delta g) - f(g)$, we have

$$\Delta T \approx -\pi\sqrt{\frac{l}{g^3}}\Delta g = -2\pi\sqrt{\frac{l}{g}}\frac{\Delta g}{2g} = \frac{-T}{2}\frac{\Delta g}{g}.$$

$$\Delta T \approx \frac{-T}{2}\frac{\Delta g}{g}.$$

(b) If g increases by 1%, we know

$$\frac{\Delta g}{g} = 0.01.$$

Thus,

$$\frac{\Delta T}{T} \approx -\frac{1}{2}\frac{\Delta g}{g} = -\frac{1}{2}(0.01) = -0.005,$$

So, T decreases by 0.5%.

17. Note that

$$[f(g(x))]' = \lim_{h \to 0} \frac{f(g(x+h)) - f(g(x))}{h}.$$

Using the local linearizations of f and g, we get that

$$f(g(x+h)) - f(g(x)) \approx f\left(g(x) + g'(x)h\right) - f(g(x))$$
$$\approx f\left(g(x)\right) + f'(g(x))g'(x)h - f(g(x))$$
$$= f'(g(x))g'(x)h.$$

Therefore,

$$[f(g(x))]' = \lim_{h \to 0} \frac{f(g(x+h)) - f(g(x))}{h}$$
$$= \lim_{h \to 0} \frac{f'(g(x))g'(x)h}{h}$$
$$= \lim_{h \to 0} f'(g(x))g'(x) = f'(g(x))g'(x).$$

A more complete derivation can be given using the error term discussed in the section on Differentiability and Linear Approximation in Chapter 2. Adapting the notation of that section to this problem, we write

$$f(z+k) = f(z) + f'(z)k + E_f(k) \quad \text{and} \quad g(x+h) = g(x) + g'(x)h + E_g(h),$$

where $\lim_{h \to 0} \dfrac{E_g(h)}{h} = \lim_{k \to 0} \dfrac{E_f(k)}{k} = 0.$

Now we let $z = g(x)$ and $k = g(x+h) - g(x)$. Then we have $k = g'(x)h + E_g(h)$. Thus,

$$\frac{f(g(x+h)) - f(g(x))}{h} = \frac{f(z+k) - f(z)}{h}$$
$$= \frac{f(z) + f'(z)k + E_f(k) - f(z)}{h} = \frac{f'(z)k + E_f(k)}{h}$$
$$= \frac{f'(z)g'(x)h + f'(z)E_g(h)}{h} + \frac{E_f(k)}{k} \cdot \left(\frac{k}{h}\right)$$
$$= f'(z)g'(x) + \frac{f'(z)E_g(h)}{h} + \frac{E_f(k)}{k}\left[\frac{g'(x)h + E_g(h)}{h}\right]$$
$$= f'(z)g'(x) + \frac{f'(z)E_g(h)}{h} + \frac{g'(x)E_f(k)}{k} + \frac{E_g(h) \cdot E_f(k)}{h \cdot k}$$

Now, if $h \to 0$ then $k \to 0$ as well, and all the terms on the right except the first go to zero, leaving us with the term $f'(z)g'(x)$. Substituting $g(x)$ for z, we obtain

$$[f(g(x))]' = \lim_{h \to 0} \frac{f(g(x+h)) - f(g(x))}{h} = f'(g(x))g'(x).$$

Solutions for Section 3.10

Exercises

1. Since $f'(a) > 0$ and $g'(a) < 0$, l'Hopital's rule tells us that

$$\lim_{x \to a} \frac{f(x)}{g(x)} = \frac{f'(a)}{g'(a)} < 0.$$

5. The denominator approaches zero as x goes to zero and the numerator goes to zero even faster, so you should expect that the limit to be 0. You can check this by substituting several values of x close to zero. Alternatively, using l'Hopital's rule, we have

$$\lim_{x \to 0} \frac{x^2}{\sin x} = \lim_{x \to 0} \frac{2x}{\cos x} = 0.$$

9. The larger power dominates. Using l'Hopital's rule

$$\lim_{x \to \infty} \frac{x^5}{0.1x^7} = \lim_{x \to \infty} \frac{5x^4}{0.7x^6} = \lim_{x \to \infty} \frac{20x^3}{4.2x^5}$$

$$= \lim_{x \to \infty} \frac{60x^2}{21x^4} = \lim_{x \to \infty} \frac{120x}{84x^3} = \lim_{x \to \infty} \frac{120}{252x^2} = 0$$

so $0.1x^7$ dominates.

Problems

13. Let $f(x) = \ln x$ and $g(x) = 1/x$ so $f'(x) = 1/x$ and $g'(x) = -1/x^2$ and

$$\lim_{x \to 0+} \frac{\ln x}{1/x} = \lim_{x \to 0+} \frac{1/x}{-1/x^2} = \lim_{x \to 0+} \frac{x}{-1} = 0.$$

17. Let $f(x) = e^{-x}$ and $g(x) = \sin x$. Observe that as x increases, $f(x)$ approaches 0 but $g(x)$ oscillates between -1 and 1. Since $g(x)$ does not approach 0 in the limit, l'Hopital's rule does not apply. Because $g(x)$ is in the denominator and oscillates through 0 forever, the limit does not exist.

Solutions for Chapter 3 Review

Exercises

1. $f'(t) = \dfrac{d}{dt}\left(2te^t - \dfrac{1}{\sqrt{t}}\right) = 2e^t + 2te^t + \dfrac{1}{2t^{3/2}}.$

5. $g'(x) = \dfrac{d}{dx}\left(x^k + k^x\right) = kx^{k-1} + k^x \ln k.$

9. $s'(\theta) = \dfrac{d}{d\theta}\sin^2(3\theta - \pi) = 6\cos(3\theta - \pi)\sin(3\theta - \pi).$

13. $g'(x) = \dfrac{d}{dx}\left(x^{\frac{1}{2}} + x^{-1} + x^{-\frac{3}{2}}\right) = \dfrac{1}{2}x^{-\frac{1}{2}} - x^{-2} - \dfrac{3}{2}x^{-\frac{5}{2}}.$

17. Using the chain rule we get:

$$k'(\alpha) = e^{\tan(\sin \alpha)}(\tan(\sin \alpha))' = e^{\tan(\sin \alpha)} \cdot \frac{1}{\cos^2(\sin \alpha)} \cdot \cos \alpha.$$

21. $\dfrac{d}{dx}xe^{\tan x} = e^{\tan x} + xe^{\tan x}\dfrac{1}{\cos^2 x}.$

25. $\dfrac{dy}{dx} = (\ln 2)2^{\sin x}\cos x \cdot \cos x + 2^{\sin x}(-\sin x) = 2^{\sin x}\left((\ln 2)\cos^2 x - \sin x\right)$

29. Using the product rule and factoring gives $f'(t) = e^{-4kt}(\cos t - 4k\sin t)$.

33. Using the quotient rule gives

$$
\begin{aligned}
f'(s) &= \frac{-2s\sqrt{a^2 + s^2} - \dfrac{s}{\sqrt{a^2+s^2}}(a^2 - s^2)}{(a^2 + s^2)} \\[2mm]
&= \frac{-2s(a^2 + s^2) - s(a^2 - s^2)}{(a^2 + s^2)^{3/2}} \\[2mm]
&= \frac{-2a^2 s - 2s^3 - a^2 s + s^3}{(a^2 + s^2)^{3/2}} \\[2mm]
&= \frac{-3a^2 s - s^3}{(a^2 + s^2)^{3/2}}.
\end{aligned}
$$

37. Using the chain rule gives $r'(t) = \dfrac{\cos(\frac{t}{k})}{\sin(\frac{t}{k})}\left(\dfrac{1}{k}\right).$

41. Using the quotient and chain rules, we have

$$
\begin{aligned}
\frac{dy}{dx} &= \frac{(ae^{ax} + ae^{-ax})(e^{ax} + e^{-ax}) - (e^{ax} - e^{-ax})(ae^{ax} - ae^{-ax})}{(e^{ax} + e^{-ax})^2} \\[2mm]
&= \frac{a(e^{ax} + e^{-ax})^2 - a(e^{ax} - e^{-ax})^2}{(e^{ax} + e^{-ax})^2} \\[2mm]
&= \frac{a[(e^{2ax} + 2 + e^{-2ax}) - (e^{2ax} - 2 + e^{-2ax})]}{(e^{ax} + e^{-ax})^2} \\[2mm]
&= \frac{4a}{(e^{ax} + e^{-ax})^2}
\end{aligned}
$$

45. Since $\tan(\arctan(k\theta)) = k\theta$, because tan and arctan are inverse functions, we have $N'(\theta) = k$.

49. Since $\cos^2 y + \sin^2 y = 1$, we have $s(y) = \sqrt[3]{1 + 3} = \sqrt[3]{4}$. Thus $s'(y) = 0$.

53. We wish to find the slope $m = dy/dx$. To do this, we can implicitly differentiate the given formula in terms of x:

$$
\begin{aligned}
x^2 + 3y^2 &= 7 \\
2x + 6y\frac{dy}{dx} &= \frac{d}{dx}(7) = 0 \\
\frac{dy}{dx} &= \frac{-2x}{6y} = \frac{-x}{3y}.
\end{aligned}
$$

Thus, at $(2, -1)$, $m = -(2)/3(-1) = 2/3$.

57. Differentiating implicitly on both sides with respect to x,

$$
\begin{aligned}
a\cos(ay)\frac{dy}{dx} - b\sin(bx) &= y + x\frac{dy}{dx} \\
(a\cos(ay) - x)\frac{dy}{dx} &= y + b\sin(bx) \\
\frac{dy}{dx} &= \frac{y + b\sin(bx)}{a\cos(ay) - x}.
\end{aligned}
$$

Problems

61. (a) $H'(2) = r'(2)s(2) + r(2)s'(2) = -1 \cdot 1 + 4 \cdot 3 = 11.$

   (b) $H'(2) = \dfrac{r'(2)}{2\sqrt{r(2)}} = \dfrac{-1}{2\sqrt{4}} = -\dfrac{1}{4}.$

   (c) $H'(2) = r'(s(2))s'(2) = r'(1) \cdot 3$, but we don't know $r'(1)$.

   (d) $H'(2) = s'(r(2))r'(2) = s'(4)r'(2) = -3.$

65. It makes sense to define the angle between two curves to be the angle between their tangent lines. (The tangent lines are the best linear approximations to the curves). See Figure 3.8. The functions $\sin x$ and $\cos x$ are equal at $x = \frac{\pi}{4}$.

$$\text{For } f_1(x) = \sin x, \quad f_1'(\frac{\pi}{4}) = \cos(\frac{\pi}{4}) = \frac{\sqrt{2}}{2}$$

$$\text{For } f_2(x) = \cos x, \quad f_2'(\frac{\pi}{4}) = -\sin(\frac{\pi}{4}) = -\frac{\sqrt{2}}{2}.$$

Using the point $(\frac{\pi}{4}, \frac{\sqrt{2}}{2})$ for each tangent line we get $y = \frac{\sqrt{2}}{2}x + \frac{\sqrt{2}}{2}(1 - \frac{\pi}{4})$ and $y = -\frac{\sqrt{2}}{2}x + \frac{\sqrt{2}}{2}(1 + \frac{\pi}{4})$, respectively.

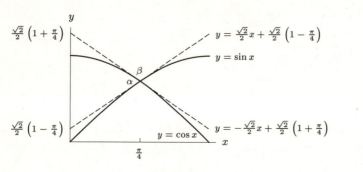

Figure 3.8

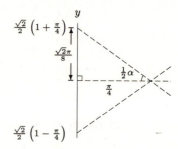

Figure 3.9

There are two possibilities of how to define the angle between the tangent lines, indicated by α and β above. The choice is arbitrary, so we will solve for both. To find the angle, α, we consider the triangle formed by these two lines and the y-axis. See Figure 3.9.

$$\tan\left(\frac{1}{2}\alpha\right) = \frac{\sqrt{2}\pi/8}{\pi/4} = \frac{\sqrt{2}}{2}$$

$$\frac{1}{2}\alpha = 0.61548 \text{ radians}$$

$$\alpha = 1.231 \text{ radians, or } 70.5°.$$

Now let us solve for β, the other possible measure of the angle between the two tangent lines. Since α and β are supplementary, $\beta = \pi - 1.231 = 1.909$ radians, or $109.4°$.

69. (a) $\frac{dg}{dr} = GM\frac{d}{dr}\left(\frac{1}{r^2}\right) = GM\frac{d}{dr}\left(r^{-2}\right) = GM(-2)r^{-3} = -\frac{2GM}{r^3}.$

(b) $\frac{dg}{dr}$ is the rate of change of acceleration due to the pull of gravity. The further away from the center of the earth, the weaker the pull of gravity is. So g is decreasing and therefore its derivative, $\frac{dg}{dr}$, is negative.

(c) By part (a),

$$\frac{dg}{dr}\bigg|_{r=6400} = -\frac{2GM}{r^3}\bigg|_{r=6400} = -\frac{2(6.67 \times 10^{-20})(6 \times 10^{24})}{(6400)^3} \approx -3.05 \times 10^{-6}.$$

(d) It is reasonable to assume that g is a constant near the surface of the earth.

73.

$$\frac{dy}{dt} = -7.5(0.507)\sin(0.507t) = -3.80\sin(0.507t)$$

(a) When $t = 6$, $\frac{dy}{dt} = -3.80\sin(0.507 \cdot 6) = -0.38$ meters/hour. So the tide is falling at 0.38 meters/hour.

(b) When $t = 9$, $\frac{dy}{dt} = -3.80\sin(0.507 \cdot 9) = 3.76$ meters/hour. So the tide is rising at 3.76 meters/hour.

(c) When $t = 12$, $\frac{dy}{dt} = -3.80\sin(0.507 \cdot 12) = 0.75$ meters/hour. So the tide is rising at 0.75 meters/hour.

(d) When $t = 18$, $\frac{dy}{dt} = -3.80\sin(0.507 \cdot 18) = -1.12$ meters/hour. So the tide is falling at 1.12 meters/hour.

77. Let r be the radius of the balloon. Then its volume, V, is

$$V = \frac{4}{3}\pi r^3.$$

We need to find the rate of change of V with respect to time, that is dV/dt. Since $V = V(r)$,

$$\frac{dV}{dr} = 4\pi r^2$$

so that by the chain rule,

$$\frac{dV}{dt} = \frac{dV}{dr}\frac{dr}{dt} = 4\pi r^2 \cdot 1.$$

When $r = 5$, $dV/dt = 100\pi$ cm^3/sec.

81. We want to find dP/dV. Solving $PV = k$ for P gives

$$P = k/V$$

so,

$$\frac{dP}{dV} = -\frac{k}{V^2}.$$

85. This problem can be solved by using either the quotient rule or the fact that

$$\frac{f'}{f} = \frac{d}{dx}(\ln f) \quad \text{and} \quad \frac{g'}{g} = \frac{d}{dx}(\ln g).$$

We use the second method. The relative rate of change of f/g is $(f/g)'/(f/g)$, so

$$\frac{(f/g)'}{f/g} = \frac{d}{dx}\ln\left(\frac{f}{g}\right) = \frac{d}{dx}(\ln f - \ln g) = \frac{d}{dx}(\ln f) - \frac{d}{dx}(\ln g) = \frac{f'}{f} - \frac{g'}{g}.$$

Thus, the relative rate of change of f/g is the difference between the relative rates of change of f and of g.

CAS Challenge Problems

89. (a) A CAS gives $h'(t) = 0$
(b) By the chain rule

$$h'(t) = \frac{\frac{d}{dt}\left(1 - \frac{1}{t}\right)}{1 - \frac{1}{t}} + \frac{\frac{d}{dt}\left(\frac{t}{t-1}\right)}{\frac{t}{t-1}} = \frac{\frac{1}{t^2}}{\frac{t-1}{t}} + \frac{\frac{1}{t-1} - \frac{t}{(t-1)^2}}{\frac{t}{t-1}}$$

$$= \frac{1}{t^2 - t} + \frac{(t-1) - t}{t^2 - t} = \frac{1}{t^2 - t} + \frac{-1}{t^2 - t} = 0.$$

(c) The expression inside the first logarithm is $1 - (1/t) = (t-1)/t$. Using the property $\log A + \log B = \log(AB)$, we get

$$\ln\left(1 - \frac{1}{t}\right) + \ln\left(\frac{t}{t-1}\right) = \ln\left(\frac{t-1}{t}\right) + \ln\left(\frac{t}{t-1}\right)$$

$$= \ln\left(\frac{t-1}{t} \cdot \frac{t}{1-t}\right) = \ln 1 = 0.$$

Thus $h(t) = 0$, so $h'(t) = 0$ also.

CHECK YOUR UNDERSTANDING

1. True. Since $d(x^n)/dx = nx^{n-1}$, the derivative of a power function is a power function, so the derivative of a polynomial is a polynomial.

5. True. Since $f'(x)$ is the limit

$$f'(x) = \lim_{h \to 0} \frac{f(x+h) - f(x)}{h},$$

the function f must be defined for all x.

9. True; differentiating the equation with respect to x, we get

$$2y\frac{dy}{dx} + y + x\frac{dy}{dx} = 0.$$

Solving for dy/dx, we get that

$$\frac{dy}{dx} = \frac{-y}{2y + x}.$$

Thus dy/dx exists where $2y + x \neq 0$. Now if $2y + x = 0$, then $x = -2y$. Substituting for x in the original equation, $y^2 + xy - 1 = 0$, we get

$$y^2 - 2y^2 - 1 = 0.$$

This simplifies to $y^2 + 1 = 0$, which has no solutions. Thus dy/dx exists everywhere.

13. False; the fourth derivative of $\cos t + C$, where C is any constant, is indeed $\cos t$. But any function of the form $\cos t + p(t)$, where $p(t)$ is a polynomial of degree less than or equal to 3, also has its fourth derivative equal to $\cos t$. So $\cos t + t^2$ will work.

17. False; for example, if both f and g are constant functions, then the derivative of $f(g(x))$ is zero, as is the derivative of $f(x)$. Another example is $f(x) = 5x + 7$ and $g(x) = x + 2$.

21. False. Let $f(x) = e^{-x}$ and $g(x) = x^2$. Let $h(x) = f(g(x)) = e^{-x^2}$. Then $h'(x) = -2xe^{-x^2}$ and $h''(x) = (-2 + 4x^2)e^{-x^2}$. Since $h''(0) < 0$, clearly h is not concave up for all x.

CHAPTER FOUR

Solutions for Section 4.1

Exercises

1. We sketch a graph which is horizontal at the two critical points. One possibility is shown in Figure 4.1.

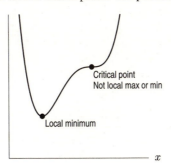

Figure 4.1

5. (a) A graph of $f(x) = e^{-x^2}$ is shown in Figure 4.2. It appears to have one critical point, at $x = 0$, and two inflection points, one between 0 and 1 and the other between 0 and -1.

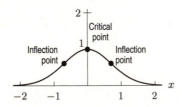

Figure 4.2

(b) To find the critical points, we set $f'(x) = 0$. Since $f'(x) = -2xe^{-x^2} = 0$, there is one solution, $x = 0$. The only critical point is at $x = 0$.

To find the inflection points, we first use the product rule to find $f''(x)$. We have

$$f''(x) = (-2x)(e^{-x^2}(-2x)) + (-2)(e^{-x^2}) = 4x^2e^{-x^2} - 2e^{-x^2}.$$

We set $f''(x) = 0$ and solve for x by factoring:

$$4x^2e^{-x^2} - 2e^{-x^2} = 0$$
$$(4x^2 - 2)e^{-x^2} = 0.$$

Since e^{-x^2} is never zero, we have

$$4x^2 - 2 = 0$$
$$x^2 = \frac{1}{2}$$
$$x = \pm 1/\sqrt{2}.$$

There are exactly two inflection points, at $x = 1/\sqrt{2} \approx 0.707$ and $x = -1/\sqrt{2} \approx -0.707$.

9. (a) A critical point occurs when $f'(x) = 0$. Since $f'(x)$ changes sign between $x = 2$ and $x = 3$, between $x = 6$ and $x = 7$, and between $x = 9$ and $x = 10$, we expect critical points at around $x = 2.5$, $x = 6.5$, and $x = 9.5$.

(b) Since $f'(x)$ goes from positive to negative at $x \approx 2.5$, a local maximum should occur there. Similarly, $x \approx 6.5$ is a local minimum and $x \approx 9.5$ a local maximum.

13.

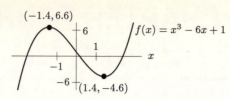

The graph of f above appears to be increasing for $x < -1.4$, decreasing for $-1.4 < x < 1.4$, and increasing for $x > 1.4$. There is a local maximum near $x = -1.4$ and local minimum near $x = 1.4$. The derivative of f is $f'(x) = 3x^2 - 6$. Thus $f'(x) = 0$ when $x^2 = 2$, that is $x = \pm\sqrt{2}$. This explains the critical points near $x = \pm 1.4$. Since $f'(x)$ changes from positive to negative at $x = -\sqrt{2}$, and from negative to positive at $x = \sqrt{2}$, there is a local maximum at $x = -\sqrt{2}$ and a local minimum at $x = \sqrt{2}$.

17.

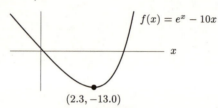

The graph of f above appears to be decreasing for $x < 2.3$ (almost like a straight line for $x < 0$), and increasing sharply for $x > 2.3$. Here $f'(x) = e^x - 10$, so $f'(x) = 0$ when $e^x = 10$, that is $x = \ln 10 = 2.302...$ This is the only place where $f'(x)$ changes sign, and it is a minimum of f. Notice that e^x is small for $x < 0$ so $f'(x) \approx -10$ for $x < 0$, which means the graph looks like a straight line of slope -10 for $x < 0$. However, e^x gets large quickly for $x > 0$, so $f'(x)$ gets large quickly for $x > \ln 10$, meaning the graph increases sharply there.

21. (13) The graph of $f(x) = x^3 - 6x + 1$ appears to be concave up for $x > 0$ and concave down for $x < 0$, with a point of inflection at $x = 0$. This is because $f''(x) = 6x$ is negative for $x < 0$ and positive for $x > 0$.

(14) Same answer as number 13.

(15) There appear to be three points of inflection at about $x = \pm 0.7$ and $x = 0$. This is because $f''(x) = 60x^3 - 30x = 30x(2x^2 - 1)$, which changes sign at $x = 0$ and $x = \pm 1/\sqrt{2}$.

(16) There appear to be points of inflection equally spaced about 3 units apart. This is because $f''(x) = -2\sin x$, which changes sign at $x = 0, \pm\pi, \pm 2\pi, \dots$.

(17) The graph appears to be concave up for all x. This is because $f''(x) = e^x > 0$ for all x.

(18) The graph appears to be concave up for all $x > 0$, and has almost periodic changes in concavity for $x < 0$. This is because for $x > 0$, $f''(x) = e^x - \sin x > 0$, and for $x < 0$, since e^x is small, $f''(x)$ changes sign at approximately the same values of x as $\sin x$.

(19) There appears to be a point of inflection for some $x < -0.71$, for $x = 0$, and for some $x > 0.71$. This is because $f'(x) = e^{-x^2}(1 - 2x^2)$ so

$$f''(x) = e^{-x^2}(-4x) + (1 - 2x^2)e^{-x^2}(-2x)$$
$$= e^{-x^2}(4x^3 - 6x).$$

Since $e^{-x^2} > 0$, this means $f''(x)$ has the same sign as $(4x^3 - 6x) = 2x(2x^2 - 3)$. Thus $f''(x)$ changes sign at $x = 0$ and $x = \pm\sqrt{3/2} \approx \pm 1.22$.

(20) The graph appears to be concave up for all x. This is because $f'(x) = 1 + \ln x$, so $f''(x) = 1/x$, which is greater than 0 for all $x > 0$.

Problems

25. We wish to have $f'(3) = 0$. Differentiating to find $f'(x)$ and then solving $f'(3) = 0$ for a gives:

$$f'(x) = x(ae^{ax}) + 1(e^{ax}) = e^{ax}(ax + 1)$$
$$f'(3) = e^{3a}(3a + 1) = 0$$
$$3a + 1 = 0$$
$$a = -\frac{1}{3}.$$

Thus, $f(x) = xe^{-x/3}$.

29. (a) Since the volume of water in the container is proportional to its depth, and the volume is increasing at a constant rate,

$$d(t) = \text{Depth at time } t = Kt,$$

where K is some positive constant. So the graph is linear, as shown in Figure 4.3. Since initially no water is in the container, we have $d(0) = 0$, and the graph starts from the origin.

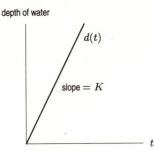

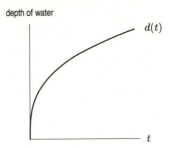

Figure 4.3

Figure 4.4

(b) As time increases, the additional volume needed to raise the water level by a fixed amount increases. Thus, although the depth, $d(t)$, of water in the cone at time t, continues to increase, it does so more and more slowly. This means $d'(t)$ is positive but decreasing, i.e., $d(t)$ is concave down. See Figure 4.4.

33. (a)   **(b)**

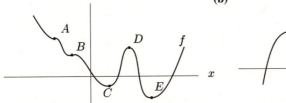

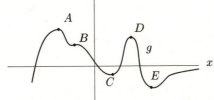

37. A has zeros where B has maxima and minima, so A could be a derivative of B. This is confirmed by comparing intervals on which B is increasing and A is positive. (They are the same.) So, C is either the derivative of A or the derivative of C is B. However, B does not have a zero at the point where C has a minimum, so B cannot be the derivative of C. Therefore, C is the derivative of A. So $B = f$, $A = f'$, and $C = f''$.

41.

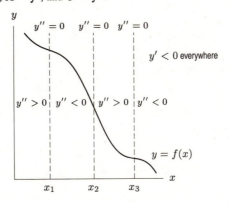

45. (a) Since $f''(x) > 0$ and $g''(x) > 0$ for all x, then $f''(x) + g''(x) > 0$ for all x, so $f(x) + g(x)$ is concave up for all x.
   (b) Nothing can be concluded about the concavity of $(f + g)(x)$. For example, if $f(x) = ax^2$ and $g(x) = bx^2$ with $a > 0$ and $b < 0$, then $(f + g)''(x) = a + b$. So $f + g$ is either always concave up, always concave down, or a straight line, depending on whether $a > |b|$, $a < |b|$, or $a = |b|$. More generally, it is even possible that $(f + g)(x)$ may have one or more changes in concavity.
   (c) It is possible to have infinitely many changes in concavity. Consider $f(x) = x^2 + \cos x$ and $g(x) = -x^2$. Since $f''(x) = 2 - \cos x$, we see that $f(x)$ is concave up for all x. Clearly $g(x)$ is concave down for all x. However, $f(x) + g(x) = \cos x$, which changes concavity an infinite number of times.

Solutions for Section 4.2

Exercises

1. We want a function of the form $y = a(x - h)^2 + k$, with $a < 0$ because the parabola opens downward. Since (h, k) is the vertex, we must take $h = 2$, $k = 5$, but we can take any negative value of a. Figure 4.5 shows the graph with $a = -1$, namely $y = -(x - 2)^2 + 5$.

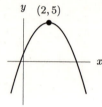

Figure 4.5: Graph of $y = -(x - 2)^2 + 5$

5. Since the vertical asymptote is $x = 2$, we have $b = -2$. The fact that the horizontal asymptote is $y = -5$ gives $a = -5$. So

$$y = \frac{-5x}{x - 2}.$$

9. Since the graph of the quartic polynomial is symmetric about the y-axis, the quartic must have only even powers and be of the form

$$y = ax^4 + bx^2 + c.$$

The y-intercept is 3, so $c = 3$. Differentiating gives

$$\frac{dy}{dx} = 4ax^3 + 2bx.$$

Since there is a maximum at $(1, 4)$, we have $dy/dx = 0$ if $x = 1$, so

$$4a(1)^3 + 2b(1) = 4a + 2b = 0 \qquad \text{so} \qquad b = -2a.$$

The fact that $dy/dx = 0$ if $x = -1$ gives us the same relationship

$$-4a - 2b = 0 \qquad \text{so} \qquad b = -2a.$$

We also know that $y = 4$ if $x = \pm 1$, so

$$a(1)^4 + b(1)^2 + 3 = a + b + 3 = 4 \qquad \text{so} \qquad a + b = 1.$$

Solving for a and b gives

$$a - 2a = 1 \qquad \text{so} \qquad a = -1 \text{ and } b = 2.$$

Finding d^2y/dx^2 so that we can check that $x = \pm 1$ are maxima, not minima, we see

$$\frac{d^2y}{dx^2} = 12ax^2 + 2b = -12x^2 + 4.$$

Thus $\dfrac{d^2y}{dx^2} = -8 < 0$ for $x = \pm 1$, so $x = \pm 1$ are maxima. See Figure 4.6.

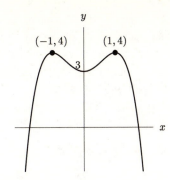

Figure 4.6: Graph of $y = -x^4 + 2x^2 + 3$

Problems

13. (a) We have $p'(x) = 3x^2 - a$, so

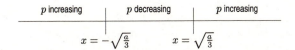

Local maximum: $p(-\sqrt{\frac{a}{3}}) = \dfrac{-a\sqrt{a}}{\sqrt{27}} + \dfrac{a\sqrt{a}}{\sqrt{3}} = +\dfrac{2a\sqrt{a}}{3\sqrt{3}}$

Local minimum: $p(\sqrt{\frac{a}{3}}) = -p(-\sqrt{\frac{a}{3}}) = -\dfrac{2a\sqrt{a}}{3\sqrt{3}}$

(b) Increasing the value of a moves the critical points of p away from the y-axis, and moves the critical values away from the x-axis. Thus, the "bumps" get further apart and higher. At the same time, increasing the value of a spreads the zeros of p further apart (while leaving the one at the origin fixed).

(c) See Figure 4.7

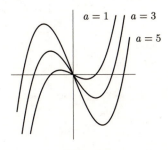

Figure 4.7

17. We begin by finding the intercepts, which occur where $f(x) = 0$, that is

$$x - k\sqrt{x} = 0$$
$$\sqrt{x}(\sqrt{x} - k) = 0$$

so $\quad x = 0 \quad$ or $\quad \sqrt{x} = k, \quad x = k^2.$

So 0 and k^2 are the x-intercepts. Now we find the location of the critical points by setting $f'(x)$ equal to 0:

$$f'(x) = 1 - k\left(\frac{1}{2}x^{-(1/2)}\right) = 1 - \frac{k}{2\sqrt{x}} = 0.$$

This means

$$1 = \frac{k}{2\sqrt{x}}, \quad \text{so} \quad \sqrt{x} = \frac{1}{2}k, \quad \text{and} \quad x = \frac{1}{4}k^2.$$

We can use the second derivative to verify that $x = \frac{k^2}{4}$ is a local minimum. $f''(x) = 1 + \frac{k}{4x^{3/2}}$ is positive for all $x > 0$. So the critical point, $x = \frac{1}{4}k^2$, is 1/4 of the way between the x-intercepts, $x = 0$ and $x = k^2$. Since $f''(x) = \frac{1}{4}kx^{-3/2}$, $f''(\frac{1}{4}k^2) = 2/k^2 > 0$, this critical point is a minimum.

21. Since $f'(x) = abe^{-bx}$, we have $f'(x) > 0$ for all x. Therefore, f is increasing for all x. Since $f''(x) = -ab^2e^{-bx}$, we have $f''(x) < 0$ for all x. Therefore, f is concave down for all x.

25. (a) Figure 4.8 suggests that each graph decreases to a local minimum and then increases sharply. The local minimum appears to move to the right as k increases. It appears to move up until $k = 1$, and then to move back down.

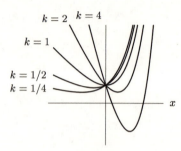

Figure 4.8

(b) $f'(x) = e^x - k = 0$ when $x = \ln k$. Since $f'(x) < 0$ for $x < \ln k$ and $f'(x) > 0$ for $x > \ln k$, f is decreasing to the left of $x = \ln k$ and increasing to the right, so f reaches a local minimum at $x = \ln k$.

(c) The minimum value of f is

$$f(\ln k) = e^{\ln k} - k(\ln k) = k - k\ln k.$$

Since we want to maximize the expression $k - k\ln k$, we can imagine a function $g(k) = k - k\ln k$. To maximize this function we simply take its derivative and find the critical points. Differentiating, we obtain

$$g'(k) = 1 - \ln k - k(1/k) = -\ln k.$$

Thus $g'(k) = 0$ when $k = 1$, $g'(k) > 0$ for $k < 1$, and $g'(k) < 0$ for $k > 1$. Thus $k = 1$ is a local maximum for $g(k)$. That is, the largest global minimum for f occurs when $k = 1$.

29.

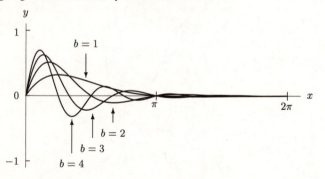

The larger the value of b, the narrower the humps and more humps per given region there are in the graph.

33. (a) The force is zero where

$$f(r) = -\frac{A}{r^2} + \frac{B}{r^3} = 0$$
$$Ar^3 = Br^2$$
$$r = \frac{B}{A}.$$

The vertical asymptote is $r = 0$ and the horizontal asymptote is the r-axis.

(b) To find critical points, we differentiate and set $f'(r) = 0$:

$$f'(r) = \frac{2A}{r^3} - \frac{3B}{r^4} = 0$$
$$2Ar^4 = 3Br^3$$
$$r = \frac{3B}{2A}.$$

Thus, $r = 3B/(2A)$ is the only critical point. Since $f'(r) < 0$ for $r < 3B/(2A)$ and $f'(r) > 0$ for $r > 3B/(2A)$, we see that $r = 3B/(2A)$ is a local minimum. At that point,

$$f\left(\frac{3B}{2A}\right) = -\frac{A}{9B^2/4A^2} + \frac{B}{27B^3/8A^3} = -\frac{4A^3}{27B^2}.$$

Differentiating again, we have

$$f''(r) = -\frac{6A}{r^4} + \frac{12B}{r^5} = -\frac{6}{r^5}(Ar - 2B).$$

So $f''(r) < 0$ where $r > 2B/A$ and $f''(r) > 0$ when $r < 2B/A$. Thus, $r = 2B/A$ is the only point of inflection. At that point

$$f\left(\frac{2B}{A}\right) = -\frac{A}{4B^2/A^2} + \frac{B}{8B^3/A^3} = -\frac{A^3}{8B^2}.$$

(c)

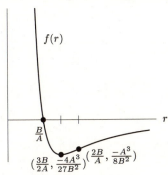

(d) (i) Increasing B means that the r-values of the zero, the minimum, and the inflection point increase, while the $f(r)$ values of the minimum and the point of inflection decrease in magnitude. See Figure 4.9.

(ii) Increasing A means that the r-values of the zero, the minimum, and the point of inflection decrease, while the $f(r)$ values of the minimum and the point of inflection increase in magnitude. See Figure 4.10.

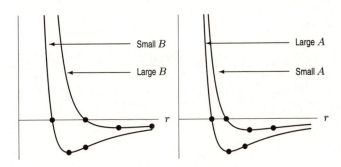

Figure 4.9: Increasing B **Figure 4.10**: Increasing A

Solutions for Section 4.3

Exercises

1.

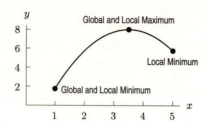

5. (a) Differentiating

$$f(x) = \sin^2 x - \cos x \quad \text{for } 0 \leq x \leq \pi$$
$$f'(x) = 2\sin x \cos x + \sin x = (\sin x)(2\cos x + 1)$$

$f'(x) = 0$ when $\sin x = 0$ or when $2\cos x + 1 = 0$. Now, $\sin x = 0$ when $x = 0$ or when $x = \pi$. On the other hand, $2\cos x + 1 = 0$ when $\cos x = -1/2$, which happens when $x = 2\pi/3$. So the critical points are $x = 0$, $x = 2\pi/3$, and $x = \pi$.

Note that $\sin x > 0$ for $0 < x < \pi$. Also, $2\cos x + 1 < 0$ if $2\pi/3 < x \leq \pi$ and $2\cos x + 1 > 0$ if $0 < x < 2\pi/3$. Therefore,

$$f'(x) < 0 \quad \text{for} \quad \frac{2\pi}{3} < x < \pi$$
$$f'(x) > 0 \quad \text{for} \quad 0 < x < \frac{2\pi}{3}.$$

Thus f has a local maximum at $x = 2\pi/3$ and local minima at $x = 0$ and $x = \pi$.

(b) We have

$$f(0) = [\sin(0)]^2 - \cos(0) = -1$$
$$f\left(\frac{2\pi}{3}\right) = \left[\sin\left(\frac{2\pi}{3}\right)\right]^2 - \cos\frac{2\pi}{3} = 1.25$$
$$f(\pi) = [\sin(\pi)]^2 - \cos(\pi) = 1.$$

Thus the global maximum is at $x = 2\pi/3$, and the global minimum is at $x = 0$.

Problems

9. We have that $v(r) = a(R - r)r^2 = aRr^2 - ar^3$, and $v'(r) = 2aRr - 3ar^2 = 2ar(R - \frac{3}{2}r)$, which is zero if $r = \frac{2}{3}R$, or if $r = 0$, and so $v(r)$ has critical points there.

$v''(r) = 2aR - 6ar$, and thus $v''(0) = 2aR > 0$, which by the second derivative test implies that v has a minimum at $r = 0$. $v''(\frac{2}{3}R) = 2aR - 4aR = -2aR < 0$, and so by the second derivative test v has a maximum at $r = \frac{2}{3}R$. In fact, this is a global max of $v(r)$ since $v(0) = 0$ and $v(R) = 0$ at the endpoints.

13. We set $f'(r) = 0$ to find the critical points:

$$\frac{2A}{r^3} - \frac{3B}{r^4} = 0$$
$$\frac{2Ar - 3B}{r^4} = 0$$
$$2Ar - 3B = 0$$
$$r = \frac{3B}{2A}.$$

The only critical point is at $r = 3B/(2A)$. If $r > 3B/(2A)$, we have $f' > 0$ and if $r < 3B/(2A)$, we have $f' < 0$. Thus, the force between the atoms is minimized at $r = 3B/(2A)$.

17. A graph of F against θ is shown below.

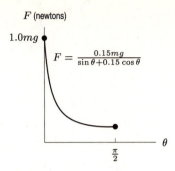

Taking the derivative:

$$\frac{dF}{d\theta} = -\frac{mg\mu(\cos\theta - \mu\sin\theta)}{(\sin\theta + \mu\cos\theta)^2}.$$

At a critical point, $dF/d\theta = 0$, so

$$\cos\theta - \mu\sin\theta = 0$$
$$\tan\theta = \frac{1}{\mu}$$
$$\theta = \arctan\left(\frac{1}{\mu}\right).$$

If $\mu = 0.15$, then $\theta = \arctan(1/0.15) = 1.422 \approx 81.5°$. To calculate the maximum and minimum values of F, we evaluate at this critical point and the endpoints:

$$\text{At } \theta = 0, \qquad F = \frac{0.15mg}{\sin 0 + 0.15\cos 0} = 1.0mg \text{ newtons.}$$

$$\text{At } \theta = 1.422, \ F = \frac{0.15mg}{\sin(1.422) + 0.15\cos(1.422)} = 0.148mg \text{ newtons.}$$

$$\text{At } \theta = \pi/2, \quad F = \frac{0.15mg}{\sin(\frac{\pi}{2}) + 0.15\cos(\frac{\pi}{2})} = 0.15mg \text{ newtons.}$$

Thus, the maximum value of F is $1.0mg$ newtons when $\theta = 0$ (her arm is vertical) and the minimum value of F is $0.148mg$ newtons is when $\theta = 1.422$ (her arm is close to horizontal). See Figure 4.11.

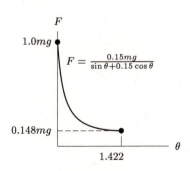

Figure 4.11

21. Let $y = e^{-x^2}$. Since $y' = -2xe^{-x^2}$, y is increasing for $x < 0$ and decreasing for $x > 0$. Hence $y = e^0 = 1$ is a global maximum.

When $x = \pm 0.3$, $y = e^{-0.09} \approx 0.9139$, which is a global minimum on the given interval. Thus $e^{-0.09} \leq y \leq 1$ for $|x| \leq 0.3$.

25. The graph of $y = x + \sin x$ in Figure 4.12 suggests that the function is nondecreasing over the entire interval. You can confirm this by looking at the derivative:

$$y' = 1 + \cos x$$

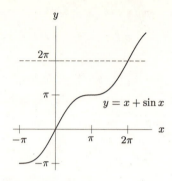

Figure 4.12: Graph of $y = x + \sin x$

Since $\cos x \geq -1$, we have $y' \geq 0$ everywhere, so y never decreases. This means that a lower bound for y is 0 (its value at the left endpoint of the interval) and an upper bound is 2π (its value at the right endpoint). That is, if $0 \leq x \leq 2\pi$:

$$0 \leq y \leq 2\pi.$$

These are the best bounds for y over the interval.

29. (a) At higher speeds, more energy is used so the graph rises to the right. The initial drop is explained by the fact that the energy it takes a bird to fly at very low speeds is greater than that needed to fly at a slightly higher speed. When it flies slightly faster, the amount of energy consumed decreases. But when it flies at very high speeds, the bird consumes a lot more energy (this is analogous to our swimming in a pool).

(b) $f(v)$ measures energy per second; $a(v)$ measures energy per meter. A bird traveling at rate v will in 1 second travel v meters, and thus will consume $v \cdot a(v)$ joules of energy in that 1 second period. Thus $v \cdot a(v)$ represents the energy consumption per second, and so $f(v) = v \cdot a(v)$.

(c) Since $v \cdot a(v) = f(v)$, $a(v) = f(v)/v$. But this ratio has the same value as the slope of a line passing from the origin through the point $(v, f(v))$ on the curve (see figure). Thus $a(v)$ is minimal when the slope of this line is minimal. To find the value of v minimizing $a(v)$, we solve $a'(v) = 0$. By the quotient rule,

$$a'(v) = \frac{vf'(v) - f(v)}{v^2}.$$

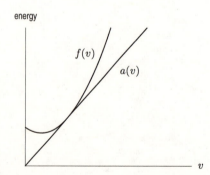

Thus $a'(v) = 0$ when $vf'(v) = f(v)$, or when $f'(v) = f(v)/v = a(v)$. Since $a(v)$ is represented by the slope of a line through the origin and a point on the curve, $a(v)$ is minimized when this line is tangent to $f(v)$, so that the slope $a(v)$ equals $f'(v)$.

(d) The bird should minimize $a(v)$ assuming it wants to go from one particular point to another, i.e. where the distance is set. Then minimizing $a(v)$ minimizes the total energy used for the flight.

33. Here is one possible graph of g:

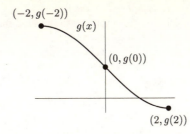

(a) From left to right, the graph of $g(x)$ starts "flat", decreases slowly at first then more rapidly, most rapidly at $x = 0$. The graph then continues to decrease but less and less rapidly until flat again at $x = 2$. The graph should exhibit symmetry about the point $(0, g(0))$.

(b) The graph has an inflection point at $(0, g(0))$ where the slope changes from negative and decreasing to negative and increasing.

(c) The function has a global maximum at $x = -2$ and a global minimum at $x = 2$.

(d) Since the function is decreasing over the interval $-2 \leq x \leq 2$

$$g(-2) = 5 > g(0) > g(2).$$

Since the function appears symmetric about $(0, g(0))$, we have

$$g(-2) - g(0) = g(0) - g(2).$$

Solutions for Section 4.4

Exercises

1. The fixed costs are $5000, the marginal cost per item is $2.40, and the price per item is $4.

5. Since fixed costs are represented by the vertical intercept, they are $1.1 million. The quantity that maximizes profit is about $q = 70$, and the profit achieved is $\$(3.7 - 2.5) = \$1.2$ million

Problems

9. (a) We know that Profit = Revenue − Cost, so differentiating with respect to q gives:

$$\text{Marginal Profit} = \text{Marginal Revenue} - \text{Marginal Cost}.$$

We see from the figure in the problem that just to the left of $q = a$, marginal revenue is less than marginal cost, so marginal profit is negative there. To the right of $q = a$ marginal revenue is greater than marginal cost, so marginal profit is positive there. At $q = a$ marginal profit changes from negative to positive. This means that profit is decreasing to the left of a and increasing to the right. The point $q = a$ corresponds to a local minimum of profit, and does not maximize profit. It would be a terrible idea for the company to set its production level at $q = a$.

(b) We see from the figure in the problem that just to the left of $q = b$ marginal revenue is greater than marginal cost, so marginal profit is positive there. Just to the right of $q = b$ marginal revenue is less than marginal cost, so marginal profit is negative there. At $q = b$ marginal profit changes from positive to negative. This means that profit is increasing to the left of b and decreasing to the right. The point $q = b$ corresponds to a local maximum of profit. In fact, since the area between the MC and MR curves in the figure in the text between $q = a$ and $q = b$ is bigger than the area between $q = 0$ and $q = a$, $q = b$ is in fact a global maximum.

13. (a) $N = 100 + 20x$, graphed in Figure 4.13.

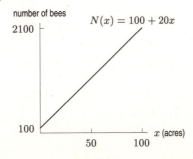

Figure 4.13

(b) $N'(x) = 20$ and its graph is just a horizontal line. This means that rate of increase of the number of bees with acres of clover is constant — each acre of clover brings 20 more bees.

On the other hand, $N(x)/x = 100/x + 20$ means that the average number of bees per acre of clover approaches 20 as more acres are put under clover. See Figure 4.14. As x increases, $100/x$ decreases to 0, so $N(x)/x$ approaches 20 (i.e. $N(x)/x \to 20$). Since the total number of bees is 20 per acre plus the original 100, the average number of bees per acre is 20 plus the 100 shared out over x acres. As x increases, the 100 are shared out over more acres, and so its contribution to the average becomes less. Thus the average number of bees per acre approaches 20 for large x.

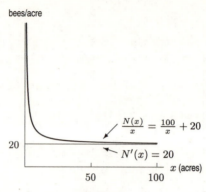

Figure 4.14

17. **(a)** Differentiating $C(q)$ gives

$$C'(q) = \frac{K}{a}q^{(1/a)-1}, \quad C''(q) = \frac{K}{a}\left(\frac{1}{a}-1\right)q^{(1/a)-2}.$$

If $a > 1$, then $C''(q) < 0$, so C is concave down.

(b) We have

$$a(q) = \frac{C(q)}{q} = \frac{Kq^{1/a} + F}{q}$$

$$C'(q) = \frac{K}{a}q^{(1/a)-1}$$

so $a(q) = C'(q)$ means

$$\frac{Kq^{1/a} + F}{q} = \frac{K}{a}q^{(1/a)-1}.$$

Solving,

$$Kq^{1/a} + F = \frac{K}{a}q^{1/a}$$

$$K\left(\frac{1}{a}-1\right)q^{1/a} = F$$

$$q = \left[\frac{Fa}{K(1-a)}\right]^a.$$

Solutions for Section 4.5

Exercises

1. We take the derivative, set it equal to 0, and solve for x:

$$\frac{dt}{dx} = \frac{1}{6} - \frac{1}{4}\cdot\frac{1}{2}\left((2000-x)^2 + 600^2\right)^{-1/2}\cdot 2(2000-x) = 0$$

$$(2000 - x) = \frac{2}{3}\left((2000 - x)^2 + 600^2\right)^{1/2}$$

$$(2000 - x)^2 = \frac{4}{9}\left((2000 - x)^2 + 600^2\right)$$

$$\frac{5}{9}(2000 - x)^2 = \frac{4}{9} \cdot 600^2$$

$$2000 - x = \sqrt{\frac{4}{5} \cdot 600^2} = \frac{1200}{\sqrt{5}}$$

$$x = 2000 - \frac{1200}{\sqrt{5}} \text{ feet.}$$

Note that $2000 - (1200/\sqrt{5}) \approx 1463$ feet, as given in the example.

Problems

5. (a) Suppose the height of the box is h. The box has six sides, four with area xh and two, the top and bottom, with area x^2. Thus,

$$4xh + 2x^2 = A.$$

So

$$h = \frac{A - 2x^2}{4x}.$$

Then, the volume, V, is given by

$$V = x^2 h = x^2 \left(\frac{A - 2x^2}{4x}\right) = \frac{x}{4}\left(A - 2x^2\right)$$

$$= \frac{A}{4}x - \frac{1}{2}x^3.$$

(b) The graph is shown in Figure 4.15. We are assuming A is a positive constant. Also, we have drawn the whole graph, but we should only consider $V > 0$, $x > 0$ as V and x are lengths.

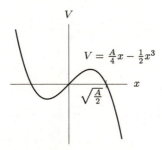

Figure 4.15

(c) To find the maximum, we differentiate, regarding A as a constant:

$$\frac{dV}{dx} = \frac{A}{4} - \frac{3}{2}x^2.$$

So $dV/dx = 0$ if

$$\frac{A}{4} - \frac{3}{2}x^2 = 0$$

$$x = \pm\sqrt{\frac{A}{6}}.$$

For a real box, we must use $x = \sqrt{A/6}$. Figure 4.15 makes it clear that this value of x gives the maximum. Evaluating at $x = \sqrt{A/6}$, we get

$$V = \frac{A}{4}\sqrt{\frac{A}{6}} - \frac{1}{2}\left(\sqrt{\frac{A}{6}}\right)^3 = \frac{A}{4}\sqrt{\frac{A}{6}} - \frac{1}{2} \cdot \frac{A}{6}\sqrt{\frac{A}{6}} = \left(\frac{A}{6}\right)^{3/2}.$$

9. Figure 4.16 shows the the pool has dimensions x by y and the deck extends 5 feet at either side and 10 feet at the ends of the pool.

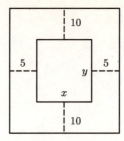

Figure 4.16

The dimensions of the plot of land containing the pool are then $(x + 5 + 5)$ by $(y + 10 + 10)$. The area of the land is then
$$A = (x + 10)(y + 20),$$
which is to be minimized. We also are told that the area of the pool is $xy = 1800$, so
$$y = 1800/x$$
and
$$A = (x + 10)\left(\frac{1800}{x} + 20\right)$$
$$= 1800 + 20x + \frac{18000}{x} + 200.$$

We find dA/dx and set it to zero to get
$$\frac{dA}{dx} = 20 - \frac{18000}{x^2} = 0$$
$$20x^2 = 18000$$
$$x^2 = 900$$
$$x = 30 \text{ feet.}$$

Since $A \to \infty$ as $x \to 0^+$ and as $x \to \infty$, this critical point must be a global minimum. Also, $y = 1800/30 = 60$ feet. The plot of land is therefore $(30 + 10) = 40$ by $(60 + 20) = 80$ feet.

13. Any point on the curve can be written (x, x^2). The distance between such a point and $(3, 0)$ is given by
$$s(x) = \sqrt{(3 - x)^2 + (0 - x^2)^2} = \sqrt{(3 - x)^2 + x^4}.$$

Plotting this function in Figure 4.17, we see that there is a minimum near $x = 1$.

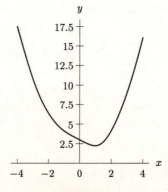

Figure 4.17

To find the value of x that minimizes the distance we can instead minimize the function $Q = s^2$ (the derivative is simpler). Then we have

$$Q(x) = (3 - x)^2 + x^4.$$

Differentiating $Q(x)$ gives

$$\frac{dQ}{dx} = -6 + 2x + 4x^3.$$

Plotting the function $4x^3 + 2x - 6$ shows that there is one real solution at $x = 1$, which can be verified by substitution; the required coordinates are therefore $(1, 1)$. Because $Q''(x) = 2 + 12x^2$ is always positive, $x = 1$ is indeed the minimum. See Figure 4.18.

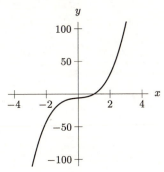

Figure 4.18

17. If v is the speed of the boat in miles per hour, then

$$\text{Cost of fuel per hour (in \$/hour)} = kv^3,$$

where k is the constant of proportionality. To find k, use the information that the boat uses $100 worth of fuel per hour when cruising at 10 miles per hour: $100 = k10^3$, so $k = 100/10^3 = 0.1$. Thus,

$$\text{Cost of fuel per hour (in \$/hour)} = 0.1v^3.$$

From the given information, we also have

$$\text{Cost of other operations (labor, maintenance, etc.) per hour (in \$/hour)} = 675.$$

So

$$\text{Total Cost per hour (in \$/hour)} = \text{Cost of fuel (in \$/hour) + Cost of other (in \$/hour)}$$
$$= 0.1v^3 + 675.$$

However, we want to find the Cost per *mile*, which is the Total Cost per *hour* divided by the number of miles that the ferry travels in one hour. Since v is the speed in miles/hour at which the ferry travels, the number of miles that the ferry travels in one hour is simply v miles. Let $C =$ Cost per *mile*. Then

$$\text{Cost per } mile \text{ (in \$/mile)} = \frac{\text{Total Cost per } hour \text{ (in \$/hour)}}{\text{Distance traveled per hour (in miles/hour)}}$$

$$C = \frac{0.1v^3 + 675}{v} = 0.1v^2 + \frac{675}{v}.$$

We also know that $0 < v < \infty$. To find the speed at which Cost per *mile* is minimized, set

$$\frac{dC}{dv} = 2(0.1)v - \frac{675}{v^2} = 0$$

so

$$2(0.1)v = \frac{675}{v^2}$$

$$v^3 = \frac{675}{2(0.1)} = 3375$$

$$v = 15 \text{ miles/hour.}$$

Since

$$\frac{d^2C}{dv^2} = 0.2 + \frac{2(675)}{v^3} > 0$$

for $v > 0$, $v = 15$ gives a local minimum for C by the second-derivative test. Since this is the only critical point for $0 < v < \infty$, it must give a global minimum.

21. Let x be as indicated in the figure in the text. Then the distance from S to Town 1 is $\sqrt{1 + x^2}$ and the distance from S to Town 2 is $\sqrt{(4 - x)^2 + 4^2} = \sqrt{x^2 - 8x + 32}$.

$$\text{Total length of pipe } = f(x) = \sqrt{1 + x^2} + \sqrt{x^2 - 8x + 32}.$$

We want to look for critical points of f. The easiest way is to graph f and see that it has a local minimum at about $x = 0.8$ miles. Alternatively, we can use the formula:

$$f'(x) = \frac{2x}{2\sqrt{1 + x^2}} + \frac{2x - 8}{2\sqrt{x^2 - 8x + 32}}$$

$$= \frac{x}{\sqrt{1 + x^2}} + \frac{x - 4}{\sqrt{x^2 - 8x + 32}}$$

$$= \frac{x\sqrt{x^2 - 8x + 32} + (x - 4)\sqrt{1 + x^2}}{\sqrt{1 + x^2}\sqrt{x^2 - 8x + 32}} = 0.$$

$f'(x)$ is equal to zero when the numerator is equal to zero.

$$x\sqrt{x^2 - 8x + 32} + (x - 4)\sqrt{1 + x^2} = 0$$

$$x\sqrt{x^2 - 8x + 32} = (4 - x)\sqrt{1 + x^2}.$$

Squaring both sides and simplifying, we get

$$x^2(x^2 - 8x + 32) = (x^2 - 8x + 16)(14x^2)$$

$$x^4 - 8x^3 + 32x^2 = x^4 - 8x^3 + 17x^2 - 8x + 16$$

$$15x^2 + 8x - 16 = 0,$$

$$(3x + 4)(5x - 4) = 0.$$

So $x = 4/5$. (Discard $x = -4/3$ since we are only interested in x between 0 and 4, between the two towns.) Using the second derivative test, we can verify that $x = 4/5$ is a local minimum.

25. (a) Since $RB' = x$ and $A'R = c - x$, we have

$$AR = \sqrt{a^2 + (c - x)^2} \quad \text{and} \quad RB = \sqrt{b^2 + x^2}.$$

See Figure 4.19.

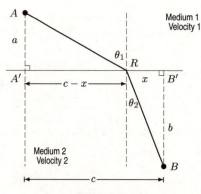

Figure 4.19

The time traveled, T, is given by

$$T = \text{Time } AR + \text{Time } RB = \frac{\text{Distance } AR}{v_1} + \frac{\text{Distance } RB}{v_2}$$

$$= \frac{\sqrt{a^2 + (c - x)^2}}{v_1} + \frac{\sqrt{b^2 + x^2}}{v_2}.$$

(b) Let us calculate dT/dx:

$$\frac{dT}{dx} = \frac{-2(c-x)}{2v_1\sqrt{a^2+(c-x)^2}} + \frac{2x}{2v_2\sqrt{b^2+x^2}}.$$

At the minimum $dT/dx = 0$, so

$$\frac{c-x}{v_1\sqrt{a^2+(c-x)^2}} = \frac{x}{v_2\sqrt{b^2+x^2}}.$$

But we have

$$\sin\theta_1 = \frac{c-x}{\sqrt{a^2+(c-x)^2}} \quad \text{and} \quad \sin\theta_2 = \frac{x}{\sqrt{b^2+x^2}}.$$

Therefore, setting $dT/dx = 0$ tells us that

$$\frac{\sin\theta_1}{v_1} = \frac{\sin\theta_2}{v_2}$$

which gives

$$\frac{\sin\theta_1}{\sin\theta_2} = \frac{v_1}{v_2}.$$

Solutions for Section 4.6

Exercises

1. Using the chain rule, $\dfrac{d}{dx}\left(\cosh(2x)\right) = (\sinh(2x)) \cdot 2 = 2\sinh(2x)$.

5. Using the chain rule,

$$\frac{d}{dt}\left(\cosh^2 t\right) = 2\cosh t \cdot \sinh t.$$

9. Substitute $x = 0$ into the formula for $\sinh x$. This yields

$$\sinh 0 = \frac{e^0 - e^{-0}}{2} = \frac{1-1}{2} = 0.$$

Problems

13. First we observe that

$$\sinh(2x) = \frac{e^{2x} - e^{-2x}}{2}.$$

Now let's calculate

$$\begin{aligned}
(\sinh x)(\cosh x) &= \left(\frac{e^x - e^{-x}}{2}\right)\left(\frac{e^x + e^{-x}}{2}\right) \\
&= \frac{(e^x)^2 - (e^{-x})^2}{4} \\
&= \frac{e^{2x} - e^{-2x}}{4} \\
&= \frac{1}{2}\sinh(2x).
\end{aligned}$$

Thus, we see that

$$\sinh(2x) = 2\sinh x \cosh x.$$

17. (a) The graph in Figure 4.20 looks like the graph of $y = \cosh x$, with the minimum at about $(0.5, 6.3)$.

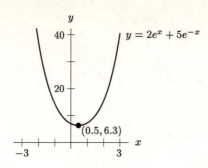

Figure 4.20

(b) We want to write

$$y = 2e^x + 5e^{-x} = A\cosh(x - c) = \frac{A}{2}e^{x-c} + \frac{A}{2}e^{-(x-c)}$$
$$= \frac{A}{2}e^x e^{-c} + \frac{A}{2}e^{-x}e^{c}$$
$$= \left(\frac{Ae^{-c}}{2}\right)e^x + \left(\frac{Ae^{c}}{2}\right)e^{-x}.$$

Thus, we need to choose A and c so that

$$\frac{Ae^{-c}}{2} = 2 \quad \text{and} \quad \frac{Ae^{c}}{2} = 5.$$

Dividing gives

$$\frac{Ae^{c}}{Ae^{-c}} = \frac{5}{2}$$
$$e^{2c} = 2.5$$
$$c = \frac{1}{2}\ln 2.5 \approx 0.458.$$

Solving for A gives

$$A = \frac{4}{e^{-c}} = 4e^{c} \approx 6.325.$$

Thus,

$$y = 6.325\cosh(x - 0.458).$$

Rewriting the function in this way shows that the graph in part (a) is the graph of $\cosh x$ shifted to the right by 0.458 and stretched vertically by a factor of 6.325.

21.

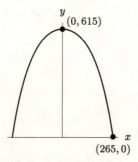

We know $x = 0$ and $y = 615$ at the top of the arch, so

$$615 = b - a\cosh(0/a) = b - a.$$

This means $b = a + 615$. We also know that $x = 265$ and $y = 0$ where the arch hits the ground, so

$$0 = b - a\cosh(265/a) = a + 615 - a\cosh(265/a).$$

We can solve this equation numerically on a calculator and get $a \approx 100$, which means $b \approx 715$. This results in the equation

$$y \approx 715 - 100\cosh\left(\frac{x}{100}\right).$$

Solutions for Section 4.7

Exercises

1. False. For example, if $f(x) = x^3$, then $f'(0) = 0$, so $x = 0$ is a critical point, but $x = 0$ is neither a local maximum nor a local minimum.

5. False. The horse that wins the race may have been moving faster for some, but not all, of the race. The Racetrack Principle guarantees the converse—that if the horses start at the same time and one moves faster throughout the race, then that horse wins.

9. No, it does not satisfy the hypotheses. The function does not appear to be differentiable. There appears to be no tangent line, and hence no derivative, at the "corner."

No, it does not satisfy the conclusion as there is no horizontal tangent.

Problems

13. Let $f(x) = \sin x$ and $g(x) = x$. Then $f(0) = 0$ and $g(0) = 0$. Also $f'(x) = \cos x$ and $g'(x) = 1$, so for all $x \geq 0$ we have $f'(x) \leq g'(x)$. So the graphs of f and g both go through the origin and the graph of f climbs slower than the graph of g. Thus the graph of f is below the graph of g for $x \geq 0$ by the Racetrack Principle. In other words, $\sin x \leq x$ for $x \geq 0$.

17. The Decreasing Function Theorem is: Suppose that f is continuous on $[a, b]$ and differentiable on (a, b). If $f'(x) < 0$ on (a, b), then f is decreasing on $[a, b]$. If $f'(x) \leq 0$ on (a, b), then f is nonincreasing on $[a, b]$.

To prove the theorem, we note that if f is decreasing then $-f$ is increasing and vice-versa. Similarly, if f is nonincreasing, then $-f$ is nondecreasing. Thus if $f'(x) < 0$, then $-f'(x) > 0$, so $-f$ is increasing, which means f is decreasing. And if $f'(x) \leq 0$, then $-f'(x) \geq 0$, so $-f$ is nondecreasing, which means f is nonincreasing.

21. By the Mean Value Theorem, Theorem 4.3, there is a number c, with $0 < c < 1$, such that

$$f'(c) = \frac{f(1) - f(0)}{1 - 0}.$$

Since $f(1) - f(0) > 0$, we have $f'(c) > 0$.

Alternatively if $f'(c) \leq 0$ for all c in $(0, 1)$, then by the Increasing Function Theorem, $f(0) \geq f(1)$.

25. If $f'(x) = 0$, then both $f'(x) \geq 0$ and $f'(x) \leq 0$. By the Increasing and Decreasing Function Theorems, f is both nondecreasing and nonincreasing, so f is constant.

29. (a) Since $f''(x) \geq 0$, $f'(x)$ is nondecreasing on (a, b). Thus $f'(c) \leq f'(x)$ for $c \leq x < b$ and $f'(x) \leq f'(c)$ for $a < x \leq c$.

(b) Let $g(x) = f(c) + f'(c)(x - c)$ and $h(x) = f(x)$. Then $g(c) = f(c) = h(c)$, and $g'(x) = f'(c)$ and $h'(x) = f'(x)$. If $c \leq x < b$, then $g'(x) \leq h'(x)$, and if $a < x \leq c$, then $g'(x) \geq h'(x)$, by (a). By the Racetrack Principle, $g(x) \leq h'(x)$ for $c \leq x < b$ and for $a < x \leq c$, as we wanted.

Solutions for Chapter 4 Review

Exercises

1.

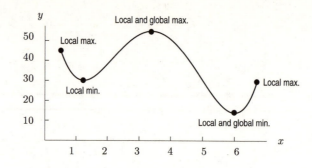

5. (a) Decreasing for $x < 0$, increasing for $0 < x < 4$, and decreasing for $x > 4$.

   (b) $f(0)$ is a local minimum, and $f(4)$ is a local maximum.

9. (a) First we find f' and f'':

$$f'(x) = -e^{-x}\sin x + e^{-x}\cos x$$
$$f''(x) = e^{-x}\sin x - e^{-x}\cos x$$
$$-e^{-x}\cos x - e^{-x}\sin x$$
$$= -2e^{-x}\cos x$$

   (b) The critical points are $x = \pi/4, 5\pi/4$, since $f'(x) = 0$ here.

   (c) The inflection points are $x = \pi/2, 3\pi/2$, since f'' changes sign at these points.

   (d) At the endpoints, $f(0) = 0$, $f(2\pi) = 0$. So we have $f(\pi/4) = (e^{-\pi/4})(\sqrt{2}/2)$ as the global maximum; $f(5\pi/4) = -e^{-5\pi/4}(\sqrt{2}/2)$ as the global minimum.

   (e)

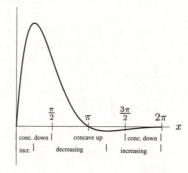

13. As $x \to -\infty$, $e^{-x} \to \infty$, so $xe^{-x} \to -\infty$. Thus $\lim_{x \to -\infty} xe^{-x} = -\infty$.

As $x \to \infty$, $\frac{x}{e^x} \to 0$, since e^x grows much more quickly than x. Thus $\lim_{x \to \infty} xe^{-x} = 0$.

Using the product rule,

$$f'(x) = e^{-x} - xe^{-x} = (1-x)e^{-x},$$

which is zero when $x = 1$, negative when $x > 1$, and positive when $x < 1$. Thus $f(1) = 1/e^1 = 1/e$ is a local maximum.

Again, using the product rule,

$$f''(x) = -e^{-x} - e^{-x} + xe^{-x}$$
$$= xe^{-x} - 2e^{-x}$$
$$= (x-2)e^{-x},$$

which is zero when $x = 2$, positive when $x > 2$, and negative when $x < 2$, giving an inflection point at $(2, \frac{2}{e^2})$. With the above, we have the following diagram:

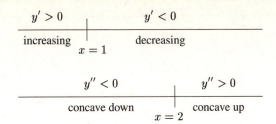

The graph of f is shown below.

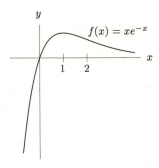

and $f(x)$ has one global maximum at $1/e$ and no local or global minima.

17. $\lim_{x \to +\infty} f(x) = +\infty$, $\lim_{x \to -\infty} f(x) = 0$.

$y = 0$ is the horizontal asymptote.

$f'(x) = 2xe^{5x} + 5x^2e^{5x} = xe^{5x}(5x + 2)$.

Thus, $x = -\frac{2}{5}$ and $x = 0$ are the critical points.

$$f''(x) = 2e^{5x} + 2xe^{5x} \cdot 5 + 10xe^{5x} + 25x^2e^{5x}$$
$$= e^{5x}(25x^2 + 20x + 2).$$

So, $x = \dfrac{-2 \pm \sqrt{2}}{5}$ are inflection points.

x		$\frac{-2-\sqrt{2}}{5}$		$-\frac{2}{5}$		$\frac{-2+\sqrt{2}}{5}$		0	
f'	+	+	+	0	−	−	−	0	+
f''	+	0	−	−	−	0	+	+	+
f	↗⌣		↗⌢		↘⌢		↘⌣		↗⌣

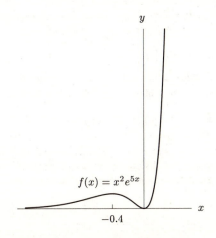

So, $f(-\frac{2}{5})$ is a local maximum; $f(0)$ is a local and global minimum.

Problems

21. We want the maximum value of $r(t) = ate^{-bt}$ to be 0.3 ml/sec and to occur at $t = 0.5$ sec. Differentiating gives

$$r'(t) = ae^{-bt} - abte^{-bt},$$

so $r'(t) = 0$ when

$$ae^{-bt}(1 - bt) = 0 \qquad \text{or} \qquad t = \frac{1}{b}.$$

Since the maximum occurs at $t = 0.5$, we have

$$\frac{1}{b} = 0.5 \qquad \text{so} \qquad b = 2.$$

Thus, $r(t) = ate^{-2t}$. The maximum value of r is given by

$$r(0.5) = a(0.5)e^{-2(0.5)} = 0.5ae^{-1}.$$

Since the maximum value of r is 0.3, we have

$$0.5ae^{-1} = 0.3 \qquad \text{so} \qquad a = \frac{0.3e}{0.5} = 1.63.$$

Thus, $r(t) = 1.63te^{-2t}$ ml/sec.

25. The local maxima and minima of f correspond to places where f' is zero and changes sign or, possibly, to the endpoints of intervals in the domain of f. The points at which f changes concavity correspond to local maxima and minima of f'. The change of sign of f', from positive to negative corresponds to a maximum of f and change of sign of f' from negative to positive corresponds to a minimum of f.

29. Since the volume is fixed at 200 ml (i.e. 200 cm^3), we can solve the volume expression for h in terms of r to get (with h and r in centimeters)

$$h = \frac{200 \cdot 3}{7\pi r^2}.$$

Using this expression in the surface area formula we arrive at

$$S = 3\pi r \sqrt{r^2 + \left(\frac{600}{7\pi r^2}\right)^2}$$

By plotting $S(r)$ we see that there is a minimum value near $r = 2.7$ cm.

33. Let $f(x) = x \sin x$. Then $f'(x) = x \cos x + \sin x$.

$f'(x) = 0$ when $x = 0$, $x \approx 2$, and $x \approx 5$. The latter two estimates we can get from the graph of $f'(x)$.

Zooming in (or using some other approximation method), we can find the zeros of $f'(x)$ with more precision. They are (approximately) 0, 2.029, and 4.913. We check the endpoints and critical points for the global maximum and minimum.

$$f(0) = 0, \qquad\qquad f(2\pi) = 0,$$
$$f(2.029) \approx 1.8197, \qquad f(4.914) \approx -4.814.$$

Thus for $0 \leq x \leq 2\pi$, $-4.81 \leq f(x) \leq 1.82$.

37. (a) We have $g'(t) = \frac{t(1/t) - \ln t}{t^2} = \frac{1 - \ln t}{t^2}$, which is zero if $t = e$, negative if $t > e$, and positive if $t < e$, since $\ln t$ is increasing. Thus $g(e) = \frac{1}{e}$ is a global maximum for g. Since $t = e$ was the only point at which $g'(t) = 0$, there is no minimum.

(b) Now $\ln t/t$ is increasing for $0 < t < e$, $\ln 1/1 = 0$, and $\ln 5/5 \approx 0.322 < \ln(e)/e$. Thus, for $1 < t < e$, $\ln t/t$ increases from 0 to above $\ln 5/5$, so there must be a t between 1 and e such that $\ln t/t = \ln 5/5$. For $t > e$, there is only one solution to $\ln t/t = \ln 5/5$, namely $t = 5$, since $\ln t/t$ is decreasing for $t > e$. For $0 < t < 1$, $\ln t/t$ is negative and so cannot equal $\ln 5/5$. Thus $\ln x/x = \ln t/t$ has exactly two solutions.

(c) The graph of $\ln t/t$ intersects the horizontal line $y = \ln 5/5$, at $x = 5$ and $x \approx 1.75$.

CAS Challenge Problems

41. (a) Since $k > 0$, we have $\lim\limits_{t \to \infty} e^{-kt} = 0$. Thus

$$\lim_{t \to \infty} P = \lim_{t \to \infty} \frac{L}{1 + Ce^{-kt}} = \frac{L}{1 + C \cdot 0} = L.$$

The constant L is called the carrying capacity of the environment because it represents the long-run population in the environment.

(b) Using a CAS, we find

$$\frac{d^2 P}{dt^2} = -\frac{LCk^2 e^{-kt}(1 - Ce^{-kt})}{(1 + Ce^{-kt})^3}.$$

Thus, $d^2 P/dt^2 = 0$ when

$$1 - Ce^{-kt} = 0$$
$$t = -\frac{\ln(1/C)}{k}.$$

Since e^{-kt} and $(1 + Ce^{-kt})$ are both always positive, the sign of $d^2 P/dt^2$ is negative when $(1 - Ce^{-kt}) > 0$, that is, for $t > -\ln(1/C)/k$. Similarly, the sign of $d^2 P/dt^2$ is positive when $(1 - Ce^{-kt}) < 0$, that is, for $t < -\ln(1/C)/k$. Thus, there is an inflection point at $t = -\ln(1/C)/k$.

For $t = -\ln(1/C)/k$,

$$P = \frac{L}{1 + Ce^{\ln(1/C)}} = \frac{L}{1 + C(1/C)} = \frac{L}{2}.$$

Thus, the inflection point occurs where $P = L/2$.

45. (a) Using a computer algebra system or differentiating by hand, we get

$$f'(x) = \frac{1}{2\sqrt{a+x}(\sqrt{a} + \sqrt{x})} - \frac{\sqrt{a+x}}{2\sqrt{x}(\sqrt{a} + \sqrt{x})^2}.$$

Simplifying gives

$$f'(x) = \frac{-a + \sqrt{a}\sqrt{x}}{2\left(\sqrt{a} + \sqrt{x}\right)^2 \sqrt{x}\sqrt{a+x}}.$$

The denominator of the derivative is always positive if $x > 0$, and the numerator is zero when $x = a$. Writing the numerator as $\sqrt{a}(\sqrt{x} - \sqrt{a})$, we see that the derivative changes from negative to positive at $x = a$. Thus, by the first derivative test, the function has a local minimum at $x = a$.

(b) As a increases, the local minimum moves to the right. See Figure 4.21. This is consistent with what we found in part (a), since the local minimum is at $x = a$.

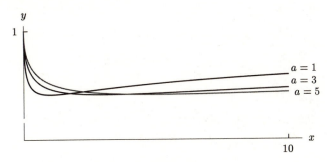

Figure 4.21

(c) Using a computer algebra system to find the second derivative when $a = 2$, we get

$$f''(x) = \frac{4\sqrt{2} + 12\sqrt{x} + 6x^{3/2} - 3\sqrt{2}x^2}{4\left(\sqrt{2} + \sqrt{x}\right)^3 x^{3/2}(2 + x)^{3/2}}.$$

Using the computer algebra system again to solve $f''(x) = 0$, we find that it has one zero at $x = 4.6477$. Graphing the second derivative, we see that it goes from positive to negative at $x = 4.6477$, so this is an inflection point.

CHECK YOUR UNDERSTANDING

1. True. Since the domain of f is all real numbers, all local minima occur at critical points.

5. False. For example, if $f(x) = x^3$, then $f'(0) = 0$, but $f(x)$ does not have either a local maximum or a local minimum at $x = 0$.

9. True, by the Increasing Function Theorem, Theorem 4.4.

13. Let $f(x) = ax^2$, with $a \neq 0$. Then $f'(x) = 2ax$, so f has a critical point only at $x = 0$.

17. Let f be defined by

$$f(x) = \begin{cases} x^2 & \text{if } 0 \leq x < 1 \\ 1/2 & \text{if } x = 1. \end{cases}$$

Then f is not continuous at $x = 1$, but f is differentiable on $(0,1)$ and $f'(x) = 2x$ for $0 < x < 1$. Thus, $c = 1/4$ satisfies

$$f'(c) = \frac{f(1) - f(0)}{1 - 0} = \frac{1}{2}, \quad \text{since} \quad f'\left(\frac{1}{4}\right) = 2 \cdot \frac{1}{4} = \frac{1}{2}.$$

21. This is impossible. If $f(a) > 0$, then the downward concavity forces the graph of f to cross the x-axis to the right or left of $x = a$, which means $f(x)$ cannot be positive for all values of x. More precisely, suppose that $f(x)$ is positive for all x and f is concave down. Thus there must be some value $x = a$ where $f(a) > 0$ and $f'(a)$ is not zero, since a constant function is not concave down. The tangent line at $x = a$ has nonzero slope and hence must cross the x-axis somewhere to the right or left of $x = a$. Since the graph of f must lie below this tangent line, it must also cross the x-axis, contradicting the assumption that $f(x)$ is positive for all x.

25. This is impossible. Since f''' exists, f'' must be continuous. By the Intermediate Value Theorem, $f''(x)$ cannot change sign, since $f''(x)$ cannot be zero. In the same way, we can show that $f'(x)$ and $f(x)$ cannot change sign. Since the product of these three with $f'''(x)$ cannot change sign, $f'''(x)$ cannot change sign. Thus $f(x)f''(x)$ and $f'(x)f'''(x)$ cannot change sign. Since their product is negative for all x, one or the other must be negative for all x. By Problem 24, this is impossible.

CHAPTER FIVE

Solutions for Section 5.1

Exercises

1. **(a)** Suppose $f(t)$ is the flowrate in m^3/hr at time t. We are only given two values of the flowrate, so in making our estimates of the flow, we use one subinterval, with $\Delta t = 3/1 = 3$:

$$\text{Left estimate} = 3[f(6 \text{ am})] = 3 \cdot 100 = 300 \text{ m}^3 \quad \text{(an underestimate)}$$
$$\text{Right estimate} = 3[f(9 \text{ am})] = 3 \cdot 280 = 840 \text{ m}^3 \quad \text{(an overestimate)}.$$

The best estimate is the average of these two estimates,

$$\text{Best estimate} = \frac{\text{Left} + \text{Right}}{2} = \frac{300 + 840}{2} = 570 \text{ m}^3/\text{hr}.$$

(b) Since the flowrate is increasing throughout, the error, i.e., the difference between over- and under-estimates, is given by

$$\text{Error} \leq \Delta t \left[f(9 \text{ am}) - f(6 \text{ am}) \right] = \Delta t [280 - 100] = 180 \Delta t.$$

We wish to choose Δt so that the the error $180\Delta t \leq 6$, or $\Delta t \leq 6/180 = 1/30$. So the flowrate guage should be read every $1/30$ of an hour, or every 2 minutes.

Problems

5. **(a)** Car A has the largest maximum velocity because the peak of car A's velocity curve is higher than the peak of B's.
   (b) Car A stops first because the curve representing its velocity hits zero (on the t-axis) first.
   (c) Car B travels farther because the area under car B's velocity curve is the larger.

9. Using $\Delta t = 0.2$, our upper estimate is

$$\frac{1}{1+0}(0.2) + \frac{1}{1+0.2}(0.2) + \frac{1}{1+0.4}(0.2) + \frac{1}{1+0.6}(0.2) + \frac{1}{1+0.8}(0.2) \approx 0.75.$$

The lower estimate is

$$\frac{1}{1+0.2}(0.2) + \frac{1}{1+0.4}(0.2) + \frac{1}{1+0.6}(0.2) + \frac{1}{1+0.8}(0.2)\frac{1}{1+1}(0.2) \approx 0.65.$$

Since v is a decreasing function, the bug has crawled more than 0.65 meters, but less than 0.75 meters. We average the two to get a better estimate:

$$\frac{0.65 + 0.75}{2} = 0.70 \text{ meters}.$$

13. **(a)** An upper estimate is $9.81 + 8.03 + 6.53 + 5.38 + 4.41 = 34.16$ m/sec. A lower estimate is $8.03 + 6.53 + 5.38 + 4.41 + 3.61 = 27.96$ m/sec.
   (b) The average is $\frac{1}{2}(34.16 + 27.96) = 31.06$ m/sec. Because the graph of acceleration is concave up, this estimate is too high, as can be seen in the figure to below. The area of the shaded region is the average of the areas of the rectangles $ABFE$ and $CDFE$.

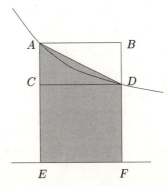

Solutions for Section 5.2

Exercises

1.

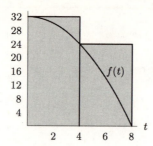

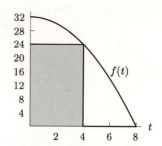

Figure 5.1: Left Sum, $\Delta t = 4$ **Figure 5.2**: Right Sum, $\Delta t = 4$

(a) Left-hand sum $= 32(4) + 24(4) = 224$.
(b) Right-hand sum $= 24(4) + 0(4) = 96$.

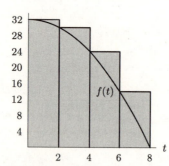

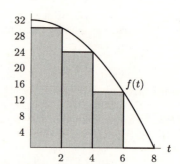

Figure 5.3: Left Sum, $\Delta t = 2$ **Figure 5.4**: Right Sum, $\Delta t = 2$

(c) Left-hand sum $= 32(2) + 30(2) + 24(2) + 14(2) = 200$.
(d) Right-hand sum $= 30(2) + 24(2) + 14(2) + 0(2) = 136$.

5. With $\Delta x = 5$, we have

$$\text{Left-hand sum} = 5(0 + 100 + 200 + 100 + 200 + 250 + 275) = 5625,$$

$$\text{Right-hand sum} = 5(100 + 200 + 100 + 200 + 250 + 275 + 300) = 7125.$$

The average of these two sums is our best guess for the value of the integral;

$$\int_{-15}^{20} f(x)\, dx \approx \frac{5625 + 7125}{2} = 6375.$$

9. We use a calculator or computer to see that $\int_{0}^{3} 2^x\, dx = 10.0989$.

13.

n	2	10	50	250
Left-hand Sum	1.34076	1.07648	1.01563	1.00314
Right-hand Sum	0.55536	0.91940	0.98421	0.99686

The sums seem to be converging to 1. Since $\cos x$ is monotone on $[0, \pi/2]$, the true value is between 1.00314 and .99686 .

17. The graph of $y = 7 - x^2$ has intercepts $x = \pm\sqrt{7}$. See Figure 5.5. Therefore we have

$$\text{Area} = \int_{-\sqrt{7}}^{\sqrt{7}} (7 - x^2)\, dx = 24.7.$$

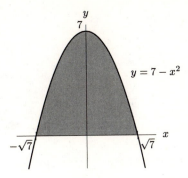

Figure 5.5

21. Since $x^{1/2} \le x^{1/3}$ for $0 \le x \le 1$, we have

$$\text{Area} = \int_0^1 (x^{1/3} - x^{1/2})\, dx = 0.0833.$$

Problems

25. The areas we computed are shaded in Figure 5.6. Since $y = x^2$ and $y = x^{1/2}$ are inverse functions, their graphs are reflections about the line $y = x$. Similarly, $y = x^3$ and $y = x^{1/3}$ are inverse functions and their graphs are reflections about the line $y = x$. Therefore, the two shaded areas in Figure 5.6 are equal.

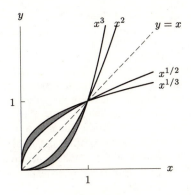

Figure 5.6

29. We have $\Delta x = 2/500 = 1/250$. The formulas for the left- and right-hand Riemann sums give us that

$$\text{Left} = \Delta x[f(-1) + f(-1 + \Delta x) + \ldots + f(1 - 2\Delta x) + f(1 - \Delta x)]$$
$$\text{Right} = \Delta x[f(-1 + \Delta x) + f(-1 + 2\Delta x) + \ldots + f(1 - \Delta x) + f(1)].$$

Subtracting these yields

$$\text{Right} - \text{Left} = \Delta x[f(1) - f(-1)] = \frac{1}{250}[6 - 2] = \frac{4}{250} = \frac{2}{125}.$$

33. We have

$$\Delta x = \frac{4}{3} = \frac{b-a}{n} \quad \text{and} \quad n = 3, \quad \text{so} \quad b-a = 4 \quad \text{or} \quad b = a + 4.$$

The function, $f(x)$, is squaring something. Since it is a left-hand sum, $f(x)$ could equal x^2 with $a = 2$ and $b = 6$ (note that $2 + 3(\frac{4}{3})$ gives the right-hand endpoint of the last interval). Or, $f(x)$ could possibly equal $(x + 2)^2$ with $a = 0$ and $b = 4$. Other answers are possible.

37.

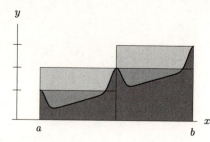

Figure 5.7: Integral vs. Left- and Right-Hand Sums

Solutions for Section 5.3

Exercises

1. Average value $= \dfrac{1}{2-0} \displaystyle\int_0^2 (1+t)\, dt = \dfrac{1}{2}(4) = 2.$

5. The units of measurement are dollars.

9. (a)

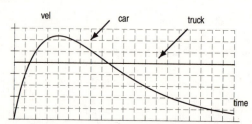

(b) The graphs intersect twice, at about 0.7 hours and 4.3 hours. At each intersection point, the velocity of the car is equal to the velocity of the truck, so $v_{car} = v_{truck}$. From the time they start until 0.7 hours later, the truck is traveling at a greater velocity than the car, so the truck is ahead of the car and is pulling farther away. At 0.7 hours they are traveling at the same velocity, and after 0.7 hours the car is traveling faster than the truck, so that the car begins to gain on the truck. Thus, at 0.7 hours the truck is farther from the car than it is immediately before or after 0.7 hours.

Similarly, because the car's velocity is greater than the truck's after 0.7 hours, it will catch up with the truck and eventually pass and pull away from the truck until 4.3 hours, at which point the two are again traveling at the same velocity. After 4.3 hours the truck travels faster than the car, so that it now gains on the car. Thus, 4.3 hours represents the point where the car is farthest ahead of the truck.

Problems

13. (a) The integral is the area above the x-axis minus the area below the x-axis. Thus, we can see that $\int_{-3}^{3} f(x)\, dx$ is about $-6 + 2 = -4$ (the negative of the area from $t = -3$ to $t = 1$ plus the area from $t = 1$ to $t = 3$.)

(b) Since the integral in part (a) is negative, the average value of $f(x)$ between $x = -3$ and $x = 3$ is negative. From the graph, however, it appears that the average value of $f(x)$ from $x = 0$ to $x = 3$ is positive. Hence (ii) is the larger quantity.

17. The time period 9am to 5pm is represented by the time $t = 0$ to $t = 8$ and $t = 24$ to $t = 32$. The area under the curve, or total number of worker-hours for these times, is about 9 boxes or $9(80) = 720$ worker-hours. The total cost for 9am to 5pm is $(720)(10) = \$7200$. The area under the rest of the curve is about 5.5 boxes, or $5.5(80) = 440$ worker-hours. The total cost for this time period is $(440)(15) = \$6600$. The total cost is about $7200 + 6600 = \$13,800$.

21. Since the average value is given by

$$\text{Average value} = \frac{1}{b-a} \int_a^b f(x)\,dx,$$

the units for dx inside the integral are canceled by the units for $1/(b-a)$ outside the integral, leaving only the units for $f(x)$. This is as it should be, since the average value of f should be measured in the same units as $f(x)$.

25. Change in income $= \int_0^{12} r(t)\,dt = \int_0^{12} 40(1.002)^t\,dt = \$485.80$

29. We know that the the integral of F, and therefore the work, can be obtained by computing the areas in Figure 5.8.

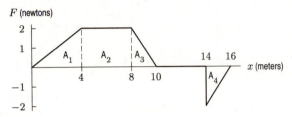

Figure 5.8

$$W = \int_0^{16} F(x)\,dx = \text{Area above } x\text{-axis} - \text{Area below } x\text{-axis}$$

$$= A_1 + A_2 + A_3 - A_4$$

$$= \frac{1}{2} \cdot 4 \cdot 2 + 4 \cdot 2 + \frac{1}{2} \cdot 2 \cdot 2 - \frac{1}{2} \cdot 2 \cdot 2$$

$$= 12 \text{ newton} \cdot \text{meters}.$$

Solutions for Section 5.4

Exercises

1. We find the changes in $f(x)$ between any two values of x by counting the area between the curve of $f'(x)$ and the x-axis. Since $f'(x)$ is linear throughout, this is quite easy to do. From $x = 0$ to $x = 1$, we see that $f'(x)$ outlines a triangle of area $1/2$ below the x-axis (the base is 1 and the height is 1). By the Fundamental Theorem,

$$\int_0^1 f'(x)\,dx = f(1) - f(0),$$

so

$$f(0) + \int_0^1 f'(x)\,dx = f(1)$$

$$f(1) = 2 - \frac{1}{2} = \frac{3}{2}$$

Similarly, between $x = 1$ and $x = 3$ we can see that $f'(x)$ outlines a rectangle below the x-axis with area -1, so $f(2) = 3/2 - 1 = 1/2$. Continuing with this procedure (note that at $x = 4$, $f'(x)$ becomes positive), we get the table below.

x	0	1	2	3	4	5	6
$f(x)$	2	3/2	1/2	−1/2	−1	−1/2	1/2

Problems

5. Note that $\int_a^b f(z)\,dz = \int_a^b f(x)\,dx$. Thus, we have

$$\int_a^b cf(z)\,dz = c\int_a^b f(z)\,dz = 8c.$$

9. The graph of $y = f(x - 5)$ is the graph of $y = f(x)$ shifted to the right by 5. Since the limits of integration have also shifted by 5 (to $a + 5$ and $b + 5$), the areas corresponding to $\int_{a+5}^{b+5} f(x - 5)\,dx$ and $\int_a^b f(x)\,dx$ are the same. Thus,

$$\int_{a+5}^{b+5} f(x - 5)\,dx = \int_a^b f(x)\,dx = 8.$$

13. **(a)** The integrand is positive, so the integral can't be negative.
   (b) The integrand ≥ 0. If the integral $= 0$, then the integrand must be identically 0, which isn't true.

17.

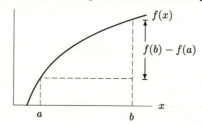

21. **(a)** $\dfrac{1}{\sqrt{2\pi}}\displaystyle\int_1^3 e^{-\frac{x^2}{2}}\,dx$

$$= \frac{1}{\sqrt{2\pi}}\int_0^3 e^{-\frac{x^2}{2}}\,dx - \frac{1}{\sqrt{2\pi}}\int_0^1 e^{-\frac{x^2}{2}}\,dx$$
$$\approx 0.4987 - 0.3413 = 0.1574.$$

   (b) $\left(\text{by symmetry of } e^{x^2/2}\right)$ $\dfrac{1}{\sqrt{2\pi}}\displaystyle\int_{-2}^3 e^{-\frac{x^2}{2}}\,dx = \frac{1}{\sqrt{2\pi}}\int_{-2}^0 e^{-\frac{x^2}{2}}\,dx + \frac{1}{\sqrt{2\pi}}\int_0^3 e^{-\frac{x^2}{2}}\,dx$

$$= \frac{1}{\sqrt{2\pi}}\int_0^2 e^{-\frac{x^2}{2}}\,dx + \frac{1}{\sqrt{2\pi}}\int_0^3 e^{-\frac{x^2}{2}}\,dx$$
$$\approx 0.4772 + 0.4987 = 0.9759.$$

Solutions for Chapter 5 Review

Exercises

1. **(a)** We calculate the right- and left-hand sums as follows:

$$\text{Left} = 2[80 + 52 + 28 + 10] = 340 \text{ ft.}$$
$$\text{Right} = 2[52 + 28 + 10 + 0] = 180 \text{ ft.}$$

Our best estimate will be the average of these two sums,

$$\text{Best} = \frac{\text{Left} + \text{Right}}{2} = \frac{340 + 180}{2} = 260 \text{ ft.}$$

   (b) Since v is decreasing throughout,

$$\text{Left} - \text{Right} = \Delta t \cdot [f(0) - f(8)]$$
$$= 80\Delta t.$$

Since our best estimate is the average of Left and Right, the maximum error is $(80)\Delta t/2$. For $(80)\Delta t/2 \leq 20$, we must have $\Delta t \leq 1/2$. In other words, we must measure the velocity every 0.5 second.

5. We take $\Delta t = 20$. Then:

$$\text{Left-hand sum} = 1.2(20) + 2.8(20) + 4.0(20) + 4.7(20) + 5.1(20)$$
$$= 356.$$
$$\text{Right-hand sum} = 2.8(20) + 4.0(20) + 4.7(20) + 5.1(20) + 5.2(20)$$
$$= 436.$$
$$\int_0^{100} f(t)\,dt \approx \text{Average} = \frac{356 + 436}{2} = 396.$$

9. The x intercepts of $y = x^2 - 9$ are $x = -3$ and $x = 3$, and since the graph is below the x axis on the interval $[-3, 3]$.

$$\text{Area} = -\int_{-3}^{3} (x^2 - 9)\,dx = 36.00.$$

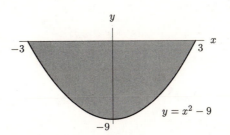

13. The graph of $y = -x^2 + 5x - 4$ is shown in Figure 5.9. We wish to find the area shaded. Since the graph crosses the x-axis at $x = 1$, we must split the integral at $x = 1$. For $x < 1$, the graph is below the x-axis, so the area is the negative of the integral. Thus

$$\text{Area shaded} = -\int_0^1 (-x^2 + 5x - 4)\,dx + \int_1^3 (-x^2 + 5x - 4)\,dx.$$

Using a calculator or computer, we find

$$\int_0^1 (-x^2 + 5x - 4)\,dx = -1.8333 \quad \text{and} \quad \int_1^3 (-x^2 + 5x - 4)\,dx = 3.3333.$$

Thus,

$$\text{Area shaded} = 1.8333 + 3.3333 = 5.1666.$$

(Notice that $\int_0^3 f(x)\,dx = -1.8333 + 3.333 = 1.5$, but the value of this integral is not the area shaded.)

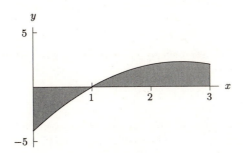

Figure 5.9

Problems

17. From $t = 0$ to $t = 3$, you are moving away from home ($v > 0$); thereafter you move back toward home. So you are the farthest from home at $t = 3$. To find how far you are then, we can measure the area under the v curve as about 9 squares, or $9 \cdot 10 \text{ km/hr} \cdot 1 \text{ hr} = 90 \text{ km}$. To find how far away from home you are at $t = 5$, we measure the area from $t = 3$ to $t = 5$ as about 25 km, except that this distance is directed toward home, giving a total distance from home during the trip of $90 - 25 = 65 \text{ km}$.

21. (a) Clearly, the points where $x = \sqrt{\pi}, \sqrt{2\pi}, \sqrt{3\pi}, \sqrt{4\pi}$ are where the graph intersects the x-axis because $f(x) = \sin(x^2) = 0$ where x is the square root of some multiple of π.

(b) Let $f(x) = \sin(x^2)$, and let A, B, C, and D be the areas of the regions indicated in the figure below. Then we see that $A > B > C > D$.

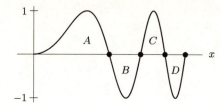

Note that

$$\int_0^{\sqrt{\pi}} f(x)\,dx = A, \qquad \int_0^{\sqrt{2\pi}} f(x)\,dx = A - B,$$

$$\int_0^{\sqrt{3\pi}} f(x)\,dx = A - B + C, \quad \text{and} \quad \int_0^{\sqrt{4\pi}} f(x)\,dx = A - B + C - D.$$

It follows that

$$\int_0^{\sqrt{\pi}} f(x)\,dx = A > \int_0^{\sqrt{3\pi}} f(x)\,dx = A - (B - C) = A - B + C >$$

$$\int_0^{\sqrt{4\pi}} f(x)\,dx = A - B + C - D > \int_0^{\sqrt{2\pi}} f(x)\,dx = (A - B) > 0.$$

And thus the ordering is $n = 1$, $n = 3$, $n = 4$, and $n = 2$ from largest to smallest. All the numbers are positive.

25. (a) We know that $\int_2^5 f(x)\,dx = \int_0^5 f(x)\,dx - \int_0^2 f(x)\,dx$. By symmetry, $\int_0^2 f(x)\,dx = \frac{1}{2}\int_{-2}^2 f(x)\,dx$, so $\int_2^5 f(x)\,dx = \int_0^5 f(x)\,dx - \frac{1}{2}\int_{-2}^2 f(x)\,dx$.

(b) $\int_2^5 f(x)\,dx = \int_{-2}^5 f(x)\,dx - \int_{-2}^2 f(x)\,dx = \int_{-2}^5 f(x)\,dx - 2\int_{-2}^0 f(x)\,dx$.

(c) Using symmetry again, $\int_0^2 f(x)\,dx = \frac{1}{2}\left(\int_{-2}^5 f(x)\,dx - \int_2^5 f(x)\,dx\right)$.

29. The change in the amount of water is the integral of rate of change, so we have

$$\text{Number of liters pumped out} = \int_0^{60} (5 - 5e^{-0.12t})\,dt = 258.4 \text{ liters}.$$

Since the tank contained 1000 liters of water initially, we see that

$$\text{Amount in tank after one hour} = 1000 - 258.4 = 741.6 \text{ liters}.$$

33. (a) About 300 meter^3/sec.

(b) About 250 meter^3/sec.

(c) Looking at the graph, we can see that the 1996 flood reached its maximum just between March and April, for a high of about 1250 meter^3/sec. Similarly, the 1957 flood reached its maximum in mid-June, for a maximum flow rate of 3500 meter^3/sec.

(d) The 1996 flood lasted about 1/3 of a month, or about 10 days. The 1957 flood lasted about 4 months.

(e) The area under the controlled flood graph is about 2/3 box. Each box represents 500 meter^3/sec for one month. Since

$$1 \text{ month} = 30\frac{\text{days}}{\text{month}} \cdot 24\frac{\text{hours}}{\text{day}} \cdot 60\frac{\text{minutes}}{\text{hour}} \cdot 60\frac{\text{seconds}}{\text{minute}}$$
$$= 2.592 \cdot 10^6 \approx 3 \cdot 10^6 \text{ seconds},$$

each box represents

$$\text{Flow} \approx (500 \text{ meter}^3/\text{sec}) \cdot (2.6 \cdot 10^6 \text{ sec}) = 13 \cdot 10^8 \text{ meter}^3 \text{ of water}.$$

So, for the artificial flood,

$$\text{Additional flow} \approx \frac{2}{3} \cdot 13 \cdot 10^8 = 9 \cdot 10^8 \text{ meter}^3 \approx 10^9 \text{ meter}^3.$$

(f) The 1957 flood released a volume of water represented by about 12 boxes above the 250 meter/sec baseline. Thus, for the natural flood,

$$\text{Additional flow} \approx 12 \cdot 15 \cdot 10^8 = 1.8 \cdot 10^{10} \approx 2 \cdot 10^{10} \text{ meter}^3.$$

So, the natural flood was nearly 20 times larger than the controlled flood and lasted much longer.

37. The graph of rate against time is the straight line shown in Figure 5.10. Since the shaded area is 270, we have

$$\frac{1}{2}(10 + 50) \cdot t = 270$$

$$t = \frac{270}{60} \cdot 2 = 9 \text{ years}$$

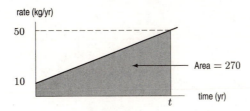

Figure 5.10

CAS Challenge Problems

41. (a) Since the length of the interval of integration is $2 - 1 = 1$, the width of each subdivision is $\Delta t = 1/n$. Thus the endpoints of the subdivision are

$$t_0 = 1, \quad t_1 = 1 + \Delta t = 1 + \frac{1}{n}, \quad t_2 = 1 + 2\Delta t = 1 + \frac{2}{n}, \dots,$$

$$t_i = 1 + i\Delta t = 1 + \frac{i}{n}, \dots, \quad t_{n-1} = 1 + (n-1)\Delta t = 1 + \frac{n-1}{n}.$$

Thus, since the integrand is $f(t) = t^2$,

$$\text{Left-hand sum} = \sum_{i=0}^{n-1} f(t_i)\Delta t = \sum_{i=0}^{n-1} t_i^2 \Delta t = \sum_{i=0}^{n-1} \left(1 + \frac{i}{n}\right)^2 \frac{1}{n} = \sum_{i=0}^{n-1} \frac{(n+i)^2}{n^3}.$$

(b) Using a CAS to find the sum, we get

$$\sum_{i=0}^{n-1} \frac{(n+i)^2}{n^3} = \frac{(-1+2n)(-1+7n)}{6n^2} = \frac{7}{3} + \frac{1}{6n^2} - \frac{3}{2n}.$$

(c) Taking the limit as $n \to \infty$

$$\lim_{n \to \infty} \left(\frac{7}{3} + \frac{1}{6n^2} - \frac{3}{2n}\right) = \lim_{n \to \infty} \frac{7}{3} + \lim_{n \to \infty} \frac{1}{6n^2} - \lim_{n \to \infty} \frac{3}{2n} = \frac{7}{3} + 0 + 0 = \frac{7}{3}.$$

(d) We have calculated $\int_1^2 t^2 \, dt$ using Riemann sums. Since t^2 is above the t-axis between $t = 1$ and $t = 2$, this integral is the area; so the area is 7/3.

CHECK YOUR UNDERSTANDING

1. True, since $\int_0^2 (f(x) + g(x))dx = \int_0^2 f(x)dx + \int_0^2 g(x)dx$.

5. False. This would be true if $h(x) = 5f(x)$. However, we cannot assume that $f(5x) = 5f(x)$, so for many functions this statement is false. For example, if f is the constant function $f(x) = 3$, then $h(x) = 3$ as well, so $\int_0^2 f(x) \, dx = \int_0^2 h(x) \, dx = 6$.

9. False. If the graph of f is symmetric about the y-axis, this is true, but otherwise it is usually not true. For example, if $f(x) = x + 1$ the area under the graph of f for $-1 \leq x \leq 0$ is less than the area under the graph of f for $0 \leq x \leq 1$, so $\int_{-1}^{1} f(x)\,dx < 2\int_{0}^{1} f(x)\,dx$. See Figure 5.11.

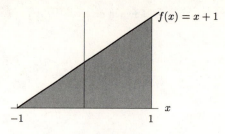

Figure 5.11

13. True, by Theorem 5.4 on Comparison of Definite Integrals:

$$\frac{1}{b-a}\int_{a}^{b} f(x)\,dx \leq \frac{1}{b-a}\int_{a}^{b} g(x)\,dx.$$

17. True. By the properties of integrals in Theorem 5.3, we have:

$$\int_{a}^{b}(f(x) + g(x))dx = \int_{a}^{b} f(x)dx + \int_{a}^{b} g(x)dx.$$

Dividing both sides of this equation through by $b - a$, we get that the average value of $f(x) + g(x)$ is average value of $f(x)$ plus the average value of $g(x)$:

$$\frac{1}{b-a}\int_{a}^{b}(f(x) + g(x))\,dx = \frac{1}{b-a}\int_{a}^{b} f(x)\,dx + \frac{1}{b-a}\int_{a}^{b} g(x)\,dx.$$

CHAPTER SIX

Solutions for Section 6.1

Exercises

1.

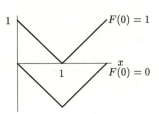

5. By the Fundamental Theorem of Calculus, we know that

$$f(2) - f(0) = \int_0^2 f'(x)dx.$$

Using a left-hand sum, we estimate $\int_0^2 f'(x)dx \approx (10)(2) = 20$. Using a right-hand sum, we estimate $\int_0^2 f'(x)dx \approx (18)(2) = 36$. Averaging, we have

$$\int_0^2 f'(x)dx \approx \frac{20 + 36}{2} = 28.$$

We know $f(0) = 100$, so

$$f(2) = f(0) + \int_0^2 f'(x)dx \approx 100 + 28 = 128.$$

Similarly, we estimate

$$\int_2^4 f'(x)dx \approx \frac{(18)(2) + (23)(2)}{2} = 41,$$

so

$$f(4) = f(2) + \int_2^4 f'(x)dx \approx 128 + 41 = 169.$$

Similarly,

$$\int_4^6 f'(x)dx \approx \frac{(23)(2) + (25)(2)}{2} = 48,$$

so

$$f(6) = f(4) + \int_4^6 f'(x)dx \approx 169 + 48 = 217.$$

The values are shown in the table.

x	0	2	4	6
$f(x)$	100	128	169	217

Problems

9. (a) Critical points of $F(x)$ are the zeros of f: $x = 1$ and $x = 3$.
  (b) $F(x)$ has a local minimum at $x = 1$ and a local maximum at $x = 3$.
  (c)

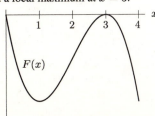

Notice that the graph could also be above or below the x-axis at $x = 3$.

13.

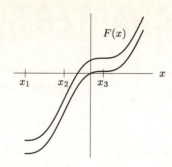

Note that since $f(x_1) = 0$, $F(x_1)$ is either a local minimum or a point of inflection; it is impossible to tell which from the graph. Since $f'(x_3) = 0$, and f' changes sign around $x = x_3$, $F(x_3)$ is an inflection point. Also, since $f'(x_2) = 0$ and f changes from increasing to decreasing about $x = x_2$, F has another inflection point at $x = x_2$.

17. The critical points are at $(0, 5)$, $(2, 21)$, $(4, 13)$, and $(5, 15)$. A graph is given below.

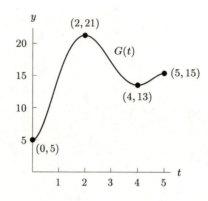

21. **(a)** The total volume emptied must increase with time and cannot decrease. The smooth graph (I) that is always increasing is therefore the volume emptied from the bladder. The jagged graph (II) that increases then decreases to zero is the flow rate.

   (b) The total change in volume is the integral of the flow rate. Thus, the graph giving total change (I) shows an antiderivative of the rate of change in graph (II).

25. **(a)** Suppose $Q(t)$ is the amount of water in the reservoir at time t. Then

$$Q'(t) = \frac{\text{Rate at which water}}{\text{in reservoir is changing}} = \frac{\text{Inflow}}{\text{rate}} - \frac{\text{Outflow}}{\text{rate}}$$

Thus the amount of water in the reservoir is increasing when the inflow curve is above the outflow, and decreasing when it is below. This means that $Q(t)$ is a maximum where the curves cross in July 1993 (as shown in Figure 6.1), and $Q(t)$ is decreasing fastest when the outflow is farthest above the inflow curve, which occurs about October 1993 (see Figure 6.1).

   To estimate values of $Q(t)$, we use the Fundamental Theorem which says that the change in the total quantity of water in the reservoir is given by

$$Q(t) - Q(\text{Jan'93}) = \int_{\text{Jan93}}^{t} (\text{inflow rate} - \text{outflow rate})\, dt$$

   or $$Q(t) = Q(\text{Jan'93}) + \int_{\text{Jan93}}^{t} (\text{inflow rate} - \text{outflow rate})\, dt.$$

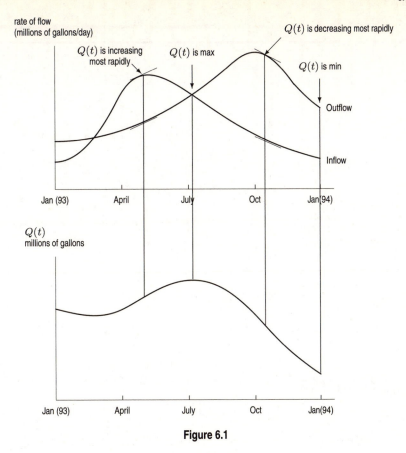

Figure 6.1

(b) See Figure 6.1. Maximum in July 1993. Minimum in Jan 1994.

(c) See Figure 6.1. Increasing fastest in May 1993. Decreasing fastest in Oct 1993.

(d) In order for the water to be the same as Jan '93 the total amount of water which has flowed into the reservoir must be 0. Referring to Figure 6.2, we have

$$\int_{\text{Jan}93}^{\text{July}94} (\text{inflow} - \text{outflow})dt = -A_1 + A_2 - A_3 + A_4 = 0$$

giving $A_1 + A_3 = A_2 + A_4$

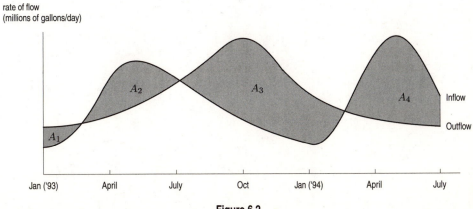

Figure 6.2

Solutions for Section 6.2

Exercises

1. $5x$

5. $\sin t$

9. $-\dfrac{1}{2z^2}$

13. $\dfrac{t^4}{4} - \dfrac{t^3}{6} - \dfrac{t^2}{2}$

17. $-\cos 2\theta$

21. $\dfrac{5}{2}x^2 - \dfrac{2}{3}x^{\frac{3}{2}}$

25. $R(t) = \displaystyle\int (t^3 + 5t - 1)\, dt = \dfrac{t^4}{4} + \dfrac{5}{2}t^2 - t + C$

29. $H(x) = \displaystyle\int (4x^3 - 7)\, dx = x^4 - 7x + C$

33. $F(x) = \displaystyle\int \dfrac{1}{x^2}\, dx = -\dfrac{1}{x} + C$

37. $f(x) = \frac{1}{4}x$, so $F(x) = \frac{x^2}{8} + C$. $F(0) = 0$ implies that $\frac{1}{8} \cdot 0^2 + C = 0$, so $C = 0$. Thus $F(x) = x^2/8$ is the only possibility.

41. $f(x) = \sin x$, so $F(x) = -\cos x + C$. $F(0) = 0$ implies that $-\cos 0 + C = 0$, so $C = 1$. Thus $F(x) = -\cos x + 1$ is the only possibility.

45. $\displaystyle\int (x^3 - 2)\, dx = \dfrac{x^4}{4} - 2x + C$

49. $\displaystyle\int \dfrac{4}{t^2}\, dt = -\dfrac{4}{t} + C$

53. $\dfrac{x^2}{2} + 2x^{1/2} + C$

57. $\sin(x + 1) + C$

61. $\displaystyle\int_0^3 (x^2 + 4x + 3)\, dx = \left(\dfrac{x^3}{3} + 2x^2 + 3x \right)\Bigg|_0^3 = (9 + 18 + 9) - 0 = 36$

65. $\displaystyle\int_2^5 (x^3 - \pi x^2)\, dx = \left(\dfrac{x^4}{4} - \dfrac{\pi x^3}{3} \right)\Bigg|_2^5 = \dfrac{609}{4} - 39\pi \approx 29.728.$

69. $\displaystyle\int_0^{\pi/4} (\sin t + \cos t)\, dt = (-\cos t + \sin t)\Bigg|_0^{\pi/4} = \left(-\dfrac{\sqrt{2}}{2} + \dfrac{\sqrt{2}}{2} \right) - (-1 + 0) = 1.$

73. $\displaystyle\int 2^x\, dx = \dfrac{1}{\ln 2} 2^x + C$, since $\dfrac{d}{dx} 2^x = \ln 2 \cdot 2^x$, so

$$\int_{-1}^1 2^x\, dx = \dfrac{1}{\ln 2} \left[2^x \Big|_{-1}^1 \right] = \dfrac{3}{2 \ln 2} \approx 2.164.$$

Problems

77. Since the graph of $y = e^x$ is above the graph of $y = \cos x$ (see the figure below), we have

$$\text{Area} = \int_0^1 (e^x - \cos x)\, dx$$

$$= \int_0^1 e^x\, dx - \int_0^1 \cos x\, dx$$

$$= e^x \Big|_0^1 - \sin x \Big|_0^1$$

$$= e^1 - e^0 - \sin 1 + \sin 0$$

$$= e - 1 - \sin 1.$$

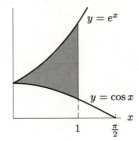

81. The average value of $v(x)$ on the interval $1 \leq x \leq c$ is

$$\frac{1}{c-1} \int_1^c \frac{6}{x^2}\, dx = \frac{1}{c-1} \left(-\frac{6}{x}\right)\Big|_1^c = \frac{1}{c-1} \left(\frac{-6}{c} + 6\right) = \frac{6}{c}.$$

Since $\dfrac{1}{c-1} \displaystyle\int_1^c \frac{6}{x^2}\, dx = 1$, we have $\dfrac{6}{c} = 1$, so $c = 6$.

Solutions for Section 6.3

Exercises

1. $y = \displaystyle\int (x^3 + 5)\, dx = \dfrac{x^4}{4} + 5x + C$

5. Since $y = x + \sin x - \pi$, we differentiate to see that $dy/dx = 1 + \cos x$, so y satisfies the differential equation. To show that it also satisfies the initial condition, we check that $y(\pi) = 0$:

$$y = x + \sin x - \pi$$
$$y(\pi) = \pi + \sin \pi - \pi = 0.$$

9. Integrating gives

$$\int \frac{dq}{dz}\, dz = \int (2 + \sin z)\, dz = 2z - \cos z + C.$$

If $q = 5$ when $z = 0$, then $2(0) - \cos(0) + C = 5$ so $C = 6$. Thus $q = 2z - \cos z + 6$.

Problems

13.

$$\frac{dy}{dt} = k\sqrt{t} = kt^{1/2}$$

$$y = \frac{2}{3} kt^{3/2} + C.$$

Since $y = 0$ when $t = 0$, we have $C = 0$, so

$$y = \frac{2}{3} kt^{3/2}.$$

17. (a)

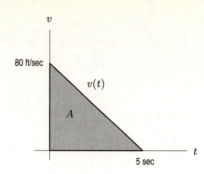

(b) The total distance is represented by the shaded region A, the area under the graph of $v(t)$.

(c) The area A, a triangle, is given by

$$A = \frac{1}{2}(\text{base})(\text{height}) = \frac{1}{2}(5\,\text{sec})(80\,\text{ft/sec}) = 200\,\text{ft}.$$

(d) Using integration and the Fundamental Theorem of Calculus, we have $A = \int_0^5 v(t)\,dt$ or $A = s(5) - s(0)$, where $s(t)$ is an antiderivative of $v(t)$.

We have that $a(t)$, the acceleration, is constant: $a(t) = k$ for some constant k. Therefore $v(t) = kt + C$ for some constant C. We have $80 = v(0) = k(0) + C = C$, so that $v(t) = kt + 80$. Putting in $t = 5$, $0 = v(5) = (k)(5) + 80$, or $k = -80/5 = -16$.

Thus $v(t) = -16t + 80$, and an antiderivative for $v(t)$ is $s(t) = -8t^2 + 80t + C$. Since the total distance traveled at $t = 0$ is 0, we have $s(0) = 0$ which means $C = 0$. Finally, $A = \int_0^5 v(t)\,dt = s(5) - s(0) = (-8(5)^2 + (80)(5)) - (-8(0)^2 + (80)(0)) = 200\,\text{ft}$, which agrees with the previous part.

21. The equation of motion is $y = -\frac{gt^2}{2} + v_0 t + y_0 = -16t^2 + 128t + 320$. Taking the first derivative, we get $v = -32t + 128$. The second derivative gives us $a = -32$.

(a) At its highest point, the stone's velocity is zero:
$v = 0 = -32t + 128$, so $t = 4$.

(b) At $t = 4$, the height is $y = -16(4)^2 + 128(4) + 320 = 576$ ft

(c) When the stone hits the beach,

$$y = 0 = -16t^2 + 128t + 320$$
$$0 = -t^2 + 8t + 20 = (10 - t)(2 + t).$$

So $t = 10$ seconds.

(d) Impact is at $t = 10$. The velocity, v, at this time is $v(10) = -32(10) + 128 = -192$ ft/sec. Upon impact, the stone's velocity is 192 ft/sec downward.

25. The first thing we should do is convert our units. We'll bring everything into feet and seconds. Thus, the initial speed of the car is

$$\frac{70\,\text{miles}}{\text{hour}}\left(\frac{1\,\text{hour}}{3600\,\text{sec}}\right)\left(\frac{5280\,\text{feet}}{1\,\text{mile}}\right) \approx 102.7\,\text{ft/sec}.$$

We assume that the acceleration is constant as the car comes to a stop. A graph of its velocity versus time is given in Figure 6.3. We know that the area under the curve represents the distance that the car travels before it comes to a stop, 157 feet. But this area is a triangle, so it is easy to find t_0, the time the car comes to rest. We solve

$$\frac{1}{2}(102.7)t_0 = 157,$$

which gives

$$t_0 \approx 3.06\,\text{sec}.$$

Since acceleration is the rate of change of velocity, the car's acceleration is given by the slope of the line in Figure 6.3. Thus, the acceleration, k, is given by

$$k = \frac{102.7 - 0}{0 - 3.06} \approx -33.56\,\text{ft/sec}^2.$$

Notice that k is negative because the car is slowing down.

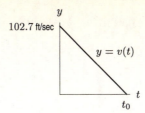

Figure 6.3: Graph of velocity versus time

Solutions for Section 6.4

Exercises

1.

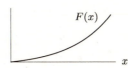

By the Fundamental Theorem, $f(x) = F'(x)$. Since f is positive and increasing, F is increasing and concave up. Since $F(0) = \int_0^0 f(t)dt = 0$, the graph of F must start from the origin.

5. Using the Fundamental Theorem, we know that the change in F between $x = 0$ and $x = 0.5$ is given by

$$F(0.5) - F(0) = \int_0^{0.5} \sin t \cos t \, dt \approx 0.115.$$

Since $F(0) = 1.0$, we have $F(0.5) \approx 1.115$. The other values are found similarly, and are given in Table 6.1.

Table 6.1

b	0	0.5	1	1.5	2	2.5	3
$F(b)$	1	1.11492	1.35404	1.4975	1.41341	1.17908	1.00996

9. If $f'(x) = \text{Si}(x)$, then $f(x)$ is of the form

$$f(x) = C + \int_a^x \text{Si}(t) \, dt.$$

Since $f(0) = 2$, we take $a = 0$ and $C = 2$, giving

$$f(x) = 2 + \int_0^x \text{Si}(t) \, dt.$$

Problems

13. Since $G'(x) = \cos(x^2)$ and $G(0) = -3$, we have

$$G(x) = G(0) + \int_0^x \cos(t^2) \, dt = -3 + \int_0^x \cos(t^2) \, dt.$$

Substituting $x = -1$ and evaluating the integral numerically gives

$$G(-1) = -3 + \int_0^{-1} \cos(t^2) \, dt = -3.905.$$

17. $\dfrac{d}{dt} \displaystyle\int_t^\pi \cos(z^3) \, dz = \dfrac{d}{dt} \left(-\int_\pi^t \cos(z^3) \, dz \right) = -\cos(t^3).$

21. **(a)** Since $\frac{d}{dt}(\cos(2t)) = -2\sin(2t)$, we have $F(\pi) = \int_0^\pi \sin(2t)\, dt = -\frac{1}{2}\cos(2t)\Big|_0^\pi = -\frac{1}{2}(1-1) = 0$.

(b) $F(\pi) = $ (Area above t-axis) $-$ (Area below t-axis) $= 0$. (The two areas are equal.)

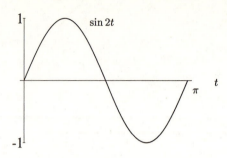

(c) $F(x) \geq 0$ everywhere. $F(x) = 0$ only at integer multiples of π. This can be seen for $x \geq 0$ by noting $F(x) = $ (Area above t-axis) $-$ (Area below t-axis), which is always non-negative and only equals zero when x is an integer multiple of π. For $x > 0$

$$F(-x) = \int_0^{-x} \sin 2t\, dt$$

$$= -\int_{-x}^0 \sin 2t\, dt$$

$$= \int_0^x \sin 2t\, dt = F(x),$$

since the area from $-x$ to 0 is the negative of the area from 0 to x. So we have $F(x) \geq 0$ for all x.

25. If we let $f(x) = \int_0^x e^{-t^2}\, dt$ and $g(x) = x^3$, then we use the chain rule because we are looking for $\frac{d}{dx}f(g(x)) = f'(g(x)) \cdot g'(x)$. Since $f'(x) = e^{-x^2}$, we have

$$\frac{d}{dx}\left(\int_0^{x^3} e^{-t^2}\, dt\right) = f'(x^3) \cdot 3x^2 = e^{-(x^3)^2} \cdot 3x^2 = 3x^2 e^{-x^6}.$$

Solutions for Section 6.5

Exercises

1. **(a)** The object is thrown from an initial height of $y = 1.5$ meters.

(b) The velocity is obtained by differentiating, which gives $v = -9.8t + 7$ m/sec. The initial velocity is $v = 7$ m/sec upward.

(c) The acceleration due to gravity is obtained by differentiating again, giving $g = -9.8$ m/sec^2, or 9.8 m/sec^2 downward.

Problems

5. $a(t) = -32$. Since $v(t)$ is the antiderivative of $a(t)$, $v(t) = -32t + v_0$. But $v_0 = 0$, so $v(t) = -32t$. Since $s(t)$ is the antiderivative of $v(t)$, $s(t) = -16t^2 + s_0$, where s_0 is the height of the building. Since the ball hits the ground in 5 seconds, $s(5) = 0 = -400 + s_0$. Hence $s_0 = 400$ feet, so the window is 400 feet high.

9. **(a)** Since $s(t) = -\frac{1}{2}gt^2$, the distance a body falls in the first second is

$$s(1) = -\frac{1}{2} \cdot g \cdot 1^2 = -\frac{g}{2}.$$

In the second second, the body travels

$$s(2) - s(1) = -\frac{1}{2}\left(g \cdot 2^2 - g \cdot 1^2\right) = -\frac{1}{2}(4g - g) = -\frac{3g}{2}.$$

In the third second, the body travels

$$s(3) - s(2) = -\frac{1}{2}\left(g \cdot 3^2 - g \cdot 2^2\right) = -\frac{1}{2}(9g - 4g) = -\frac{5g}{2},$$

and in the fourth second, the body travels

$$s(4) - s(3) = -\frac{1}{2}\left(g \cdot 4^2 - g \cdot 3^2\right) = -\frac{1}{2}(16g - 9g) = -\frac{7g}{2}.$$

(b) Galileo seems to have been correct. His observation follows from the fact that the differences between consecutive squares are consecutive odd numbers. For, if n is any number, then $n^2 - (n-1)^2 = 2n - 1$, which is the nth odd number (where 1 is the first).

Solutions for Chapter 6 Review

Exercises

1. $\frac{5}{2}x^2 + 7x + C$

5. $\displaystyle\int (3e^x + 2\sin x)\, dx = 3e^x - 2\cos x + C$

9. $e^x + 5x + C$

13. $\displaystyle\int (x+1)^2\, dx = \frac{(x+1)^3}{3} + C.$

Another way to work the problem is to expand $(x+1)^2$ to $x^2 + 2x + 1$ as follows:

$$\int (x+1)^2\, dx = \int (x^2 + 2x + 1)\, dx = \frac{x^3}{3} + x^2 + x + C.$$

These two answers are the same, since $\dfrac{(x+1)^3}{3} = \dfrac{x^3 + 3x^2 + 3x + 1}{3} = \dfrac{x^3}{3} + x^2 + x + \dfrac{1}{3}$, which is $\dfrac{x^3}{3} + x^2 + x$, plus a constant.

17. Since $f(x) = x + 1 + \dfrac{1}{x}$, the indefinite integral is $\dfrac{1}{2}x^2 + x + \ln|x| + C$

21. $2e^x - 8\sin x + C$

25. $G(x) = \displaystyle\int \sin x\, dx = -\cos x + C$

29. $F(x) = \displaystyle\int (e^x - 1)\, dx = e^x - x + C$

33. $F(x) = \displaystyle\int e^x\, dx = e^x + C.$ If $F(0) = 4$, then $F(0) = 1 + C = 4$ and thus $C = 3$. So $F(x) = e^x + 3.$

Problems

37. $\displaystyle\int_0^3 x^2\, dx = \left.\frac{x^3}{3}\right|_0^3 = 9 - 0 = 9.$

41. **(a)** See Figure 6.4. Since $f(x) > 0$ for $0 < x < 2$ and $f(x) < 0$ for $2 < x < 5$, we have

$$\text{Area} = \int_0^2 f(x)\, dx - \int_2^5 f(x)\, dx$$

$$= \int_0^2 (x^3 - 7x^2 + 10x)\, dx - \int_2^5 (x^3 - 7x^2 + 10x)\, dx$$

$$= \left(\frac{x^4}{4} - \frac{7x^3}{3} + 5x^2 \right) \Big|_0^2 - \left(\frac{x^4}{4} - \frac{7x^3}{3} + 5x^2 \right) \Big|_2^5$$

$$= \left[\left(4 - \frac{56}{3} + 20 \right) - (0 - 0 + 0) \right] - \left[\left(\frac{625}{4} - \frac{875}{3} + 125 \right) - \left(4 - \frac{56}{3} + 20 \right) \right]$$

$$= \frac{253}{12}.$$

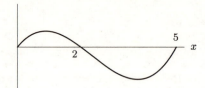

Figure 6.4: Graph of $f(x) = x^3 - 7x^2 + 10x$

(b) Calculating $\int_0^5 f(x)\,dx$ gives

$$\int_0^5 f(x)\,dx = \int_0^5 (x^3 - 7x^2 + 10x)\,dx$$

$$= \left(\frac{x^4}{4} - \frac{7x^3}{3} + 5x^2 \right) \Big|_0^5$$

$$= \left(\frac{625}{4} - \frac{875}{3} + 125 \right) - (0 - 0 + 0)$$

$$= -\frac{125}{12}.$$

This integral measures the difference between the area above the x-axis and the area below the x-axis. Since the definite integral is negative, the graph of $f(x)$ lies more below the x-axis than above it. Since the function crosses the axis at $x = 2$,

$$\int_0^5 f(x)\,dx = \int_0^2 f(x)\,dx + \int_2^5 f(x)\,dx = \frac{16}{3} - \frac{63}{4} = \frac{-125}{12},$$

whereas

$$\text{Area} = \int_0^2 f(x)\,dx - \int_2^5 f(x)\,dx = \frac{16}{3} + \frac{64}{4} = \frac{253}{12}.$$

45. See Figure 6.5. The average value of $f(x)$ is given by

$$\text{Average} = \frac{1}{9 - 0} \int_0^9 \sqrt{x}\,dx = \frac{1}{9} \left(\frac{2}{3} x^{3/2} \Big|_0^9 \right) = \frac{1}{9} \left(\frac{2}{3} 9^{3/2} - 0 \right) = \frac{1}{9} 18 = 2.$$

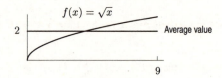

Figure 6.5

49.

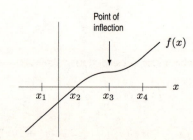

53. A function whose derivative is e^{x^2} is of the form

$$f(x) = C + \int_a^x e^{t^2}\,dt \qquad \text{for some value of } C.$$

(a) To ensure that the function goes through the point $(0, 3)$, we take $a = 0$ and $C = 3$:

$$f(x) = 3 + \int_0^x e^{t^2}\,dt.$$

(b) To ensure that the function goes through $(-1, 5)$, we take $a = -1$ and $C = 5$:

$$f(x) = 5 + \int_{-1}^x e^{t^2}\,dt.$$

57. (a) Since 6 sec $= 1/10$ min,

$$\text{Angular acceleration } = \frac{2500 - 1100}{1/10} = 14{,}000 \text{ revs/min}^2.$$

(b) We know angular acceleration is the derivative of angular velocity. Since

$$\text{Angular acceleration } = 14{,}000,$$

we have

$$\text{Angular velocity } = 14{,}000t + C.$$

Measuring time from the moment at which the angular velocity is 1100 revs/min, we have $C = 1100$. Thus,

$$\text{Angular velocity } = 14{,}000t + 1100.$$

Thus the total number of revolutions performed during the period from $t = 0$ to $t = 1/10$ min is given by

$$\begin{array}{l} \text{Number of} \\ \text{revolutions} \end{array} = \int_0^{1/10} (14000t + 1100)\,dt = 7000t^2 + 1100t \Big|_0^{1/10} = 180 \text{ revolutions.}$$

61. (a) In the beginning, both birth and death rates are small; this is consistent with a very small population. Both rates begin climbing, the birth rate faster than the death rate, which is consistent with a growing population. The birth rate is then high, but it begins to decrease as the population increases.

(b)

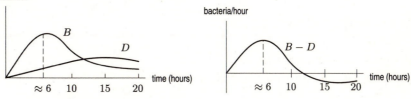

Figure 6.6: Difference between B and D is greatest at $t \approx 6$

The bacteria population is growing most quickly when $B - D$, the rate of change of population, is maximal; that happens when B is farthest above D, which is at a point where the slopes of both graphs are equal. That point is $t \approx 6$ hours.

(c) Total number born by time t is the area under the B graph from $t = 0$ up to time t. See Figure 6.7.

Total number alive at time t is the number born minus the number that have died, which is the area under the B graph minus the area under the D graph, up to time t. See Figure 6.8.

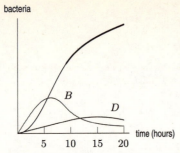

Figure 6.7: Number born by time t is $\int_0^t B(x)\,dx$

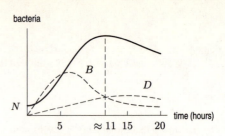

Figure 6.8: Number alive at time t is $\int_0^t (B(x) - D(x))\,dx$

From Figure 6.8, we see that the population is at a maximum when $B = D$, that is, after about 11 hours. This stands to reason, because $B - D$ is the rate of change of population, so population is maximized when $B - D = 0$, that is, when $B = D$.

CAS Challenge Problems

65. **(a)** A CAS gives

$$\int e^{2x}\,dx = \frac{1}{2}e^{2x} \qquad \int e^{3x}\,dx = \frac{1}{3}e^{3x} \qquad \int e^{3x+5}\,dx = \frac{1}{3}e^{3x+5}.$$

(b) The three integrals in part (a) obey the rule

$$\int e^{ax+b}\,dx = \frac{1}{a}e^{ax+b}.$$

(c) Checking the formula by calculating the derivative

$$
\begin{aligned}
\frac{d}{dx}\left(\frac{1}{a}e^{ax+b}\right) &= \frac{1}{a}\frac{d}{dx}e^{ax+b} \quad \text{by the constant multiple rule} \\
&= \frac{1}{a}e^{ax+b}\frac{d}{dx}(ax+b) \quad \text{by the chain rule} \\
&= \frac{1}{a}e^{ax+b}\cdot a = e^{ax+b}.
\end{aligned}
$$

CHECK YOUR UNDERSTANDING

1. True. A function can have only one derivative.

5. False. Differentiating using the product and chain rules gives

$$\frac{d}{dx}\left(\frac{-1}{2x}e^{-x^2}\right) = \frac{1}{2x^2}e^{-x^2} + e^{-x^2}.$$

9. True. If $y = F(x)$ is a solution to the differential equation $dy/dx = f(x)$, then $F'(x) = f(x)$, so $F(x)$ is an antiderivative of $f(x)$.

13. False. If f is positive then F is increasing, but if f is negative then F is decreasing.

CHAPTER SEVEN

Solutions for Section 7.1

Exercises

1. (a) $\frac{d}{dx} \sin(x^2 + 1) = 2x \cos(x^2 + 1);$ $\frac{d}{dx} \sin(x^3 + 1) = 3x^2 \cos(x^3 + 1)$
  (b) (i) $\frac{1}{2} \sin(x^2 + 1) + C$ (ii) $\frac{1}{3} \sin(x^3 + 1) + C$
  (c) (i) $-\frac{1}{2} \cos(x^2 + 1) + C$ (ii) $-\frac{1}{3} \cos(x^3 + 1) + C$

5. We use the substitution $w = 2x$, $dw = 2\,dx$.

$$\int \sin(2x)dx = \frac{1}{2} \int \sin(w)dw = -\frac{1}{2}\cos(w) + C = -\frac{1}{2}\cos(2x) + C.$$

Check: $\frac{d}{dx}\left(-\frac{1}{2}\cos(2x) + C\right) = \frac{1}{2}\sin(2x)(2) = \sin(2x).$

9. In this case, it seems easier not to substitute.

$$\int y^2(1+y)^2\,dy = \int y^2(y^2 + 2y + 1)\,dy = \int (y^4 + 2y^3 + y^2)\,dy$$

$$= \frac{y^5}{5} + \frac{y^4}{2} + \frac{y^3}{3} + C.$$

Check: $\frac{d}{dy}\left(\frac{y^5}{5} + \frac{y^4}{2} + \frac{y^3}{3} + C\right) = y^4 + 2y^3 + y^2 = y^2(y+1)^2.$

13. We use the substitution $w = 4 - x$, $dw = -dx$.

$$\int \frac{1}{\sqrt{4-x}}\,dx = -\int \frac{1}{\sqrt{w}}\,dw = -2\sqrt{w} + C = -2\sqrt{4-x} + C.$$

Check: $\frac{d}{dx}(-2\sqrt{4-x} + C) = -2 \cdot \frac{1}{2} \cdot \frac{1}{\sqrt{4-x}} \cdot -1 = \frac{1}{\sqrt{4-x}}.$

17. We use the substitution $w = \cos\theta + 5$, $dw = -\sin\theta\,d\theta$.

$$\int \sin\theta(\cos\theta + 5)^7\,d\theta = -\int w^7\,dw = -\frac{1}{8}w^8 + C$$

$$= -\frac{1}{8}(\cos\theta + 5)^8 + C.$$

Check:

$$\frac{d}{d\theta}\left[-\frac{1}{8}(\cos\theta + 5)^8 + C\right] = -\frac{1}{8} \cdot 8(\cos\theta + 5)^7 \cdot (-\sin\theta)$$

$$= \sin\theta(\cos\theta + 5)^7$$

21. We use the substitution $w = x^3 + 1$, $dw = 3x^2\,dx$, to get

$$\int x^2 e^{x^3+1}\,dx = \frac{1}{3} \int e^w\,dw = \frac{1}{3}e^w + C = \frac{1}{3}e^{x^3+1} + C.$$

Check: $\frac{d}{dx}\left(\frac{1}{3}e^{x^3+1} + C\right) = \frac{1}{3}e^{x^3+1} \cdot 3x^2 = x^2 e^{x^3+1}.$

25. We use the substitution $w = \ln z$, $dw = \frac{1}{z}\,dz$.

$$\int \frac{(\ln z)^2}{z}\,dz = \int w^2\,dw = \frac{w^3}{3} + C = \frac{(\ln z)^3}{3} + C.$$

Check: $\frac{d}{dz}\left[\frac{(\ln z)^3}{3} + C\right] = 3 \cdot \frac{1}{3}(\ln z)^2 \cdot \frac{1}{z} = \frac{(\ln z)^2}{z}.$

29. We use the substitution $w = \sqrt{y}$, $dw = \dfrac{1}{2\sqrt{y}}\,dy$.

$$\int \frac{e^{\sqrt{y}}}{\sqrt{y}}\,dy = 2\int e^w\,dw = 2e^w + C = 2e^{\sqrt{y}} + C.$$

Check: $\dfrac{d}{dy}(2e^{\sqrt{y}} + C) = 2e^{\sqrt{y}} \cdot \dfrac{1}{2\sqrt{y}} = \dfrac{e^{\sqrt{y}}}{\sqrt{y}}.$

33. We use the substitution $w = 1 + 3t^2$, $dw = 6t\,dt$.

$$\int \frac{t}{1 + 3t^2}\,dt = \int \frac{1}{w}(\tfrac{1}{6}\,dw) = \frac{1}{6}\ln|w| + C = \frac{1}{6}\ln(1 + 3t^2) + C.$$

(We can drop the absolute value signs since $1 + 3t^2 > 0$ for all t).

Check: $\dfrac{d}{dt}\left[\dfrac{1}{6}\ln(1 + 3t^2) + C\right] = \dfrac{1}{6}\dfrac{1}{1 + 3t^2}(6t) = \dfrac{t}{1 + 3t^2}.$

37. Since $d(\sinh x)/dx = \cosh x$, we have

$$\int \cosh x\,dx = \sinh x + C.$$

41. The general antiderivative is $\int (\pi t^3 + 4t)\,dt = (\pi/4)t^4 + 2t^2 + C.$

45. Make the substitution $w = 2 - 5x$, then $dw = -5dx$. We have

$$\int \sin(2 - 5x)dx = \int \sin w \left(-\frac{1}{5}\right)dw = -\frac{1}{5}(-\cos w) + C = \frac{1}{5}\cos(2 - 5x) + C.$$

49. $\displaystyle\int_0^{\pi} \cos(x + \pi)\,dx = \sin(x + \pi)\big|_0^{\pi} = \sin(2\pi) - \sin(\pi) = 0 - 0 = 0$

53. We substitute $w = \sqrt[3]{x} = x^{\frac{1}{3}}$. Then $dw = \dfrac{1}{3}x^{-\frac{2}{3}}\,dx = \dfrac{1}{3\sqrt[3]{x^2}}\,dx.$

$$\int_1^8 \frac{e^{\sqrt[3]{x}}}{\sqrt[3]{x^2}}dx = \int_{x=1}^{x=8} e^w (3\,dw) = 3e^w \bigg|_{x=1}^{x=8} = 3e^{\sqrt[3]{x}}\bigg|_1^8 = 3(e^2 - e).$$

57.

$$\int_{-1}^3 (x^3 + 5x)\,dx = \frac{x^4}{4}\bigg|_{-1}^3 + \frac{5x^2}{2}\bigg|_{-1}^3 = 40.$$

61. $\displaystyle\int_{-1}^2 \sqrt{x + 2}\,dx = \frac{2}{3}(x + 2)^{3/2}\bigg|_{-1}^2 = \frac{2}{3}\left[(4)^{3/2} - (1)^{3/2}\right] = \frac{2}{3}(7) = \frac{14}{3}$

Problems

65. Since $f(x) = 1/(x + 1)$ is positive on the interval $x = 0$ to $x = 2$, we have

$$\text{Area} = \int_0^2 \frac{1}{x + 1}\,dx = \ln(x + 1)\bigg|_0^2 = \ln 3 - \ln 1 = \ln 3.$$

The area is $\ln 3 \approx 1.0986$.

69. If $f(x) = \dfrac{1}{x + 1}$, the average value of f on the interval $0 \le x \le 2$ is defined to be

$$\frac{1}{2 - 0}\int_0^2 f(x)\,dx = \frac{1}{2}\int_0^2 \frac{dx}{x + 1}.$$

We'll integrate by substitution. We let $w = x + 1$ and $dw = dx$, and we have

$$\int_{x=0}^{x=2} \frac{dx}{x + 1} = \int_{w=1}^{w=3} \frac{dw}{w} = \ln w \bigg|_1^3 = \ln 3 - \ln 1 = \ln 3.$$

Thus, the average value of $f(x)$ on $0 \le x \le 2$ is $\frac{1}{2}\ln 3 \approx 0.5493$. See Figure 7.1.

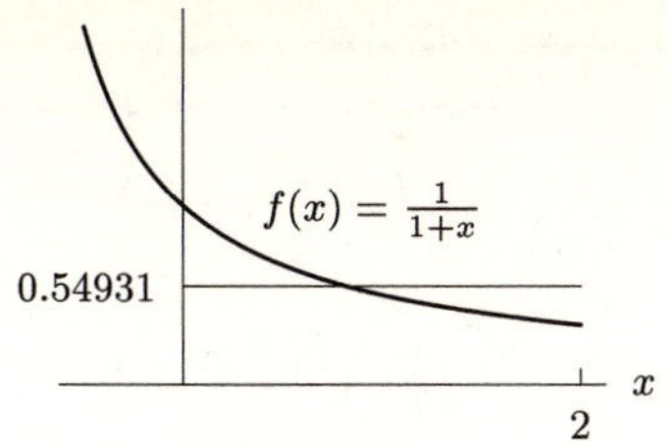

Figure 7.1

73. (a) In 1990, we have $P = 5.3e^{0.014(0)} = 5.3$ billion people.
In 2000, we have $P = 5.3e^{0.014(10)} = 6.1$ billion people.

(b) We have

$$\text{Average population} = \frac{1}{10-0}\int_0^{10} 5.3e^{0.014t}\,dt = \frac{1}{10}\cdot\frac{5.3}{0.014}e^{0.014t}\Big|_0^{10}$$

$$= \frac{1}{10}\left(\frac{5.3}{0.014}(e^{0.14} - e^0)\right) = 5.7.$$

The average population of the world during the 1990s was 5.7 billion people.

77. Since v is given as the velocity of a falling body, the height h is decreasing, so $v = -\frac{dh}{dt}$, and it follows that $h(t) = -\int v(t)\,dt$ and $h(0) = h_0$. Let $w = e^{t\sqrt{gk}} + e^{-t\sqrt{gk}}$. Then

$$dw = \sqrt{gk}\left(e^{t\sqrt{gk}} - e^{-t\sqrt{gk}}\right)dt,$$

so $\dfrac{dw}{\sqrt{gk}} = (e^{t\sqrt{gk}} - e^{-t\sqrt{gk}})\,dt$. Therefore,

$$-\int v(t)dt = -\int \sqrt{\frac{g}{k}}\left(\frac{e^{t\sqrt{gk}} - e^{-t\sqrt{gk}}}{e^{t\sqrt{gk}} + e^{-t\sqrt{gk}}}\right)dt$$

$$= -\sqrt{\frac{g}{k}}\int \frac{1}{e^{t\sqrt{gk}} + e^{-t\sqrt{gk}}}\left(e^{t\sqrt{gk}} - e^{-t\sqrt{gk}}\right)dt$$

$$= -\sqrt{\frac{g}{k}}\int \left(\frac{1}{w}\right)\frac{dw}{\sqrt{gk}}$$

$$= -\sqrt{\frac{g}{gk^2}}\ln|w| + C$$

$$= -\frac{1}{k}\ln\left(e^{t\sqrt{gk}} + e^{-t\sqrt{gk}}\right) + C.$$

Since

$$h(0) = -\frac{1}{k}\ln(e^0 + e^0) + C = -\frac{\ln 2}{k} + C = h_0,$$

we have $C = h_0 + \dfrac{\ln 2}{k}$. Thus,

$$h(t) = -\frac{1}{k}\ln\left(e^{t\sqrt{gk}} + e^{-t\sqrt{gk}}\right) + \frac{\ln 2}{k} + h_0 = -\frac{1}{k}\ln\left(\frac{e^{t\sqrt{gk}} + e^{-t\sqrt{gk}}}{2}\right) + h_0.$$

Solutions for Section 7.2

Exercises

1. Let $u = \arctan x$, $v' = 1$. Then $v = x$ and $u' = \dfrac{1}{1 + x^2}$. Integrating by parts, we get:

$$\int 1 \cdot \arctan x \, dx = x \cdot \arctan x - \int x \cdot \frac{1}{1 + x^2} \, dx.$$

To compute the second integral use the substitution, $z = 1 + x^2$.

$$\int \frac{x}{1 + x^2} \, dx = \frac{1}{2} \int \frac{dz}{z} = \frac{1}{2} \ln |z| + C = \frac{1}{2} \ln(1 + x^2) + C.$$

Thus,

$$\int \arctan x \, dx = x \cdot \arctan x - \frac{1}{2} \ln(1 + x^2) + C.$$

5. Let $u = t$, $v' = \sin t$. Thus, $v = -\cos t$ and $u' = 1$. With this choice of u and v, integration by parts gives:

$$\int t \sin t \, dt = -t \cos t - \int (-\cos t) \, dt$$
$$= -t \cos t + \sin t + C.$$

9. Let $u = t^2$, $v' = \sin t$ implying $v = -\cos t$ and $u' = 2t$. Integrating by parts, we get:

$$\int t^2 \sin t \, dt = -t^2 \cos t - \int 2t(-\cos t) \, dt.$$

Again, applying integration by parts with $u = t$, $v' = \cos t$, we have:

$$\int t \cos t \, dt = t \sin t + \cos t + C.$$

Thus

$$\int t^2 \sin t \, dt = -t^2 \cos t + 2t \sin t + 2 \cos t + C.$$

13. Let $u = \ln 5q$, $v' = q^5$. Then $v = \frac{1}{6}q^6$ and $u' = \dfrac{1}{q}$. Integrating by parts, we get:

$$\int q^5 \ln 5q \, dq = \frac{1}{6}q^6 \ln 5q - \int \left(5 \cdot \frac{1}{5q}\right) \cdot \frac{1}{6}q^6 \, dq$$
$$= \frac{1}{6}q^6 \ln 5q - \frac{1}{36}q^6 + C.$$

17. Let $u = \theta + 1$ and $v' = \sin(\theta + 1)$, so $u' = 1$ and $v = -\cos(\theta + 1)$.

$$\int (\theta + 1) \sin(\theta + 1) \, d\theta = -(\theta + 1) \cos(\theta + 1) + \int \cos(\theta + 1) \, d\theta$$
$$= -(\theta + 1) \cos(\theta + 1) + \sin(\theta + 1) + C.$$

21. $\displaystyle \int \frac{t + 7}{\sqrt{5 - t}} \, dt = \int \frac{t}{\sqrt{5 - t}} \, dt + 7 \int (5 - t)^{-1/2} \, dt.$

To calculate the first integral, we use integration by parts. Let $u = t$ and $v' = \frac{1}{\sqrt{5-t}}$, so $u' = 1$ and $v = -2(5-t)^{1/2}$. Then

$$\int \frac{t}{\sqrt{5 - t}} \, dt = -2t(5 - t)^{1/2} + 2 \int (5 - t)^{1/2} \, dt = -2t(5 - t)^{1/2} - \frac{4}{3}(5 - t)^{3/2} + C.$$

We can calculate the second integral directly: $7 \displaystyle \int (5 - t)^{-1/2} = -14(5 - t)^{1/2} + C_1$. Thus

$$\int \frac{t + 7}{\sqrt{5 - t}} \, dt = -2t(5 - t)^{1/2} - \frac{4}{3}(5 - t)^{3/2} - 14(5 - t)^{1/2} + C_2.$$

25. This integral can first be simplified by making the substitution $w = x^2$, $dw = 2x\,dx$. Then

$$\int x \arctan x^2 \, dx = \frac{1}{2} \int \arctan w \, dw.$$

To evaluate $\int \arctan w \, dw$, we'll use integration by parts. Let $u = \arctan w$ and $v' = 1$, so $u' = \frac{1}{1+w^2}$ and $v = w$. Then

$$\int \arctan w \, dw = w \arctan w - \int \frac{w}{1+w^2} \, dw = w \arctan w - \frac{1}{2} \ln|1 + w^2| + C.$$

Since $1 + w^2$ is never negative, we can drop the absolute value signs. Thus, we have

$$\int x \arctan x^2 \, dx = \frac{1}{2} \left(x^2 \arctan x^2 - \frac{1}{2} \ln(1 + (x^2)^2) + C \right)$$

$$= \frac{1}{2} x^2 \arctan x^2 - \frac{1}{4} \ln(1 + x^4) + C.$$

29. $\displaystyle\int_3^5 x \cos x \, dx = (\cos x + x \sin x) \Big|_3^5 = \cos 5 + 5 \sin 5 - \cos 3 - 3 \sin 3 \approx -3.944.$

33. $\displaystyle\int_0^5 \ln(1+t) \, dt = ((1+t)\ln(1+t) - (1+t)) \Big|_0^5 = 6 \ln 6 - 5 \approx 5.751.$

Problems

37.

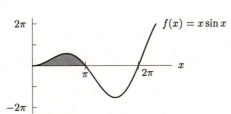

The graph of $f(x) = x \sin x$ is shown above. The first positive zero is at $x = \pi$, so, using integration by parts,

$$\text{Area} = \int_0^\pi x \sin x \, dx$$

$$= -x \cos x \Big|_0^\pi + \int_0^\pi \cos x \, dx$$

$$= -x \cos x \Big|_0^\pi + \sin x \Big|_0^\pi$$

$$= -\pi \cos \pi - (-0 \cos 0) + \sin \pi - \sin 0 = \pi.$$

41. Let $u = e^\theta$ and $v' = \cos \theta$, so $u' = e^\theta$ and $v = \sin \theta$. Then $\int e^\theta \cos \theta \, d\theta = e^\theta \sin \theta - \int e^\theta \sin \theta \, d\theta$.

In Problem 40 we found that $\displaystyle\int e^x \sin x \, dx = \frac{1}{2} e^x (\sin x - \cos x) + C.$

$$\int e^\theta \cos \theta \, d\theta = e^\theta \sin \theta - \left[\frac{1}{2} e^\theta (\sin \theta - \cos \theta) \right] + C$$

$$= \frac{1}{2} e^\theta (\sin \theta + \cos \theta) + C.$$

45. We integrate by parts. Let $u = x^n$ and $v' = \cos ax$, so $u' = nx^{n-1}$ and $v = \frac{1}{a} \sin ax$. Then

$$\int x^n \cos ax \, dx = \frac{1}{a} x^n \sin ax - \int (nx^{n-1})(\frac{1}{a} \sin ax) \, dx$$

$$= \frac{1}{a} x^n \sin ax - \frac{n}{a} \int x^{n-1} \sin ax \, dx.$$

49. Since $f'(x) = 2x$, integration by parts tells us that

$$\int_0^{10} f(x)g'(x)\,dx = f(x)g(x)\Big|_0^{10} - \int_0^{10} f'(x)g(x)\,dx$$

$$= f(10)g(10) - f(0)g(0) - 2\int_0^{10} xg(x)\,dx.$$

We can use left and right Riemann Sums with $\Delta x = 2$ to approximate $\int_0^{10} xg(x)\,dx$:

$$\text{Left sum} \approx 0 \cdot g(0)\Delta x + 2 \cdot g(2)\Delta x + 4 \cdot g(4)\Delta x + 6 \cdot g(6)\Delta x + 8 \cdot g(8)\Delta x$$

$$= (0(2.3) + 2(3.1) + 4(4.1) + 6(5.5) + 8(5.9))\,2 = 205.6.$$

$$\text{Right sum} \approx 2 \cdot g(2)\Delta x + 4 \cdot g(4)\Delta x + 6 \cdot g(6)\Delta x + 8 \cdot g(8)\Delta x + 10 \cdot g(10)\Delta x$$

$$= (2(3.1) + 4(4.1) + 6(5.5) + 8(5.9) + 10(6.1))\,2 = 327.6.$$

A good estimate for the integral is the average of the left and right sums, so

$$\int_0^{10} xg(x)\,dx \approx \frac{205.6 + 327.6}{2} = 266.6.$$

Substituting values for f and g, we have

$$\int_0^{10} f(x)g'(x)\,dx = f(10)g(10) - f(0)g(0) - 2\int_0^{10} xg(x)\,dx$$

$$\approx 10^2(6.1) - 0^2(2.3) - 2(266.6) = 76.8 \approx 77.$$

53. (a) Increasing V_0 increases the maximum value of V, since this maximum is V_0. Increasing ω or ϕ does not affect the maximum of V.

(b) Since

$$\frac{dV}{dt} = -\omega V_0 \sin(\omega t + \phi),$$

the maximum of dV/dt is ωV_0. Thus, the maximum of dV/dt is increased if V_0 or ω is increased, and is unaffected if ϕ is increased.

(c) The period of $V = V_0 \cos(\omega t + \phi)$ is $2\pi/\omega$, so

$$\text{Average value} = \frac{1}{2\pi/\omega}\int_0^{2\pi/\omega}(V_0\cos(\omega t + \phi))^2\,dt.$$

Substituting $x = \omega t + \phi$, we have $dx = \omega dt$. When $t = 0$, $x = \phi$, and when $t = 2\pi/\omega$, $x = 2\pi + \phi$. Thus,

$$\text{Average value} = \frac{\omega}{2\pi}\int_\phi^{2\pi+\phi}V_0^2(\cos x)^2\frac{1}{\omega}\,dx$$

$$= \frac{V_0^2}{2\pi}\int_\phi^{2\pi+\phi}(\cos x)^2\,dx.$$

Using integration by parts and the fact that $\sin^2 x = 1 - \cos^2 x$, we see that

$$\text{Average value} = \frac{V_0^2}{2\pi}\left[\frac{1}{2}(\cos x \sin x + x)\right]_\phi^{2\pi+\phi}$$

$$= \frac{V_0^2}{4\pi}[\cos(2\pi + \phi)\sin(2\pi + \phi) + (2\pi + \phi) - \cos\phi\sin\phi - \phi]$$

$$= \frac{V_0^2}{4\pi} \cdot 2\pi = \frac{V_0^2}{2}.$$

Thus, increasing V_0 increases the average value; increasing ω or ϕ has no effect.

However, it is not in fact necessary to compute the integral to see that ω does not affect the average value, since all ω's dropped out of the average value expression when we made the substitution $x = \omega t + \phi$.

Solutions for Section 7.3

Exercises

1. $\dfrac{1}{10}e^{(-3\theta)}(-3\cos\theta + \sin\theta) + C.$
(Let $a = -3, b = 1$ in II-9.)

5. Note that you can't use substitution here: letting $w = x^3 + 5$ doesn't work, since there is no $dw = 3x^2\,dx$ in the integrand. What will work is simply multiplying out the square: $(x^3 + 5)^2 = x^6 + 10x^3 + 25$. Then use I-1:

$$\int (x^3 + 5)^2\,dx = \int x^6\,dx + 10\int x^3\,dx + 25\int 1\,dx = \frac{1}{7}x^7 + 10\cdot\frac{1}{4}x^4 + 25x + C.$$

9. Let $m = 3$ in IV-21.

$$\int \frac{1}{\cos^3 x}\,dx = \frac{1}{2}\frac{\sin x}{\cos^2 x} + \frac{1}{2}\int \frac{1}{\cos x}\,dx$$
$$= \frac{1}{2}\frac{\sin x}{\cos^2 x} + \frac{1}{4}\ln\left|\frac{\sin x + 1}{\sin x - 1}\right| + C \text{ by IV-22.}$$

13. $\left(\dfrac{1}{3}x^2 - \dfrac{2}{9}x + \dfrac{2}{27}\right)e^{3x} + C.$
(Let $a = 3, p(x) = x^2$ in III-14.)

17. Use long division to reorganize the integral:

$$\int \frac{t^2 + 1}{t^2 - 1}\,dt = \int \left(1 + \frac{2}{t^2 - 1}\right)dt = \int dt + \int \frac{2}{(t-1)(t+1)}\,dt.$$

To get this second integral, let $a = 1, b = -1$ in V-26, so

$$\int \frac{t^2 + 1}{t^2 - 1}\,dt = t + \ln|t - 1| - \ln|t + 1| + C.$$

21. $\dfrac{1}{34}e^{5x}(5\sin 3x - 3\cos 3x) + C.$
(Let $a = 5, b = 3$ in II-8.)

25. Substitute $w = 7x$, $dw = 7\,dx$. Then use IV-21.

$$\int \frac{1}{\cos^4 7x}\,dx = \frac{1}{7}\int \frac{1}{\cos^4 w}\,dw = \frac{1}{7}\left[\frac{1}{3}\frac{\sin w}{\cos^3 w} + \frac{2}{3}\int \frac{1}{\cos^2 w}\,dw\right]$$
$$= \frac{1}{21}\frac{\sin w}{\cos^3 w} + \frac{2}{21}\left[\frac{\sin w}{\cos w} + C\right]$$
$$= \frac{1}{21}\frac{\tan w}{\cos^2 w} + \frac{2}{21}\tan w + C$$
$$= \frac{1}{21}\frac{\tan 7x}{\cos^2 7x} + \frac{2}{21}\tan 7x + C.$$

29.

$$\int \frac{dy}{4 - y^2} = -\int \frac{dy}{(y+2)(y-2)} = -\frac{1}{4}(\ln|y - 2| - \ln|y + 2|) + C.$$

(Let $a = 2, b = -2$ in V-26.)

33. We use the Pythagorean Identity to change the integrand in the following manner:

$$\sin^3 x = \left(\sin^2 x\right)\sin x = \left(1 - \cos^2 x\right)\sin x = \sin x - \cos^2 x \sin x.$$

Thus, we have

$$\int \sin^3 x \, dx = \int \left(\sin x - \cos^2 x \sin x \right) dx$$

$$= \int \sin x \, dx - \int \cos^2 x \sin x \, dx.$$

The first of these new integrals can be easily found. The second can be found using the substitution $w = \cos x$ so $dw = -\sin x \, dx$. The second integral becomes

$$\int \cos^2 x \sin x \, dx = -\int w^2 \, dw$$

$$= -\frac{1}{3} w^3 + C$$

$$= -\frac{1}{3} \cos^3 x + C$$

and so our final answer is

$$\int \sin^3 x \, dx = \int \sin x \, dx - \int \cos^2 x \sin x \, dx$$

$$= -\cos x + (1/3) \cos^3 x + C.$$

Problems

37. Using formula II-11, if $m \neq \pm n$, then

$$\int_{-\pi}^{\pi} \cos m\theta \cos n\theta \, d\theta = \frac{1}{n^2 - m^2} \left(n \cos m\theta \sin n\theta - m \sin m\theta \cos n\theta \right) \Big|_{-\pi}^{\pi}.$$

We see that in the evaluation, each term will have a $\sin k\pi$ term, so the expression reduces to 0.

Solutions for Section 7.4 ━━━━━━━━━━━━━━

Exercises

1. Since $25 - x^2 = (5 - x)(5 + x)$, we take

$$\frac{20}{25 - x^2} = \frac{A}{5 - x} + \frac{B}{5 + x}.$$

So,

$$20 = A(5 + x) + B(5 - x)$$
$$20 = (A - B)x + 5A + 5B,$$

giving

$$A - B = 0$$
$$5A + 5B = 20.$$

Thus $A = B = 2$ and

$$\frac{20}{25 - x^2} = \frac{2}{5 - x} + \frac{2}{5 + x}.$$

5. Using the result of Problem 1, we have

$$\int \frac{20}{25 - x^2} \, dx = \int \frac{2}{5 - x} \, dx + \int \frac{2}{5 + x} \, dx = -2 \ln |5 - x| + 2 \ln |5 + x| + C.$$

9. (a) Yes, use $x = 3 \sin \theta$.
   (b) No; better to substitute $w = 9 - x^2$, so $dw = -2x \, dx$.

13. Division gives

$$\frac{x^4 + 12x^3 + 15x^2 + 25x + 11}{x^3 + 12x^2 + 11x} = x + \frac{4x^2 + 25x + 11}{x^3 + 12x^2 + 11x}.$$

Since $x^3 + 12x^2 + 11x = x(x+1)(x+11)$, we write

$$\frac{4x^2 + 25x + 11}{x^3 + 12x^2 + 11x} = \frac{A}{x} + \frac{B}{x+1} + \frac{C}{x+11}$$

giving

$$4x^2 + 25x + 11 = A(x+1)(x+11) + Bx(x+11) + Cx(x+1)$$
$$4x^2 + 25x + 11 = (A+B+C)x^2 + (12A + 11B + C)x + 11A$$

so

$$A + B + C = 4$$
$$12A + 11B + C = 25$$
$$11A = 11.$$

Thus, $A = B = 1, C = 2$ so

$$\int \frac{x^4 + 12x^3 + 15x^2 + 25x + 11}{x^3 + 12x^2 + 11x}\, dx = \int x\, dx + \int \frac{dx}{x} + \int \frac{dx}{x+1} + \int \frac{2\, dx}{x+11}$$

$$= \frac{x^2}{2} + \ln|x| + \ln|x+1| + 2\ln|x+11| + K.$$

17. Since $x = \sin t + 2$, we have

$$4x - 3 - x^2 = 4(\sin t + 2) - 3 - (\sin t + 2)^2 = 1 - \sin^2 t = \cos^2 t$$

and $dx = \cos t\, dt$, so substitution gives

$$\int \frac{1}{\sqrt{4x - 3 - x^2}} = \int \frac{1}{\sqrt{\cos^2 t}} \cos t\, dt = \int dt = t + C = \arcsin(x - 2) + C.$$

Problems

21. Since $x^2 + 2x + 2 = (x+1)^2 + 1$, we have

$$\int \frac{1}{x^2 + 2x + 2}\, dx = \int \frac{1}{(x+1)^2 + 1}\, dx.$$

Substitute $x + 1 = \tan\theta$, so $x = (\tan\theta) - 1$.

25. Since $t^2 + 4t + 7 = (t+2)^2 + 3$, we have

$$\int (t+2)\sin(t^2 + 4t + 7)\, dt = \int (t+2)\sin((t+2)^2 + 3)\, dt.$$

Substitute $w = (t+2)^2 + 3$, so $dw = 2(t+2)\, dt$.

This integral can also be computed without completing the square, by substituting $w = t^2 + 4t + 7$, so $dw = (2t + 4)\, dt$.

29. Since $x^3 + x = x(x^2 + 1)$ cannot be factored further, we write

$$\frac{x+1}{x^3 + x} = \frac{A}{x} + \frac{Bx + C}{x^2 + 1}.$$

Multiplying by $x(x^2 + 1)$ gives

$$x + 1 = A(x^2 + 1) + (Bx + C)x$$
$$x + 1 = (A + B)x^2 + Cx + A,$$

so

$$A + B = 0$$
$$C = 1$$
$$A = 1.$$

Thus, $A = C = 1$, $B = -1$, and we have

$$\int \frac{x+1}{x^3+x}\,dx = \int \left(\frac{1}{x} + \frac{-x+1}{x^2+1}\right) = \int \frac{dx}{x} - \int \frac{x\,dx}{x^2+1} + \int \frac{dx}{x^2+1}$$

$$= \ln|x| - \frac{1}{2}\ln\left|x^2+1\right| + \arctan x + K.$$

33. Let $t = \tan\theta$ so $dt = (1/\cos^2\theta)d\theta$. Since $\sqrt{1 + \tan^2\theta} = 1/\cos\theta$, we have

$$\int \frac{dt}{t^2\sqrt{1+t^2}} = \int \frac{1/\cos^2\theta}{\tan^2\theta\sqrt{1+\tan^2\theta}}\,d\theta = \int \frac{\cos\theta}{\tan^2\theta\cos^2\theta}\,d\theta = \int \frac{\cos\theta}{\sin^2\theta}\,d\theta.$$

The last integral can be evaluated by guess-and-check or by substituting $w = \sin\theta$. The result is

$$\int \frac{dt}{t^2\sqrt{1+t^2}} = \int \frac{\cos\theta}{\sin^2\theta}\,d\theta = -\frac{1}{\sin\theta} + C.$$

Since $t = \tan\theta$ and $1/\cos^2\theta = 1 + \tan^2\theta$, we have

$$\cos\theta = \frac{1}{\sqrt{1 + \tan^2\theta}} = \frac{1}{\sqrt{1+t^2}}.$$

In addition, $\tan\theta = \sin\theta/\cos\theta$ so

$$\sin\theta = \tan\theta\cos\theta = \frac{t}{\sqrt{1+t^2}}.$$

Thus

$$\int \frac{dt}{t^2\sqrt{1+t^2}} = -\frac{\sqrt{1+t^2}}{t} + C.$$

37. Notice that because $\frac{3x}{(x-1)(x-4)}$ is negative for $2 \leq x \leq 3$,

$$\text{Area} = -\int_2^3 \frac{3x}{(x-1)(x-4)}\,dx.$$

Using partial fractions gives

$$\frac{3x}{(x-1)(x-4)} = \frac{A}{x-1} + \frac{B}{x-4} = \frac{(A+B)x - B - 4A}{(x-1)(x-4)}.$$

Multiplying through by $(x-1)(x-4)$ gives

$$3x = (A+B)x - B - 4A$$

so $A = -1$ and $B = 4$. Thus

$$-\int_2^3 \frac{3x}{(x-1)(x-4)}\,dx = -\int_2^3 \left(\frac{-1}{x-1} + \frac{4}{x-4}\right)dx = (\ln|x-1| - 4\ln|x-4|)\Big|_2^3 = 5\ln 2.$$

41. We have

$$\text{Area} = \int_0^3 \frac{1}{\sqrt{x^2+9}}\,dx.$$

Let $x = 3\tan\theta$ so $dx = (3/\cos^2\theta)d\theta$ and

$$\sqrt{x^2+9} = \sqrt{\frac{9\sin^2\theta}{\cos^2\theta} + 9} = \frac{3}{\cos\theta}.$$

When $x = 0, \theta = 0$ and when $x = 3, \theta = \pi/4$. Thus

$$\int_0^3 \frac{1}{\sqrt{x^2+9}}\,dx = \int_0^{\pi/4} \frac{1}{\sqrt{9\tan^2\theta + 9}} \frac{3}{\cos^2\theta}\,d\theta = \int_0^{\pi/4} \frac{1}{3/\cos\theta}\cdot\frac{3}{\cos^2\theta}\,d\theta = \int_0^{\pi/4} \frac{1}{\cos\theta}\,d\theta$$

$$= \frac{1}{2}\ln\left|\frac{\sin\theta+1}{\sin\theta-1}\right|\Bigg|_0^{\pi/4} = \frac{1}{2}\ln\left|\frac{1/\sqrt{2}+1}{1/\sqrt{2}-1}\right| = \frac{1}{2}\ln\left(\frac{1+\sqrt{2}}{\sqrt{2}-1}\right).$$

This answer can be simplified to $\ln(1+\sqrt{2})$ by multiplying the numerator and denominator of the fraction by $(\sqrt{2}+1)$ and using the properties of logarithms. The integral $\int(1/\cos\theta)d\theta$ is done using the Table of Integrals.

45. **(a)** We want to evaluate the integral

$$T = \int_0^{a/2} \frac{k\,dx}{(a-x)(b-x)}.$$

Using partial fractions, we have

$$\frac{k}{(a-x)(b-x)} = \frac{C}{a-x} + \frac{D}{b-x}$$
$$k = C(b-x) + D(a-x)$$
$$k = -(C+D)x + Cb + Da$$

so

$$0 = -(C+D)$$
$$k = Cb + Da,$$

giving

$$C = -D = \frac{k}{b-a}.$$

Thus, the time is given by

$$T = \int_0^{a/2} \frac{k\,dx}{(a-x)(b-x)} = \frac{k}{b-a}\int_0^{a/2}\left(\frac{1}{a-x} - \frac{1}{b-x}\right)dx$$

$$= \frac{k}{b-a}\left(-\ln|a-x| + \ln|b-x|\right)\Bigg|_0^{a/2}$$

$$= \frac{k}{b-a}\ln\left|\frac{b-x}{a-x}\right|\Bigg|_0^{a/2}$$

$$= \frac{k}{b-a}\left(\ln\left(\frac{2b-a}{a}\right) - \ln\left(\frac{b}{a}\right)\right)$$

$$= \frac{k}{b-a}\ln\left(\frac{2b-a}{b}\right).$$

(b) A similar calculation with x_0 instead of $a/2$ leads to the following expression for the time

$$T = \int_0^{x_0} \frac{k\,dx}{(a-x)(b-x)} = \frac{k}{b-a}\ln\left|\frac{b-x}{a-x}\right|\Bigg|_0^{x_0}$$

$$= \frac{k}{b-a}\left(\ln\left|\frac{b-x_0}{a-x_0}\right| - \ln\left(\frac{b}{a}\right)\right).$$

As $x_0 \to a$, the value of $|a - x_0| \to 0$, so $|b - x_0|/|a - x_0| \to \infty$. Thus, $T \to \infty$ as $x_0 \to a$. In other words, the time taken tends to infinity.

Solutions for Section 7.5

Exercises

1. (a) The approximation LEFT(2) uses two rectangles, with the height of each rectangle determined by the left-hand endpoint. See Figure 7.2. We see that this approximation is an underestimate.

Figure 7.2

(b) The approximation RIGHT(2) uses two rectangles, with the height of each rectangle determined by the right-hand endpoint. See Figure 7.3. We see that this approximation is an overestimate.

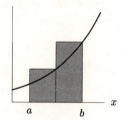

Figure 7.3

(c) The approximation TRAP(2) uses two trapezoids, with the height of each trapezoid given by the secant line connecting the two endpoints. See Figure 7.4. We see that this approximation is an overestimate.

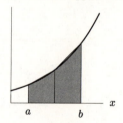

Figure 7.4

(d) The approximation MID(2) uses two rectangles, with the height of each rectangle determined by the height at the midpoint. Alternately, we can view MID(2) as a trapezoid rule where the height is given by the tangent line at the midpoint. Both interpretations are shown in Figure 7.5. We see from the tangent line interpretation that this approximation is an underestimate

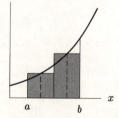 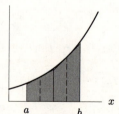

Figure 7.5

5. (a) Since two rectangles are being used, the width of each rectangle is 3. The height is given by the left-hand endpoint so we have

$$\text{LEFT}(2) = f(0) \cdot 3 + f(3) \cdot 3 = 0^2 \cdot 3 + 3^2 \cdot 3 = 27.$$

(b) Since two rectangles are being used, the width of each rectangle is 3. The height is given by the right-hand endpoint so we have
$$\text{RIGHT}(2) = f(3) \cdot 3 + f(6) \cdot 3 = 3^2 \cdot 3 + 6^2 \cdot 3 = 135.$$

(c) We know that TRAP is the average of LEFT and RIGHT and so
$$\text{TRAP}(2) = \frac{27 + 135}{2} = 81.$$

(d) Since two rectangles are being used, the width of each rectangle is 3. The height is given by the height at the midpoint so we have
$$\text{MID}(2) = f(1.5) \cdot 3 + f(4.5) \cdot 3 = (1.5)^2 \cdot 3 + (4.5)^2 \cdot 3 = 67.5.$$

Problems

9. Let $s(t)$ be the distance traveled at time t and $v(t)$ be the velocity at time t. Then the distance traveled during the interval $0 \le t \le 6$ is

$$s(6) - s(0) = s(t) \Big|_0^6$$

$$= \int_0^6 s'(t)\, dt \quad \text{(by the Fundamental Theorem)}$$

$$= \int_0^6 v(t)\, dt.$$

We estimate the distance by estimating this integral.

From the table, we find: $\text{LEFT}(6) = 31$, $\text{RIGHT}(6) = 39$, $\text{TRAP}(6) = 35$.

13. $f(x)$ is concave down, so MID gives an overestimate and TRAP gives an underestimate.

17. **(a)** TRAP(4) gives probably the best estimate of the integral. We cannot calculate MID(4).
$$\text{LEFT}(4) = 3 \cdot 100 + 3 \cdot 97 + 3 \cdot 90 + 3 \cdot 78 = 1095$$
$$\text{RIGHT}(4) = 3 \cdot 97 + 3 \cdot 90 + 3 \cdot 78 + 3 \cdot 55 = 960$$
$$\text{TRAP}(4) = \frac{1095 + 960}{2} = 1027.5.$$

(b) Because there are no points of inflection, the graph is either concave down or concave up. By plotting points, we see that it is concave down. So TRAP(4) is an underestimate.

21.

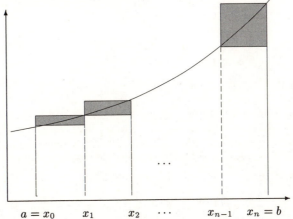

From the diagram, the difference between $\text{RIGHT}(n)$ and $\text{LEFT}(n)$ is the area of the shaded rectangles.
$$\text{RIGHT}(n) = f(x_1)\Delta x + f(x_2)\Delta x + \cdots + f(x_n)\Delta x$$
$$\text{LEFT}(n) = f(x_0)\Delta x + f(x_1)\Delta x + \cdots + f(x_{n-1})\Delta x$$
Notice that the terms in these two sums are the same, except that $\text{RIGHT}(n)$ contains $f(x_n)\Delta x$ $(= f(b)\Delta x)$, and $\text{LEFT}(n)$ contains $f(x_0)\Delta x$ $(= f(a)\Delta x)$. Thus

$$\text{RIGHT}(n) = \text{LEFT}(n) + f(x_n)\Delta x - f(x_0)\Delta x$$
$$= \text{LEFT}(n) + f(b)\Delta x - f(a)\Delta x$$

25. First, we compute:

$$(f(b) - f(a))\Delta x = (f(b) - f(a))\left(\frac{b-a}{n}\right)$$

$$= (f(5) - f(2))\left(\frac{3}{n}\right)$$

$$= (21 - 13)\left(\frac{3}{n}\right)$$

$$= \frac{24}{n}$$

RIGHT(10) = LEFT(10) + 24 = 3.156 + 2.4 = 5.556.
TRAP(10) = LEFT(10) + $\frac{1}{2}$(2.4) = 3.156 + 1.2 = 4.356.
LEFT(20) = $\frac{1}{2}$(LEFT(10) + MID(10)) = $\frac{1}{2}$(3.156 + 3.242) = 3.199.
RIGHT(20) = LEFT(20) + 2.4 = 3.199 + 1.2 = 4.399.
TRAP(20) = LEFT(20) + $\frac{1}{2}$(1.2) = 3.199 + 0.6 = 3.799.

Solutions for Section 7.6

Exercises

1. We saw in Problem 5 in Section 7.5 that, for this definite integral, we have LEFT(2) = 27, RIGHT(2) = 135, TRAP(2) = 81, and MID(2) = 67.5. Thus,

$$SIMP(2) = \frac{2MID(2) + TRAP(2)}{3} = \frac{2(67.5) + 81}{3} = 72.$$

Notice that

$$\int_0^6 x^2 dx = \left.\frac{x^3}{3}\right|_0^6 = \frac{6^3}{3} - \frac{0^3}{3} = 72,$$

and so SIMP(2) gives the exact value of the integral in this case.

Problems

5. **(a)** $\displaystyle\int_0^4 e^x \, dx = e^x \bigg|_0^4 = e^4 - e^0 \approx 53.598 \ldots .$

    (b) Computing the sums directly, since $\Delta x = 2$, we have
        LEFT(2)= $2 \cdot e^0 + 2 \cdot e^2 \approx 2(1) + 2(7.389) = 16.778$;   error = 36.820.
        RIGHT(2)= $2 \cdot e^2 + 2 \cdot e^4 \approx 2(7.389) + 2(54.598) = 123.974$;   error = -70.376.
        TRAP(2)= $\dfrac{16.778 + 123.974}{2} = 70.376$;   error = 16.778.
        MID(2)= $2 \cdot e^1 + 2 \cdot e^3 \approx 2(2.718) + 2(20.086) = 45.608$;   error = 7.990.
        SIMP(2)= $\dfrac{2(45.608) + 70.376}{3} = 53.864$;   error = -0.266.

    (c) Similarly, since $\Delta x = 1$, we have LEFT(4)= 31.193;   error = 22.405
        RIGHT(4)= 84.791;   error = -31.193
        TRAP(4)= 57.992;   error = -4.394
        MID(4)= 51.428;   error = 2.170
        SIMP(4)= 53.616;   error = -0.018

    (d) For LEFT and RIGHT, we expect the error to go down by 1/2, and this is very roughly what we see. For MID and TRAP, we expect the error to go down by 1/4, and this is approximately what we see. For SIMP, we expect the error to go down by $1/2^4 = 1/16$, and this is approximately what we see.

9. Since the midpoint rule is sensitive to f'', the simplifying assumption should be that f'' does not change sign in the interval of integration. Thus MID(10) and MID(20) will both be overestimates or will both be underestimates. Since the larger number, MID(10) is less accurate than the smaller number, they must both be overestimates. Then the information that ERROR(10) $= 4 \times$ ERROR(20) means that the the value of the integral and the two sums are arranged as follows:

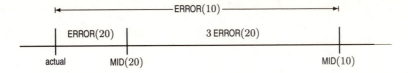

Thus

$$3 \times \text{ERROR}(20) = \text{MID}(10) - \text{MID}(20) = 35.619 - 35.415 = 0.204,$$

so ERROR(20) $= 0.068$ and ERROR(10) $= 4 \times$ ERROR(20) $= 0.272$.

Solutions for Section 7.7

Exercises

1. **(a)** See Figure 7.6. The area extends out infinitely far along the positive x-axis.

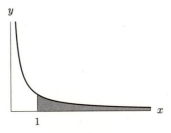

Figure 7.6

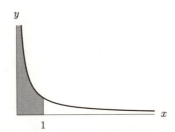

Figure 7.7

(b) See Figure 7.7. The area extends up infinitely far along the positive y-axis.

5. We have

$$\int_1^\infty \frac{1}{5x+2}\,dx = \lim_{b \to \infty} \int_1^b \frac{1}{5x+2}\,dx = \lim_{b \to \infty} \left(\frac{1}{5}\ln\left(5x+2\right)\right)\Big|_1^b = \lim_{b \to \infty} \left(\frac{1}{5}\ln\left(5b+2\right) - \frac{1}{5}\ln\left(7\right)\right).$$

As $b \leftarrow \infty$, we know that $\ln\left(5b+2\right) \to \infty$, and so this integral diverges.

9. Using integration by parts with $u = x$ and $v' = e^{-x}$, we find that

$$\int xe^{-x}\,dx = -xe^{-x} - \int -e^{-x}\,dx = -(1+x)e^{-x}$$

so

$$\int_0^\infty \frac{x}{e^x}\,dx = \lim_{b \to \infty} \int_0^b \frac{x}{e^x}\,dx$$

$$= \lim_{b \to \infty} -1(1+x)e^{-x}\Big|_0^b$$

$$= \lim_{b \to \infty} \left[1 - (1+b)e^{-b}\right]$$

$$= 1.$$

13. This is an improper integral because $\sqrt{16 - x^2} = 0$ at $x = 4$. So

$$\int_0^4 \frac{dx}{\sqrt{16 - x^2}} = \lim_{b \to 4^-} \int_0^b \frac{dx}{\sqrt{16 - x^2}}$$

$$= \lim_{b \to 4^-} (\arcsin x/4) \Big|_0^b$$

$$= \lim_{b \to 4^-} [\arcsin(b/4) - \arcsin(0)] = \pi/2 - 0 = \pi/2.$$

17.

$$\int_1^\infty \frac{1}{x^2 + 1} \, dx = \lim_{b \to \infty} \int_1^b \frac{1}{x^2 + 1} \, dx$$

$$= \lim_{b \to \infty} \arctan(x) \Big|_1^b$$

$$= \lim_{b \to \infty} [\arctan(b) - \arctan(1)]$$

$$= \pi/2 - \pi/4 = \pi/4.$$

21. With the substitution $w = \ln x$, $dw = \frac{1}{x} dx$,

$$\int \frac{dx}{x \ln x} = \int \frac{1}{w} \, dw = \ln |w| + C = \ln |\ln x| + C$$

so

$$\int_2^\infty \frac{dx}{x \ln x} = \lim_{b \to \infty} \int_2^b \frac{dx}{x \ln x}$$

$$= \lim_{b \to \infty} \ln |\ln x| \Big|_2^b$$

$$= \lim_{b \to \infty} [\ln |\ln b| - \ln |\ln 2|].$$

As $b \to \infty$, the limit goes to ∞ and hence the integral diverges.

25. Using the substitution $w = -x^{\frac{1}{2}}$, $-2dw = x^{-\frac{1}{2}} \, dx$,

$$\int e^{-x^{\frac{1}{2}}} x^{-\frac{1}{2}} \, dx = -2 \int e^w \, dw = -2e^{-x^{\frac{1}{2}}} + C.$$

So

$$\int_0^\pi \frac{1}{\sqrt{x}} e^{-\sqrt{x}} \, dx = \lim_{b \to 0^+} \int_b^\pi \frac{1}{\sqrt{x}} e^{-\sqrt{x}} \, dx$$

$$= \lim_{b \to 0^+} -2e^{-\sqrt{x}} \Big|_b^\pi$$

$$= 2 - 2e^{-\sqrt{\pi}}.$$

29. $\displaystyle\int \frac{dx}{x^2 - 1} = \int \frac{dx}{(x-1)(x+1)} = \frac{1}{2}(\ln |x-1| - \ln |x+1|) + C = \frac{1}{2} \left(\ln \frac{|x-1|}{|x+1|} \right) + C$, so

$$\int_4^\infty \frac{dx}{x^2 - 1} = \lim_{b \to \infty} \int_4^b \frac{dx}{x^2 - 1}$$

$$= \lim_{b \to \infty} \frac{1}{2} \left(\ln \frac{|x-1|}{|x+1|} \right) \Big|_4^b$$

$$= \lim_{b \to \infty} \left[\frac{1}{2} \ln \left(\frac{b-1}{b+1} \right) - \frac{1}{2} \ln \frac{3}{5} \right]$$

$$= -\frac{1}{2} \ln \frac{3}{5} = \frac{1}{2} \ln \frac{5}{3}.$$

Problems

33. Since the graph is above the x-axis for $x \geq 0$, we have

$$\text{Area} = \int_0^\infty xe^{-x}\,dx = \lim_{b\to\infty} \int_0^b xe^{-x}\,dx$$

$$= \lim_{b\to\infty}\left(-xe^{-x}\Big|_0^b + \int_0^b e^{-x}\,dx \right)$$

$$= \lim_{b\to\infty}\left(-be^{-b} - e^{-x}\Big|_0^b \right)$$

$$= \lim_{b\to\infty}(-be^{-b} - e^{-b} + e^0) = 1.$$

37. The factor $\ln x$ grows slowly enough (as $x \to 0^+$) not to change the convergence or divergence of the integral, although it will change what it converges or diverges to.

The integral is always improper, because $\ln x$ is not defined for $x = 0$. Integrating by parts (or, alternatively, the integral table) yields

$$\int_0^e x^p \ln x\,dx = \lim_{a\to 0^+} \int_a^e x^p \ln x\,dx$$

$$= \lim_{a\to 0^+}\left(\frac{1}{p+1}x^{p+1}\ln x - \frac{1}{(p+1)^2}x^{p+1} \right)\Bigg|_a^e$$

$$= \lim_{a\to 0^+}\left[\left(\frac{1}{p+1}e^{p+1} - \frac{1}{(p+1)^2}e^{p+1} \right) \right.$$

$$\left. - \left(\frac{1}{p+1}a^{p+1}\ln a - \frac{1}{(p+1)^2}a^{p+1} \right) \right].$$

If $p < -1$, then $(p+1)$ is negative, so as $a \to 0^+$, $a^{p+1} \to \infty$ and $\ln a \to -\infty$, and therefore the limit does not exist.

If $p > -1$, then $(p+1)$ is positive and it's easy to see that $a^{p+1} \to 0$ as $a \to 0$. Looking at graphs of $x^{p+1}\ln x$ (for different values of p) shows that $a^{p+1}\ln a \to 0$ as $a \to 0$. This isn't so easy to see analytically. It's true because if we let $t = \frac{1}{a}$ then

$$\lim_{a\to 0^+} a^{p+1}\ln a = \lim_{t\to\infty}\left(\frac{1}{t}\right)^{p+1}\ln\left(\frac{1}{t}\right) = \lim_{t\to\infty} -\frac{\ln t}{t^{p+1}}.$$

This last limit is zero because $\ln t$ grows very slowly, much more slowly than t^{p+1}. So if $p > -1$, the integral converges and equals $e^{p+1}[1/(p+1) - 1/(p+1)^2] = pe^{p+1}/(p+1)^2$.

What happens if $p = -1$? Then we get

$$\int_0^e \frac{\ln x}{x}\,dx = \lim_{a\to 0^+}\int_a^e \frac{\ln x}{x}\,dx$$

$$= \lim_{a\to 0^+} \frac{(\ln x)^2}{2}\Bigg|_a^e$$

$$= \lim_{a\to 0^+}\left(\frac{1 - (\ln a)^2}{2} \right).$$

Since $\ln a \to -\infty$ as $a \to 0^+$, this limit does not exist.

To summarize, $\int_0^e x^p \ln x$ converges for $p > -1$ to the value $pe^{p+1}/(p+1)^2$.

Solutions for Section 7.8

Exercises

1. For large x, the integrand behaves like $1/x^2$ because

$$\frac{x^2}{x^4+1} \approx \frac{x^2}{x^4} = \frac{1}{x^2}.$$

Since $\displaystyle\int_1^\infty \frac{dx}{x^2}$ converges, we expect our integral to converge. More precisely, since $x^4 + 1 > x^4$, we have

$$\frac{x^2}{x^4+1} < \frac{x^2}{x^4} = \frac{1}{x^2}.$$

Since $\displaystyle\int_1^\infty \frac{dx}{x^2}$ is convergent, the comparison test tells us that $\displaystyle\int_1^\infty \frac{x^2}{x^4+1}\,dx$ converges also.

5. The integrand is continuous for all $x \geq 1$, so whether the integral converges or diverges depends only on the behavior of the function as $x \to \infty$. As $x \to \infty$, polynomials behave like the highest powered term. Thus, as $x \to \infty$, the integrand $\dfrac{x}{x^2+2x+4}$ behaves like $\dfrac{x}{x^2}$ or $\dfrac{1}{x}$. Since $\displaystyle\int_1^\infty \frac{1}{x}\,dx$ diverges, we predict that the given integral will diverge.

9. The integrand is continuous for all $x \geq 1$, so whether the integral converges or diverges depends only on the behavior of the function as $x \to \infty$. As $x \to \infty$, polynomials behave like the highest powered term. Thus, as $x \to \infty$, the integrand $\dfrac{x^2+4}{x^4+3x^2+11}$ behaves like $\dfrac{x^2}{x^4}$ or $\dfrac{1}{x^2}$. Since $\displaystyle\int_1^\infty \frac{1}{x^2}\,dx$ converges, we predict that the given integral will converge.

13. The integrand is unbounded as $t \to 5$. We substitute $w = t - 5$, so $dw = dt$. When $t = 5$, $w = 0$ and when $t = 8$, $w = 3$.

$$\int_5^8 \frac{6}{\sqrt{t-5}}\,dt = \int_0^3 \frac{6}{\sqrt{w}}\,dw.$$

Since

$$\int_0^3 \frac{6}{\sqrt{w}}\,dw = \lim_{a\to 0^+} 6\int_a^3 \frac{1}{\sqrt{w}}\,dw = 6\lim_{a\to 0^+} 2w^{1/2}\Big|_a^3 = 12\lim_{a\to 0^+}(\sqrt{3}-\sqrt{a}) = 12\sqrt{3},$$

our integral converges.

17. Since $\dfrac{1}{u+u^2} < \dfrac{1}{u^2}$ for $u \geq 1$, and since $\displaystyle\int_1^\infty \frac{du}{u^2}$ converges, $\displaystyle\int_1^\infty \frac{du}{u+u^2}$ converges.

21. Since $\dfrac{1}{1+e^y} \leq \dfrac{1}{e^y} = e^{-y}$ and $\displaystyle\int_0^\infty e^{-y}\,dy$ converges, the integral $\displaystyle\int_0^\infty \frac{dy}{1+e^y}$ converges.

25. Since $\dfrac{3+\sin\alpha}{\alpha} \geq \dfrac{2}{\alpha}$ for $\alpha \geq 4$, and since $\displaystyle\int_4^\infty \frac{2}{\alpha}\,d\alpha$ diverges, then $\displaystyle\int_4^\infty \frac{3+\sin\alpha}{\alpha}\,d\alpha$ diverges.

Problems

29. First let's calculate the indefinite integral $\displaystyle\int \frac{dx}{x(\ln x)^p}$. Let $\ln x = w$, then $\dfrac{dx}{x} = dw$. So

$$\int \frac{dx}{x(\ln x)^p} = \int \frac{dw}{w^p}$$

$$= \begin{cases} \ln|w| + C, & \text{if } p = 1 \\ \frac{1}{1-p}w^{1-p} + C, & \text{if } p \neq 1 \end{cases}$$

$$= \begin{cases} \ln|\ln x| + C, & \text{if } p = 1 \\ \frac{1}{1-p}(\ln x)^{1-p} + C, & \text{if } p \neq 1. \end{cases}$$

Notice that $\displaystyle\lim_{x\to\infty} \ln x = +\infty$.

(a) $p = 1$:

$$\int_2^\infty \frac{dx}{x \ln x} = \lim_{b \to \infty} \left(\ln|\ln b| - \ln|\ln 2| \right) = +\infty.$$

(b) $p < 1$:

$$\int_2^\infty \frac{dx}{x(\ln x)^p} = \frac{1}{1-p} \left(\lim_{b \to \infty} (\ln b)^{1-p} - (\ln 2)^{1-p} \right) = +\infty.$$

(c) $p > 1$:

$$\int_2^\infty \frac{dx}{x(\ln x)^p} = \frac{1}{1-p} \left(\lim_{b \to \infty} (\ln b)^{1-p} - (\ln 2)^{1-p} \right)$$

$$= \frac{1}{1-p} \left(\lim_{b \to \infty} \frac{1}{(\ln b)^{p-1}} - (\ln 2)^{1-p} \right)$$

$$= -\frac{1}{1-p} (\ln 2)^{1-p}.$$

Thus, $\displaystyle \int_2^\infty \frac{dx}{x(\ln x)^p}$ is convergent for $p > 1$, divergent for $p \leq 1$.

33. (a) Since $e^{-x^2} \leq e^{-3x}$ for $x \geq 3$,

$$\int_3^\infty e^{-x^2} \, dx \leq \int_3^\infty e^{-3x} \, dx$$

Now

$$\int_3^\infty e^{-3x} \, dx = \lim_{b \to \infty} \int_3^b e^{-3x} \, dx = \lim_{b \to \infty} \left. -\frac{1}{3} e^{-3x} \right|_3^b$$

$$= \lim_{b \to \infty} \frac{e^{-9}}{3} - \frac{e^{-3b}}{3} = \frac{e^{-9}}{3}.$$

Thus

$$\int_3^\infty e^{-x^2} \, dx \leq \frac{e^{-9}}{3}.$$

(b) By reasoning similar to part (a),

$$\int_n^\infty e^{-x^2} \, dx \leq \int_n^\infty e^{-nx} \, dx,$$

and

$$\int_n^\infty e^{-nx} \, dx = \frac{1}{n} e^{-n^2},$$

so

$$\int_n^\infty e^{-x^2} \, dx \leq \frac{1}{n} e^{-n^2}.$$

Solutions for Chapter 7 Review

Exercises

1. Since $\dfrac{d}{dt} \cos t = -\sin t$, we have

$$\int \sin t \, dt = -\cos t + C, \quad \text{where } C \text{ is a constant.}$$

5. Since $\displaystyle \int \sin w \, d\theta = -\cos w + C$, the substitution $w = 2\theta$, $dw = 2 \, d\theta$ gives $\displaystyle \int \sin 2\theta \, d\theta = -\frac{1}{2} \cos 2\theta + C$.

9. Either expand $(r+1)^3$ or use the substitution $w = r + 1$. If $w = r + 1$, then $dw = dr$ and

$$\int (r+1)^3 \, dr = \int w^3 \, dw = \frac{1}{4} w^4 + C = \frac{1}{4}(r+1)^4 + C.$$

13. Substitute $w = t^2$, so $dw = 2t\,dt$.

$$\int te^{t^2}\,dt = \frac{1}{2}\int e^{t^2}2t\,dt = \frac{1}{2}\int e^w\,dw = \frac{1}{2}e^w + C = \frac{1}{2}e^{t^2} + C.$$

Check:

$$\frac{d}{dt}\left(\frac{1}{2}e^{t^2} + C\right) = 2t\left(\frac{1}{2}e^{t^2}\right) = te^{t^2}.$$

17. Integration by parts with $u = \ln x$, $v' = x$ gives

$$\int x\ln x\,dx = \frac{x^2}{2}\ln x - \int \frac{1}{2}x\,dx = \frac{1}{2}x^2\ln x - \frac{1}{4}x^2 + C.$$

Or use the integral table, III-13, with $n = 1$.

21. Using the exponent rules and the chain rule, we have

$$\int e^{0.5-0.3t}\,dt = e^{0.5}\int e^{-0.3t}\,dt = -\frac{e^{0.5}}{0.3}e^{-0.3t} + C = -\frac{e^{0.5-0.3t}}{0.3} + C.$$

25. Substitute $w = \sqrt{y}$, $dw = 1/(2\sqrt{y})\,dy$. Then

$$\int \frac{\cos\sqrt{y}}{\sqrt{y}}\,dy = 2\int \cos w\,dw = 2\sin w + C = 2\sin\sqrt{y} + C.$$

Check:

$$\frac{d}{dy}2\sin\sqrt{y} + C = \frac{2\cos\sqrt{y}}{2\sqrt{y}} = \frac{\cos\sqrt{y}}{\sqrt{y}}.$$

29. Substitute $w = 2x - 6$. Then $dw = 2\,dx$ and

$$\int \tan(2x - 6)\,dx = \frac{1}{2}\int \tan w\,dw = \frac{1}{2}\int \frac{\sin w}{\cos w}\,dw$$

$$= -\frac{1}{2}\ln|\cos w| + C \text{ by substitution or by I-7 of the integral table.}$$

$$= -\frac{1}{2}\ln|\cos(2x - 6)| + C.$$

33.

$$\int_0^{10} ze^{-z}\,dz = [-ze^{-z}]\Big|_0^{10} - \int_0^{10} -e^{-z}\,dz \qquad (\text{let } z = u, e^{-z} = v', -e^{-z} = v)$$

$$= -10e^{-10} - [e^{-z}]\Big|_0^{10}$$

$$= -10e^{-10} - e^{-10} + 1$$

$$= -11e^{-10} + 1.$$

37.

$$\int_0^1 \frac{dx}{x^2 + 1} = \tan^{-1}x\Big|_0^1 = \tan^{-1}1 - \tan^{-1}0 = \frac{\pi}{4} - 0 = \frac{\pi}{4}.$$

41. Dividing and then integrating, we obtain

$$\int \frac{t+1}{t^2}\,dt = \int \frac{1}{t}\,dt + \int \frac{1}{t^2}\,dt = \ln|t| - \frac{1}{t} + C, \text{ where } C \text{ is a constant.}$$

45. Using substitution,

$$\int \frac{x}{x^2 + 1}\, dx = \int \frac{1/2}{w}\, dw \qquad (x^2 + 1 = w, 2x\, dx = dw, x\, dx = \frac{1}{2}\, dw)$$

$$= \frac{1}{2}\int \frac{1}{w}\, dw = \frac{1}{2}\ln|w| + C = \frac{1}{2}\ln|x^2 + 1| + C,$$

where C is a constant.

49. Let $\cos 5\theta = w$, then $-5\sin 5\theta\, d\theta = dw$, $\sin 5\theta\, d\theta = -\frac{1}{5}dw$. So

$$\int \sin 5\theta \cos^3 5\theta\, d\theta = \int w^3 \cdot \left(-\frac{1}{5}\right) dw = -\frac{1}{5}\int w^3\, dw = -\frac{1}{20}w^4 + C$$

$$= -\frac{1}{20}\cos^4 5\theta + C,$$

where C is a constant.

53.

$$\int xe^x\, dx = xe^x - \int e^x\, dx \qquad (\text{let } x = u, e^x = v', e^x = v)$$

$$= xe^x - e^x + C,$$

where C is a constant.

57. Rewrite $9 + u^2$ as $9[1 + (u/3)^2]$ and let $w = u/3$, then $dw = du/3$ so that

$$\int \frac{du}{9 + u^2} = \frac{1}{3}\int \frac{dw}{1 + w^2} = \frac{1}{3}\arctan w + C = \frac{1}{3}\arctan\left(\frac{u}{3}\right) + C.$$

61. Let $u = 2x$, then $du = 2\, dx$ so that

$$\int \frac{dx}{\sqrt{1 - 4x^2}} = \frac{1}{2}\int \frac{du}{\sqrt{1 - u^2}} = \frac{1}{2}\arcsin u + C = \frac{1}{2}\arcsin(2x) + C.$$

65. Let $w = \ln x$. Then $dw = (1/x)dx$ which gives

$$\int \frac{dx}{x \ln x} = \int \frac{dw}{w} = \ln|w| + C = \ln|\ln x| + C.$$

69. Using integration by parts, let $r = u$ and $dt = e^{ku}du$, so $dr = du$ and $t = (1/k)e^{ku}$. Thus

$$\int ue^{ku}\, du = \frac{u}{k}e^{ku} - \frac{1}{k}\int e^{ku}\, du = \frac{u}{k}e^{ku} - \frac{1}{k^2}e^{ku} + C.$$

73. $\int (e^x + x)^2 dx = \int (e^{2x} + 2xe^x + x^2)dy$. Separating into three integrals, we have

$$\int e^{2x}\, dx = \frac{1}{2}\int e^{2x}2\, dx = \frac{1}{2}e^{2x} + C_1,$$

$$\int 2xe^x\, dx = 2\int xe^x\, dx = 2xe^x - 2e^x + C_2$$

from Formula II-13 of the integral table or integration by parts, and

$$\int x^2\, dx = \frac{x^3}{3} + C_3.$$

Combining the results and writing $C = C_1 + C_2 + C_3$, we get

$$\frac{1}{2}e^{2x} + 2xe^x - 2e^x + \frac{x^3}{3} + C.$$

77. We can factor $r^2 - 100 = (r - 10)(r + 10)$ so we can use Table V-26 (with $a = 10$ and $b = -10$) to get

$$\int \frac{dr}{r^2 - 100} = \frac{1}{20}\left[\ln|r - 10| + \ln|r + 10|\right] + C.$$

81. Since $\int (x^{\sqrt{k}} + (\sqrt{k})^x)dx = \int x^{\sqrt{k}}dx + \int (\sqrt{k})^x\,dx$, for the first integral, use Formula I-1 with $n = \sqrt{k}$. For the second integral, use Formula I-3 with $a = \sqrt{k}$. The result is

$$\int (x^{\sqrt{x}} + (\sqrt{k})^x)\,dx = \frac{x^{(\sqrt{k})+1}}{(\sqrt{k})+1} + \frac{(\sqrt{k})^x}{\ln\sqrt{k}} + C.$$

85. First divide $x^2 + 3x + 2$ into x^3 to obtain

$$\frac{x^3}{x^2 + 3x + 2} = x - 3 + \frac{7x + 6}{x^2 + 3x + 2}.$$

Since $x^2 + 3x + 2 = (x+1)(x+2)$, we can use V-27 of the integral table (with $c = 7$, $d = 6$, $a = -1$, and $b = -2$) to get

$$\int \frac{7x + 6}{x^2 + 3x + 2}\,dx = -\ln|x+1| + 8\ln|x+2| + C.$$

Including the terms $x - 3$ from the long division and integrating them gives

$$\int \frac{x^3}{x^2 + 3x + 2}\,dx = \int \left(x - 3 + \frac{7x + 6}{x^2 + 3x + 6} \right)\,dx = \frac{1}{2}x^2 - 3x - \ln|x+1| + 8\ln|x+2| + C.$$

89. This can be done by formula V–26 in the integral table or by partial fractions

$$\int \frac{dz}{z^2 + z} = \int \frac{dz}{z(z+1)} = \int \left(\frac{1}{z} - \frac{1}{z+1} \right)\,dz = \ln|z| - \ln|z+1| + C.$$

Check:

$$\frac{d}{dz}\left(\ln|z| - \ln|z+1| + C \right) = \frac{1}{z} - \frac{1}{z+1} = \frac{1}{z^2 + z}.$$

93. Multiplying out and integrating term by term gives

$$\int (x^2 + 5)^3 dx = \int (x^6 + 15x^4 + 75x^2 + 125)dx = \frac{1}{7}x^7 + 15\frac{x^5}{5} + 75\frac{x^3}{3} + 125x + C$$

$$= \frac{1}{7}x^7 + 3x^5 + 25x^3 + 125x + C.$$

97. If $u = 1 + \cos^2 w$, $du = 2(\cos w)^1(-\sin w)\,dw$, so

$$\int \frac{\sin w \cos w}{1 + \cos^2 w}\,dw = -\frac{1}{2}\int \frac{-2\sin w \cos w}{1 + \cos^2 w}\,dw = -\frac{1}{2}\int \frac{1}{u}\,du = -\frac{1}{2}\ln|u| + C$$

$$= -\frac{1}{2}\ln|1 + \cos^2 w| + C.$$

101. If $u = \sqrt{x+1}$, $u^2 = x + 1$ with $x = u^2 - 1$ and $dx = 2u\,du$. Substituting, we get

$$\int \frac{x}{\sqrt{x+1}}\,dx = \int \frac{(u^2 - 1)2u\,du}{u} = \int (u^2 - 1)2\,du = 2\int (u^2 - 1)\,du$$

$$= \frac{2u^3}{3} - 2u + C = \frac{2(\sqrt{x+1})^3}{3} - 2\sqrt{x+1} + C.$$

105. Letting $u = z - 5$, $z = u + 5$, $dz = du$, and substituting, we have

$$\int \frac{z}{(z-5)^3}dz = \int \frac{u+5}{u^3}du = \int (u^{-2} + 5u^{-3})du = \frac{u^{-1}}{-1} + 5\left(\frac{u^{-2}}{-2} \right) + C$$

$$= \frac{-1}{(z-5)} + \frac{-5}{2(z-5)^2} + C.$$

109. Let $w = 2 + 3\cos x$, so $dw = -3\sin x\,dx$, giving $-\dfrac{1}{3}\,dw = \sin x\,dx$. Then

$$\int \sin x \left(\sqrt{2 + 3\cos x}\right) dx = \int \sqrt{w}\left(-\frac{1}{3}\right) dw = -\frac{1}{3}\int \sqrt{w}\,dw$$

$$= \left(-\frac{1}{3}\right)\frac{w^{\frac{3}{2}}}{\frac{3}{2}} + C = -\frac{2}{9}(2 + 3\cos x)^{\frac{3}{2}} + C.$$

113. Let $w = x + \sin x$, then $dw = (1 + \cos x)\,dx$ which gives

$$\int (x + \sin x)^3(1 + \cos x)\,dx = \int w^3\,dw = \frac{1}{4}w^4 + C = \frac{1}{4}(x + \sin x)^4 + C.$$

117. Splitting the integrand into partial fractions with denominators x and $(x + 5)$, we have

$$\frac{1}{x(x + 5)} = \frac{A}{x} + \frac{B}{x + 5}.$$

Multiplying by $x(x + 5)$ gives the identity

$$1 = A(x + 5) + Bx$$

so

$$1 = (A + B)x + 5A.$$

Since this equation holds for all x, the constant terms on both sides must be equal. Similarly, the coefficient of x on both sides must be equal. So

$$5A = 1$$
$$A + B = 0.$$

Solving these equations gives $A = 1/5$, $B = -1/5$ and the integral becomes

$$\int \frac{1}{x(x + 5)}\,dx = \frac{1}{5}\int \frac{1}{x}\,dx - \frac{1}{5}\int \frac{1}{x + 5}\,dx = \frac{1}{5}\left(\ln|x| - \ln|x + 5|\right) + C.$$

121. The denominator can be factored to give $x(x - 1)(x + 1)$. Splitting the integrand into partial fractions with denominators x, $x - 1$, and $x + 1$, we have

$$\frac{3x + 1}{x(x - 1)(x + 1)} = \frac{A}{x - 1} + \frac{B}{x + 1} + \frac{C}{x}.$$

Multiplying by $x(x - 1)(x + 1)$ gives the identity

$$3x + 1 = Ax(x + 1) + Bx(x - 1) + C(x - 1)(x + 1)$$

so

$$3x + 1 = (A + B + C)x^2 + (A - B)x - C.$$

Since this equation holds for all x, the constant terms on both sides must be equal. Similarly, the coefficient of x and x^2 on both sides must be equal. So

$$-C = 1$$
$$A - B = 3$$
$$A + B + C = 0.$$

Solving these equations gives $A = 2$, $B = -1$ and $C = -1$. The integral becomes

$$\int \frac{3x + 1}{x(x + 1)(x - 1)}\,dx = \int \frac{2}{x - 1}\,dx - \int \frac{1}{x + 1}\,dx - \int \frac{1}{x}\,dx$$
$$= 2\ln|x - 1| - \ln|x + 1| - \ln|x| + C.$$

125. Using partial fractions, we have:

$$\frac{3x + 1}{x^2 - 3x + 2} = \frac{3x + 1}{(x - 1)(x - 2)} = \frac{A}{x - 1} + \frac{B}{x - 2}.$$

Multiplying by $(x - 1)$ and $(x - 2)$, this becomes

$$3x + 1 = A(x - 2) + B(x - 1)$$
$$= (A + B)x - 2A - B$$

which produces the system of equations

$$\begin{cases} A + B = 3 \\ -2A - B = 1. \end{cases}$$

Solving this system yields $A = -4$ and $B = 7$. So,

$$\int \frac{3x + 1}{x^2 - 3x + 2} dx = \int \left(-\frac{4}{x - 1} + \frac{7}{x - 2} \right) dx$$
$$= -4 \int \frac{dx}{x - 1} + 7 \int \frac{dx}{x - 2}$$
$$= -4 \ln |x - 1| + 7 \ln |x - 2| + C.$$

129. To find $\int we^{-w} \, dw$, integrate by parts, with $u = w$ and $v' = e^{-w}$. Then $u' = 1$ and $v = -e^{-w}$.
Then

$$\int we^{-w} \, dw = -we^{-w} + \int e^{-w} \, dw = -we^{-w} - e^{-w} + C.$$

Thus

$$\int_0^\infty we^{-w} \, dw = \lim_{b \to \infty} \int_0^b we^{-w} \, dw = \lim_{b \to \infty} \left(-we^{-w} - e^{-w} \right) \Big|_0^b = 1.$$

133. We find the exact value:

$$\int_{10}^\infty \frac{1}{z^2 - 4} dz = \int_{10}^\infty \frac{1}{(z + 2)(z - 2)} dz$$
$$= \lim_{b \to \infty} \int_{10}^b \frac{1}{(z + 2)(z - 2)} dz$$
$$= \lim_{b \to \infty} \frac{1}{4} \left(\ln |z - 2| - \ln |z + 2| \right) \Big|_{10}^b$$
$$= \frac{1}{4} \lim_{b \to \infty} \left[(\ln |b - 2| - \ln |b + 2|) - (\ln 8 - \ln 12) \right]$$
$$= \frac{1}{4} \lim_{b \to \infty} \left[\left(\ln \frac{b - 2}{b + 2} \right) + \ln \frac{3}{2} \right]$$
$$= \frac{1}{4} (\ln 1 + \ln 3/2) = \frac{\ln 3/2}{4}.$$

137. The integrand $\frac{x}{x + 1} \to 1$ as $x \to \infty$, so there's no way $\int_1^\infty \frac{x}{x + 1} dx$ can converge.

Problems

141. Since the definition of f is different on $0 \leq t \leq 1$ than it is on $1 \leq t \leq 2$, break the definite integral at $t = 1$.

$$\int_0^2 f(t) \, dt = \int_0^1 f(t) \, dt + \int_1^2 f(t) \, dt$$

$$= \int_0^1 t^2 \, dt + \int_1^2 (2-t) \, dt$$

$$= \frac{t^3}{3}\bigg|_0^1 + \left(2t - \frac{t^2}{2}\right)\bigg|_1^2$$

$$= 1/3 + 1/2 = 5/6 \approx 0.833$$

145. (a) i. 0   ii. $\frac{2}{\pi}$ iii. $\frac{1}{2}$

(b) Average value of $f(t)$ < Average value of $k(t)$ < Average value of $g(t)$

We can look at the three functions in the range $-\frac{\pi}{2} \le x \le \frac{3\pi}{2}$, since they all have periods of 2π ($|\cos t|$ and $(\cos t)^2$ also have a period of π, but that doesn't hurt our calculation). It is clear from the graphs of the three functions below that the average value for $\cos t$ is 0 (since the area above the x-axis is equal to the area below it), while the average values for the other two are positive (since they are everywhere positive, except where they are 0).

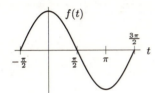
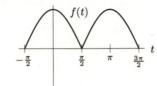
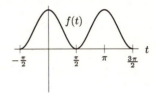

It is also fairly clear from the graphs that the average value of $g(t)$ is greater than the average value of $k(t)$; it is also possible to see this algebraically, since

$$(\cos t)^2 = |\cos t|^2 \le |\cos t|$$

because $|\cos t| \le 1$ (and both of these $\le$'s are <'s at all the points where the functions are not 0 or 1).

149. (a) Since the rate is given by $r(t) = 2te^{-2t}$ ml/sec, by the Fundamental Theorem of Calculus, the total quantity is given by the definite integral:

$$\text{Total quantity} \approx \int_0^\infty 2te^{-2t} \, dt = 2 \lim_{b \to \infty} \int_0^b te^{-2t} \, dt.$$

Integration by parts with $u = t$, $v' = e^{-2t}$ gives

$$\text{Total quantity} \approx 2 \lim_{b \to \infty} \left(-\frac{t}{2}e^{-2t} - \frac{1}{4}e^{-2t}\right)\bigg|_0^b$$

$$= 2 \lim_{b \to \infty} \left(\frac{1}{4} - \left(\frac{b}{2} + \frac{1}{4}\right)e^{-2b}\right) = 2 \cdot \frac{1}{4} = 0.5 \text{ ml}.$$

(b) At the end of 5 seconds,

$$\text{Quantity received} = \int_0^5 2te^{-2t} \, dt \approx 0.49975 \text{ ml}.$$

Since $0.49975/0.5 = 0.9995 = 99.95\%$, the patient has received 99.95% of the dose in the first 5 seconds.

CAS Challenge Problems

153. (a) A possible answer from the CAS is

$$\int \sin^3 x \, dx = \frac{-9 \cos(x) + \cos(3x)}{12}.$$

(b) Differentiating

$$\frac{d}{dx}\left(\frac{-9 \cos(x) + \cos(3x)}{12}\right) = \frac{9 \sin(x) - 3 \sin(3x)}{12} = \frac{3 \sin x - \sin(3x)}{4}.$$

(c) Using the identities, we get

$$\begin{aligned}\sin(3x) &= \sin(x + 2x) = \sin x \cos 2x + \cos x \sin 2x \\ &= \sin x(1 - 2\sin^2 x) + \cos x(2\sin x \cos x) \\ &= \sin x - 2\sin^3 x + 2\sin x(1 - \sin^2 x) \\ &= 3\sin x - 4\sin^3 x.\end{aligned}$$

Thus,

$$3\sin x - \sin(3x) = 3\sin x - (3\sin x - 4\sin^3 x) = 4\sin^3 x,$$

so

$$\frac{3\sin x - \sin(3x)}{4} = \sin^3 x.$$

CHECK YOUR UNDERSTANDING

1. False. The subdivision size $\Delta x = (1/10)(6 - 2) = 4/10$.

5. True. We have

$$\text{LEFT}(n) - \text{RIGHT}(n) = (f(x_0) + f(x_1) + \cdots + f(x_{n-1}))\Delta x - (f(x_1) + f(x_2) + \cdots + f(x_n))\Delta x.$$

On the right side of the equation, all terms cancel except the first and last, so:

$$\text{LEFT}(n) - \text{RIGHT}(n) = (f(x_0) - f(x_n))\Delta x = (f(2) - f(6))\Delta x.$$

This is also discussed in Section 5.1.

9. False. This is true if f is an increasing function or if f is a decreasing function, but it is not true in general. For example, suppose that $f(2) = f(6)$. Then $\text{LEFT}(n) = \text{RIGHT}(n)$ for all n, which means that if $\int_2^6 f(x)dx$ lies between $\text{LEFT}(n)$ and $\text{RIGHT}(n)$, then it must equal $\text{LEFT}(n)$, which is not always the case.

For example, if $f(x) = (x - 4)^2$ and $n = 1$, then $f(2) = f(6) = 4$, so

$$\text{LEFT}(1) = \text{RIGHT}(1) = 4 \cdot (6 - 2) = 16.$$

However

$$\int_2^6 (x - 4)^2 dx = \frac{(x - 4)^3}{3}\bigg|_2^6 = \frac{2^3}{3} - \left(-\frac{2^3}{3}\right) = \frac{16}{3}.$$

In this example, since $\text{LEFT}(n) = \text{RIGHT}(n)$, we have $\text{TRAP}(n) = \text{LEFT}(n)$. However trapezoids overestimate the area, since the graph of f is concave up. This is also discussed in Section 7.5.

13. True. Rewrite $\sin^7 \theta = \sin \theta \sin^6 \theta = \sin \theta(1 - \cos^2 \theta)^3$. Expanding, substituting $w = \cos \theta, dw = -\sin \theta \, d\theta$, and integrating gives a polynomial in w, which is a polynomial in $\cos \theta$.

17. True. Let $u = t, v' = \sin(5 - t)$, so $u' = 1, v = \cos(5 - t)$. Then the integral $\int 1 \cdot \cos(5 - t) \, dt$ can be done by guess-and-check or by substituting $w = 5 - t$.

21. False. Let $f(x) = x + 1$. Then

$$\int_0^\infty \frac{1}{x + 1} dx = \lim_{b \to \infty} \ln|x + 1|\bigg|_0^b = \lim_{b \to \infty} \ln(b + 1),$$

but $\lim_{b \to \infty} \ln(b + 1)$ does not exist.

25. True. Make the substitution $w = ax$. Then $dw = a \, dx$, so

$$\int_0^b f(ax) \, dx = \frac{1}{a} \int_0^c f(w) \, dw,$$

where $c = ab$. As b approaches infinity, so does c, since a is constant. Thus the limit of the left side of the equation as b approaches infinity is finite exactly when the limit of the right side of the equation as c approaches infinity is finite. That is, $\int_0^\infty f(ax) \, dx$ converges exactly when $\int_0^\infty f(x) \, dx$ converges.

CHAPTER EIGHT

Solutions for Section 8.1

Exercises

1. Each strip is a rectangle of length 3 and width Δx, so

$$\text{Area of strip } = 3\Delta x, \quad \text{so}$$

$$\text{Area of region } = \int_0^5 3 \, dx = 3x \Big|_0^5 = 15.$$

Check: This area can also be computed using Length $\times$ Width $= 5 \cdot 3 = 15$.

5. The strip has width Δy, so the variable of integration is y. The length of the strip is x. Since $x^2 + y^2 = 10$ and the region is in the first quadrant, solving for x gives $x = \sqrt{10 - y^2}$. Thus

$$\text{Area of strip } \approx x\Delta y = \sqrt{10 - y^2} \, dy.$$

The region stretches from $y = 0$ to $y = \sqrt{10}$, so

$$\text{Area of region } = \int_0^{\sqrt{10}} \sqrt{10 - y^2} \, dy.$$

Evaluating using VI-30 from the Table of Integrals, we have

$$\text{Area } = \frac{1}{2}\left(y\sqrt{10 - y^2} + 10\arcsin\left(\frac{y}{\sqrt{10}}\right)\right)\Big|_0^{\sqrt{10}} = 5(\arcsin 1 - \arcsin 0) = \frac{5}{2}\pi.$$

Check: This area can also be computed using the formula $\frac{1}{4}\pi r^2 = \frac{1}{4}\pi(\sqrt{10})^2 = \frac{5}{2}\pi$.

9. Each slice is a circular disk with radius $r = 2$ cm.

$$\text{Volume of disk } = \pi r^2 \Delta x = 4\pi \Delta x \text{ cm}^3.$$

Summing over all disks, we have

$$\text{Total volume } \approx \sum 4\pi \Delta x \text{ cm}^3.$$

Taking a limit as $\Delta x \to 0$, we get

$$\text{Total volume } = \lim_{\Delta x \to 0} \sum 4\pi \Delta x = \int_0^9 4\pi \, dx \text{ cm}^3.$$

Evaluating gives

$$\text{Total volume } = 4\pi x \Big|_0^9 = 36\pi \text{ cm}^3.$$

Check: The volume of the cylinder can also be calculated using the formula $V = \pi r^2 h = \pi 2^2 \cdot 9 = 36\pi \text{ cm}^3$.

13. Each slice is a circular disk. See Figure 8.1. The radius of the sphere is 5 mm, and the radius r at height y is given by the Pythagorean Theorem

$$y^2 + r^2 = 5^2.$$

Solving gives $r = \sqrt{5^2 - y^2}$ mm. Thus,

$$\text{Volume of disk } \approx \pi r^2 \Delta y = \pi(5^2 - y^2)\Delta y \text{ mm}^3.$$

Summing over all disks, we have

$$\text{Total volume} \approx \sum \pi(5^2 - y^2)\Delta y \text{ mm}^3.$$

Taking the limit as $\Delta y \to 0$, we get

$$\text{Total volume} = \lim_{\Delta y \to 0} \sum \pi(5^2 - y^2)\Delta y = \int_0^5 \pi(5^2 - y^2)\, dy \text{ mm}^3.$$

Evaluating gives

$$\text{Total volume} = \pi \left(25y - \frac{y^3}{3} \right)\Big|_0^5 = \frac{250}{3}\pi \text{ mm}^3.$$

Check: The volume of a hemisphere can be calculated using the formula $V = \frac{2}{3}\pi r^3 = \frac{2}{3}\pi 5^3 = \frac{250}{3}\pi \text{ mm}^3$.

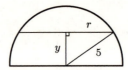

Figure 8.1

Problems

17. Triangle of base and height 1 and 3. See Figure 8.2. (Either 1 or 3 can be the base. A non-right triangle is also possible.)

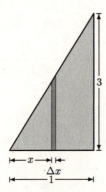

Figure 8.2

21. Hemisphere with radius 12. See Figure 8.3.

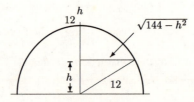

Figure 8.3

25.

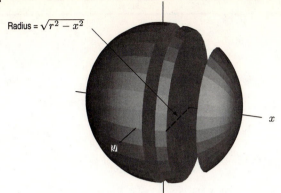

Radius = $\sqrt{r^2 - x^2}$

We slice up the sphere in planes perpendicular to the x-axis. Each slice is a circle, with radius $y = \sqrt{r^2 - x^2}$; that's the radius because $x^2 + y^2 = r^2$ when $z = 0$. Then the volume is

$$V \approx \sum \pi(y^2)\,\Delta x = \sum \pi(r^2 - x^2)\,\Delta x.$$

Therefore, as Δx tends to zero, we get

$$V = \int_{x=-r}^{x=r} \pi(r^2 - x^2)\,dx$$

$$= 2\int_{x=0}^{x=r} \pi(r^2 - x^2)\,dx$$

$$= 2\left(\pi r^2 x - \frac{\pi x^3}{3}\right)\Bigg|_0^r$$

$$= \frac{4\pi r^3}{3}.$$

29. To calculate the volume of material, we slice the dam horizontally. See Figure 8.4. The slices are rectangular, so

$$\text{Volume of slice} \approx 1400 w \Delta h \text{ m}^3.$$

Since w is a linear function of h, and $w = 160$ when $h = 0$, and $w = 10$ when $h = 150$, this function has slope $= (10 - 160)/150 = -1$. Thus

$$w = 160 - h \text{ meters},$$

so

$$\text{Volume of slice} \approx 1400(160 - h)\Delta h \text{ m}^3.$$

Summing over all slices and taking the limit as $\Delta h \to 0$ gives

$$\text{Total volume} = \lim_{\Delta h \to 0} \sum 1400(160 - h)\Delta h = \int_0^{150} 1400(160 - h)\,dh \text{ m}^3.$$

Evaluating the integral gives

$$\text{Total volume} = 1400\left(160h - \frac{h^2}{2}\right)\Bigg|_0^{150} = 1.785 \cdot 10^7 \text{ m}^3.$$

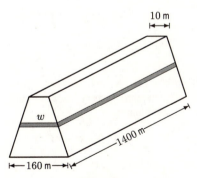

Figure 8.4

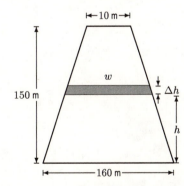

Figure 8.5

Solutions for Section 8.2

Exercises

1. The volume is given by

$$V = \int_0^1 \pi y^2 dx = \int_0^1 \pi x^4 dx = \pi \frac{x^5}{5}\Big|_0^1 = \frac{\pi}{5}.$$

5. The volume is given by

$$V = \int_{-2}^0 \pi(4 - x^2)^2 \, dx = \pi \int_{-2}^0 (16 - 8x^2 + x^4) \, dx = \pi \left(16x - \frac{8x^3}{3} + \frac{x^5}{5} \right)\Big|_{-2}^0 = \frac{256\pi}{15}.$$

9. Since the graph of $y = e^{3x}$ is above the graph of $y = e^x$ for $0 \le x \le 1$, the volume is given by

$$V = \int_0^1 \pi(e^{3x})^2 \, dx - \int_0^1 \pi(e^x)^2 \, dx = \int_0^1 \pi(e^{6x} - e^{2x}) \, dx = \pi \left(\frac{e^{6x}}{6} - \frac{e^{2x}}{2} \right)\Big|_0^1 = \pi \left(\frac{e^6}{6} - \frac{e^2}{2} + \frac{1}{3} \right).$$

13. We have

$$D = \int_0^1 \sqrt{(-e^t \sin(e^t))^2 + (e^t \cos(e^t))^2} \, dt$$

$$= \int_0^1 \sqrt{e^{2t}} \, dt = \int_0^1 e^t \, dt$$

$$= e - 1.$$

This is the length of the arc of a unit circle from the point $(\cos 1, \sin 1)$ to $(\cos e, \sin e)$—in other words between the angles $\theta = 1$ and $\theta = e$. The length of this arc is $(e - 1)$.

Problems

17.

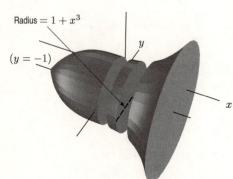

Radius $= 1 + x^3$

$(y = -1)$

We slice the region perpendicular to the x–axis. The Riemann sum we get is $\sum \pi(x^3 + 1)^2 \Delta x$. So the volume V is the integral

$$V = \int_{-1}^1 \pi(x^3 + 1)^2 \, dx$$

$$= \pi \int_{-1}^1 (x^6 + 2x^3 + 1) \, dx$$

$$= \pi \left(\frac{x^7}{7} + \frac{x^4}{2} + x \right)\Big|_{-1}^1$$

$$= (16/7)\pi \approx 7.18.$$

21.

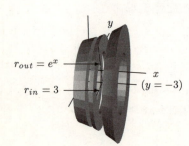

$r_{out} = e^x$

$r_{in} = 3$

$(y = -3)$

We slice the volume with planes perpendicular to the line $y = -3$. This divides the curve into thin washers, as in Example 3 on page 354 of the text, whose volumes are

$$\pi r_{out}^2 dx - \pi r_{in}^2 dx = \pi(3 + y)^2 dx - \pi 3^2 dx.$$

So the integral we get from adding all these washers up is

$$V = \int_{x=0}^{x=1} [\pi(3 + y)^2 - \pi 3^2] \, dx$$

$$= \pi \int_0^1 [(3 + e^x)^2 - 9] \, dx$$

$$= \pi \int_0^1 [e^{2x} + 6e^x] \, dx = \pi[\frac{e^{2x}}{2} + 6e^x]\Big|_0^1$$

$$= \pi[(e^2/2 + 6e) - (1/2 + 6)] \approx 42.42.$$

25. (a) We can begin by slicing the pie into horizontal slabs of thickness Δh located at height h. To find the radius of each slice, we note that radius increases linearly with height. Since $r = 4.5$ when $h = 3$ and $r = 3.5$ when $h = 0$, we should have $r = 3.5 + h/3$. Then the volume of each slab will be $\pi r^2 \, \Delta h = \pi(3.5 + h/3)^2 \, \Delta h$. To find the total volume of the pie, we integrate this from $h = 0$ to $h = 3$:

$$V = \pi \int_0^3 \left(3.5 + \frac{h}{3}\right)^2 \, dh$$

$$= \pi \left[\frac{h^3}{27} + \frac{7h^2}{6} + \frac{49h}{4}\right]\Big|_0^3$$

$$= \pi \left[\frac{3^3}{27} + \frac{7(3^2)}{6} + \frac{49(3)}{4}\right] \approx 152 \text{ in}^3.$$

(b) We use 1.5 in as a rough estimate of the radius of an apple. This gives us a volume of $(4/3)\pi(1.5)^3 \approx 10 \text{ in}^3$. Since $152/10 \approx 15$, we would need about 15 apples to make a pie.

29. (a) The volume, V, contained in the bowl when the surface has height h is

$$V = \int_0^h \pi x^2 \, dy.$$

However, since $y = x^4$, we have $x^2 = \sqrt{y}$ so that

$$V = \int_0^h \pi\sqrt{y} \, dy = \frac{2}{3}\pi h^{3/2}.$$

Differentiating gives $dV/dh = \pi h^{1/2} = \pi\sqrt{h}$. We are given that $dV/dt = -6\sqrt{h}$, where the negative sign reflects the fact that V is decreasing. Using the chain rule we have

$$\frac{dh}{dt} = \frac{dh}{dV} \cdot \frac{dV}{dt} = \frac{1}{dV/dh} \cdot \frac{dV}{dt} = \frac{1}{\pi\sqrt{h}} \cdot (-6\sqrt{h}) = -\frac{6}{\pi}.$$

Thus, $dh/dt = -6/\pi$, a constant.

(b) Since $dh/dt = -6/\pi$ we know that $h = -6t/\pi + C$. However, when $t = 0$, $h = 1$, therefore $h = 1 - 6t/\pi$. The bowl is empty when $h = 0$, that is when $t = \pi/6$ units.

33. Since $y = (e^x + e^{-x})/2$, $y' = (e^x - e^{-x})/2$. The length of the catenary is

$$\int_{-1}^1 \sqrt{1 + (y')^2} \, dx = \int_{-1}^1 \sqrt{1 + \left[\frac{e^x - e^{-x}}{2}\right]^2} \, dx = \int_{-1}^1 \sqrt{1 + \frac{e^{2x}}{4} - \frac{1}{2} + \frac{e^{-2x}}{4}} \, dx$$

$$= \int_{-1}^1 \sqrt{\left[\frac{e^x + e^{-x}}{2}\right]^2} \, dx = \int_{-1}^1 \frac{e^x + e^{-x}}{2} \, dx$$

$$= \left[\frac{e^x - e^{-x}}{2}\right]\Big|_{-1}^1 = e - e^{-1}.$$

Solutions for Section 8.3

Exercises

1. Since density is e^{-x} gm/cm,

$$\text{Mass} = \int_0^{10} e^{-x} \, dx = -e^{-x}\Big|_0^{10} = 1 - e^{-10} \text{ gm}.$$

5. **(a)** Figure 8.6 shows a graph of the density function.

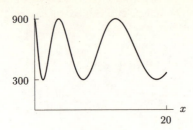

Figure 8.6

(b) Suppose we choose an x, $0 \leq x \leq 20$. We approximate the density of the number of the cars between x and $x + \Delta x$ miles as $\delta(x)$ cars per mile. Therefore, the number of cars between x and $x + \Delta x$ is approximately $\delta(x)\Delta x$. If we slice the 20 mile strip into N slices, we get that the total number of cars is

$$C \approx \sum_{i=1}^{N} \delta(x_i)\Delta x = \sum_{i=1}^{N} \left[600 + 300\sin(4\sqrt{x_i + 0.15})\right] \Delta x,$$

where $\Delta x = 20/N$. (This is a right-hand approximation; the corresponding left-hand approximation is $\sum_{i=0}^{N-1} \delta(x_i)\Delta x$.)

(c) As $N \to \infty$, the Riemann sum above approaches the integral

$$C = \int_0^{20} \left(600 + 300\sin 4\sqrt{x + 0.15}\right) dx.$$

If we calculate the integral numerically, we find $C \approx 11513$. We can also find the integral exactly as follows:

$$C = \int_0^{20} \left(600 + 300\sin 4\sqrt{x + 0.15}\right) dx$$

$$= \int_0^{20} 600 \, dx + \int_0^{20} 300\sin 4\sqrt{x + 0.15} \, dx$$

$$= 12000 + 300\int_0^{20} \sin 4\sqrt{x + 0.15} \, dx.$$

Let $w = \sqrt{x + 0.15}$, so $x = w^2 - 0.15$ and $dx = 2w \, dw$. Then

$$\int_{x=0}^{x=20} \sin 4\sqrt{x + 0.15} \, dx = 2\int_{w=\sqrt{0.15}}^{w=\sqrt{20.15}} w \sin 4w \, dw, \text{ (using integral table III-15)}$$

$$= 2\left[-\frac{1}{4}w \cos 4w + \frac{1}{16}\sin 4w\right]\Bigg|_{\sqrt{0.15}}^{\sqrt{20.15}}$$

$$\approx -1.624.$$

Using this, we have $C \approx 12000 + 300(-1.624) \approx 11513$, which matches our numerical approximation.

Problems

9. Since the density varies with x, the region must be sliced perpendicular to the x-axis. This has the effect of making the density approximately constant on each strip. See Figure 8.7. Since a strip is of height y, its area is approximately $y\Delta x$. The density on the strip is $\delta(x) = 1 + x$ gm/cm^2. Thus

$$\text{Mass of strip} \approx \text{Density} \cdot \text{Area} \approx (1 + x)y\Delta x \text{ gm.}$$

Because the tops of the strips end on two different lines, one for $x \geq 0$ and the other for $x < 0$, the mass is calculated as the sum of two integrals. See Figure 8.7. For the left part of the region, $y = x + 1$, so

$$\text{Mass of left part} = \lim_{\Delta x \to 0} \sum (1 + x) y \Delta x = \int_{-1}^{0} (1 + x)(x + 1)\, dx$$

$$= \int_{-1}^{0} (1 + x)^2\, dx = \left. \frac{(x + 1)^3}{3} \right|_{-1}^{0} = \frac{1}{3}\ \text{gm.}$$

From Figure 8.7, we see that for the right part of the region, $y = -x + 1$, so

$$\text{Mass of right part} = \lim_{\Delta x \to 0} \sum (1 + x) y \Delta x = \int_{0}^{1} (1 + x)(-x + 1)\, dx$$

$$= \int_{0}^{1} (1 - x^2)\, dx = \left. x - \frac{x^3}{3} \right|_{0}^{1} = \frac{2}{3}\ \text{gm.}$$

$$\text{Total mass} = \frac{1}{3} + \frac{2}{3} = 1\ \text{gm.}$$

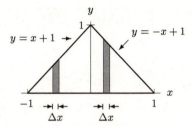

Figure 8.7

13. (a) Use the formula for the volume of a cylinder:

$$\text{Volume} = \pi r^2 l.$$

Since it is only a half cylinder

$$\text{Volume of shed} = \frac{1}{2} \pi r^2 l.$$

(b) Set up the axes as shown in Figure 8.8. The density can be defined as

$$\text{Density} = ky.$$

Now slice the sawdust horizontally into slabs of thickness Δy as shown in Figure 8.9, and calculate

$$\text{Volume of slab} \approx 2xl\Delta y = 2l(\sqrt{r^2 - y^2})\Delta y.$$

$$\text{Mass of slab} = \text{Density} \cdot \text{Volume} \approx 2kly\sqrt{r^2 - y^2}\,\Delta y.$$

Finally, we compute the total mass of sawdust:

$$\text{Total mass of sawdust} = \int_{0}^{r} 2kly\sqrt{r^2 - y^2}\, dy = \left. -\frac{2}{3} kl(r^2 - y^2)^{3/2} \right|_{0}^{r} = \frac{2klr^3}{3}.$$

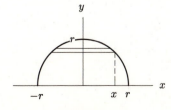

Figure 8.8

Figure 8.9

17. We need the numerator of $\bar{x}$, to be zero, i.e. $\sum x_i m_i = 0$. Since all of the masses are the same, we can factor them out and write $4 \sum x_i = 0$. Thus the fourth mass needs to be placed so that all of the positions sum to zero. The first three positions sum to $(-6 + 1 + 3) = -2$, so the fourth mass needs to be placed at $x = 2$.

21. **(a)** The density is minimum at $x = -1$ and increases as x increases, so more of the mass of the rod is in the right half of the rod. We thus expect the balancing point to be to the right of the origin.

   (b) We need to compute

$$\int_{-1}^{1} x(3 - e^{-x}) \, dx = \left(\frac{3}{2} x^2 + x e^{-x} + e^{-x} \right) \Big|_{-1}^{1} \quad \text{(using integration by parts)}$$

$$= \frac{3}{2} + e^{-1} + e^{-1} - \left(\frac{3}{2} - e^1 + e^1 \right) = \frac{2}{e}.$$

We must divide this result by the total mass, which is given by

$$\int_{-1}^{1} (3 - e^{-x}) \, dx = (3x + e^{-x}) \Big|_{-1}^{1} = 6 - e + \frac{1}{e}.$$

We therefore have

$$\bar{x} = \frac{2/e}{6 - e + (1/e)} = \frac{2}{1 + 6e - e^2} \approx 0.2.$$

25. Stand the cone with the base horizontal, with center at the origin. Symmetry gives us that $\bar{x} = \bar{y} = 0$. Since the cone is fatter near its base we expect the center of mass to be nearer to the base.

   Slice the cone into disks parallel to the xy-plane.

   As we saw in Example 2 on page 347, a disk of thickness Δz at height z above the base has

$$\text{Volume of disk} = A_z(z) \Delta z \approx \pi (5 - z)^2 \Delta z \ \text{cm}^3.$$

Thus, since the density is δ,

$$\bar{z} = \frac{\int z \delta A_z(z) \, dz}{\text{Mass}} = \frac{\int_0^5 z \cdot \delta \pi (5 - z)^2 \, dz}{\text{Mass}} \ \text{cm}.$$

To evaluate the integral in the numerator, we factor out the constant density δ and π to get

$$\int_0^5 z \cdot \delta \pi (5 - z)^2 \, dz = \delta \pi \int_0^5 z(25 - 10z + z^2) \, dz = \delta \pi \left(\frac{25z^2}{2} - \frac{10z^3}{3} + \frac{z^4}{4} \right) \Big|_0^5 = \frac{625}{12} \delta \pi.$$

We divide this result by the total mass of the cone, which is $\left(\frac{1}{3} \pi 5^2 \cdot 5 \right) \delta$:

$$\bar{z} = \frac{\frac{625}{12} \delta \pi}{\frac{1}{3} \pi 5^3 \delta} = \frac{5}{4} = 1.25 \ \text{cm}.$$

As predicted, the center of mass is closer to the base of the cone than its top.

Solutions for Section 8.4

Exercises

1. The work done is given by

$$W = \int_1^2 3x \, dx = \frac{3}{2} x^2 \Big|_1^2 = \frac{9}{2} \ \text{joules}.$$

5. The force exerted on the satellite by the earth (and vice versa!) is GMm/r^2, where r is the distance from the center of the earth to the center of the satellite, m is the mass of the satellite, M is the mass of the earth, and G is the gravitational constant. So the total work done is

$$\int_{6.4\cdot 10^6}^{8.4\cdot 10^6} F\, dr = \int_{6.4\cdot 10^6}^{8.4\cdot 10^6} \frac{GMm}{r^2}\, dr = \left(\frac{-GMm}{r}\right)\Bigg|_{6.4\cdot 10^6}^{8.4\cdot 10^6} \approx 1.489\cdot 10^{10} \text{ joules.}$$

Problems

9. Consider lifting a rectangular slab of water h feet from the top up to the top. The area of such a slab is $(10)(20) = 200$ square feet; if the thickness is dh, then the volume of such a slab is $200\, dh$ cubic feet. This much water weighs 62.4 pounds per ft^3, so the weight of such a slab is $(200\, dh)(62.4) = 12480\, dh$ pounds. To lift that much water h feet requires $12480h\, dh$ foot-pounds of work. To lift the whole tank, we lift one plate at a time; integrating over the slabs yields

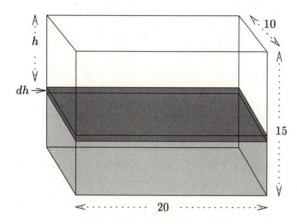

$$\int_0^{15} 12480h\, dh = \frac{12480h^2}{2}\Bigg|_0^{15} = \frac{12480\cdot 15^2}{2} = 1{,}404{,}000 \text{ foot-pounds.}$$

13.

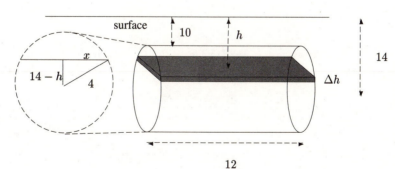

Let h represent distance below the surface in feet. We slice the tank up into horizontal slabs of thickness Δh. From looking at the figure, we can see that the slabs will be rectangular. The length of any slab is 12 feet. The width w of a slab h units below the ground will equal $2x$, where $(14 - h)^2 + x^2 = 16$, so $w = 2\sqrt{4^2 - (14 - h)^2}$. The volume of such a slab is therefore $12w\, \Delta h = 24\sqrt{16 - (14 - h)^2}\, \Delta h$ cubic feet; the slab weighs $42\cdot 24\sqrt{16 - (14 - h)^2}\, \Delta h = 1008\sqrt{16 - (14 - h)^2}\, \Delta h$ pounds. So the total work done in pumping out all the gasoline is

$$\int_{10}^{18} 1008h\sqrt{16 - (14 - h)^2}\, dh = 1008\int_{10}^{18} h\sqrt{16 - (14 - h)^2}\, dh.$$

Substitute $s = 14 - h$, $ds = -dh$. We get

$$1008 \int_{10}^{18} h\sqrt{16 - (14 - h)^2}\, dh = -1008 \int_{4}^{-4} (14 - s)\sqrt{16 - s^2}\, ds$$

$$= 1008 \cdot 14 \int_{-4}^{4} \sqrt{16 - s^2}\, ds - 1008 \int_{-4}^{4} s\sqrt{16 - s^2}\, ds.$$

The first integral represents the area of a semicircle of radius 4, which is 8π. The second is the integral of an odd function, over the interval $-4 \le s \le 4$, and is therefore 0. Hence, the total work is $1008 \cdot 14 \cdot 8\pi \approx 354{,}673$ foot-pounds.

17. Bottom:

$$\text{Water force } = 62.4(2)(12) = 1497.6 \text{ lbs.}$$

Front and back:

$$\text{Water force } = (62.4)(4) \int_{0}^{2} (2 - x)\, dx = (62.4)(4)\left(2x - \frac{1}{2}x^2\right)\Big|_{0}^{2}$$
$$= (62.4)(4)(2) = 499.2 \text{ lbs.}$$

Both sides:

$$\text{Water force } = (62.4)(3) \int_{0}^{2} (2 - x)\, dx = (62.4)(3)(2) = 374.4 \text{ lbs.}$$

21. We divide the dam into horizontal strips since the pressure is then approximately constant on each one. See Figure 8.10.

$$\text{Area of strip } \approx w\Delta h \text{ m}^2.$$

Since w is a linear function of h, and $w = 3600$ when $h = 0$, and $w = 3000$ when $h = 100$, the function has slope $(3000 - 3600)/100 = -6$. Thus,

$$w = 3600 - 6h,$$

so

$$\text{Area of strip } \approx (3600 - 6h)\Delta h \text{ m}^2.$$

The density of water is $\delta = 1000$ kg/m^3, so the pressure at depth h meters $= \delta g h = 1000 \cdot 9.8h = 9800h$ nt/m^2. Thus,

$$\text{Total force } = \lim_{\Delta h \to 0} \sum 9800h(3600 - 6h)\Delta h = 9800 \int_{0}^{100} h(3600 - 6h)\, dh \text{ newtons.}$$

Evaluating the integral gives

$$\text{Total force } = 9800(1800h^2 - 2h^3)\Big|_{0}^{100} = 1.6 \cdot 10^{11} \text{ newtons.}$$

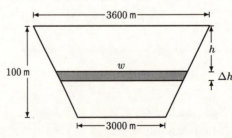

Figure 8.10

25.

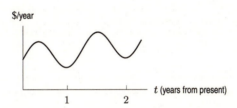

The density of the rod, in mass per unit length, is M/l (see above). So a slice of size dr has mass $\frac{M\,dr}{l}$. It pulls the small mass m with force $Gm\frac{M\,dr}{l}/r^2 = \frac{GmM\,dr}{lr^2}$. So the total gravitational attraction between the rod and point is

$$\int_a^{a+l} \frac{GmM\,dr}{lr^2} = \frac{GmM}{l}\left(-\frac{1}{r}\right)\Bigg|_a^{a+l}$$
$$= \frac{GmM}{l}\left(\frac{1}{a} - \frac{1}{a+l}\right)$$
$$= \frac{GmM}{l}\frac{l}{a(a+l)} = \frac{GmM}{a(a+l)}.$$

Solutions for Section 8.5

Exercises

1. At any time t, in a time interval Δt, an amount of $1000\Delta t$ is deposited into the account. This amount earns interest for $(10-t)$ years giving a future value of $1000e^{(0.08)(10-t)}$. Summing all such deposits, we have

$$\text{Future value} = \int_0^{10} 1000e^{0.08(10-t)}\,dt = \$15{,}319.30.$$

Problems

5.

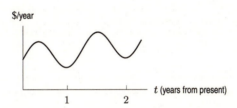

The graph reaches a peak each summer, and a trough each winter. The graph shows sunscreen sales increasing from cycle to cycle. This gradual increase may be due in part to inflation and to population growth.

9. You should choose the payment which gives you the highest present value. The immediate lump-sum payment of $2800 obviously has a present value of exactly $2800, since you are getting it now. We can calculate the present value of the installment plan as:

$$\text{PV} = 1000e^{-0.06(0)} + 1000e^{-0.06(1)} + 1000e^{-0.06(2)}$$
$$\approx \$2828.68.$$

Since the installment payments offer a (slightly) higher present value, you should accept this option.

13. (a) Suppose the oil extracted over the time period $[0, M]$ is S. (See Figure 8.11.) Since $q(t)$ is the rate of oil extraction, we have:

$$S = \int_0^M q(t)\,dt = \int_0^M (a - bt)\,dt = \int_0^M (10 - 0.1t)\,dt.$$

To calculate the time at which the oil is exhausted, set $S = 100$ and try different values of M. We find $M = 10.6$ gives

$$\int_0^{10.6} (10 - 0.1t)\,dt = 100,$$

so the oil is exhausted in 10.6 years.

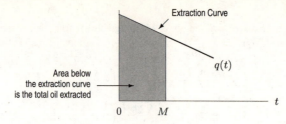

Figure 8.11

(b) Suppose p is the oil price, C is the extraction cost per barrel, and r is the interest rate. We have the present value of the profit as

$$\text{Present value of profit} = \int_0^M (p - C)q(t)e^{-rt}dt$$
$$= \int_0^{10.6} (20 - 10)(10 - 0.1t)e^{-0.1t}\, dt$$
$$= 624.9 \text{ million dollars.}$$

17.

$$\int_0^{q^*} (p^* - S(q))\, dq = \int_0^{q^*} p^*\, dq - \int_0^{q^*} S(q)\, dq$$
$$= p^* q^* - \int_0^{q^*} S(q)\, dq.$$

Using Problem 16, this integral is the extra amount consumers pay (i.e., suppliers earn over and above the minimum they would be willing to accept for supplying the good). It results from charging the equilibrium price.

Solutions for Section 8.6

Exercises

1.

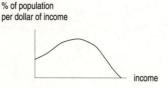

Figure 8.12: Density function

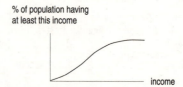

Figure 8.13: Cumulative distribution function

5. Since the function is decreasing, it cannot be a cdf (whose values never decrease). Thus, the function is a pdf.
The area under a pdf is 1, so, using the formula for the area of a triangle, we have

$$\frac{1}{2}4c = 1, \quad \text{giving} \quad c = \frac{1}{2}.$$

The pdf is

$$p(x) = \frac{1}{2} - \frac{1}{8}x \quad \text{for} \quad 0 \le x \le 4,$$

so the cdf is given in Figure 8.14 by

$$P(x) = \begin{cases} 0 & \text{for} \quad x < 0 \\ \dfrac{x}{2} - \dfrac{x^2}{16} & \text{for} \quad 0 \leq x \leq 4 \\ 1 & \text{for} \quad x > 4. \end{cases}$$

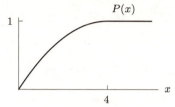

Figure 8.14

9. This function does not level off to 1, and it is not always increasing. Thus, the function is a pdf. Since the area under the curve must be 1, using the formula for the area of a triangle,

$$\frac{1}{2} \cdot c \cdot 1 = 1 \quad \text{so} \quad c = 2.$$

Thus, the pdf is given by

$$p(x) = \begin{cases} 0 & \text{for} \quad x < 0 \\ 4x & \text{for} \quad 0 \leq x \leq 0.5 \\ 2 - 4(x - 0.5) = 4 - 4x & \text{for} \quad 0.5 < x \leq 1 \\ 0 & \text{for} \quad x > 0. \end{cases}$$

To find the cdf, we integrate each part of the function separately, making sure that the constants of integration are arranged so that the cdf is continuous.

Since $\int 4x\,dx = 2x^2 + C$ and $P(0) = 0$, we have $2(0)^2 + C = 0$ so $C = 0$. Thus $P(x) = 2x^2$ on $0 \leq x \leq 0.5$. At $x = 0.5$, the cdf has value $P(0.5) = 2(0.5)^2 = 0.5$. Thus, we arrange that the integral of $4 - 4x$ goes through the point $(0.5, 0.5)$. Since $\int (4 - 4x)\,dx = 4x - 2x^2 + C$, we have

$$4(0.5) - 2(0.5)^2 + C = 0.5 \quad \text{giving} \quad C = -1.$$

Thus

$$P(x) = \begin{cases} 0 & \text{for} \quad x < 0 \\ 2x^2 & \text{for} \quad 0 \leq x \leq 0.5 \\ 4x - 2x^2 - 1 & \text{for} \quad 0.5 < x \leq 1 \\ 1 & \text{for} \quad x > 1. \end{cases}$$

See Figure 8.15.

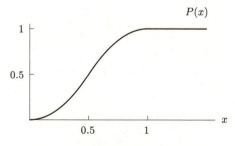

Figure 8.15

Problems

13. For a small interval Δx around 68, the fraction of the population of American men with heights in this interval is about $(0.2)\Delta x$. For example, taking $\Delta x = 0.1$, we can say that approximately $(0.2)(0.1) = 0.02 = 2\%$ of American men have heights between 68 and 68.1 inches.

17. (a) The percentage of calls lasting from 1 to 2 minutes is given by the integral

$$\int_1^2 p(x)\,dx \int_1^2 0.4e^{-0.4x}\,dx = e^{-0.4} - e^{-0.8} \approx 22.1\%.$$

(b) A similar calculation (changing the limits of integration) gives the percentage of calls lasting 1 minute or less as

$$\int_0^1 p(x)\,dx = \int_0^1 0.4e^{-0.4x}\,dx = 1 - e^{-0.4} \approx 33.0\%.$$

(c) The percentage of calls lasting 3 minutes or more is given by the improper integral

$$\int_3^\infty p(x)\,dx = \lim_{b\to\infty}\int_3^b 0.4e^{-0.4x}\,dx = \lim_{b\to\infty}(e^{-1.2} - e^{-0.4b}) = e^{-1.2} \approx 30.1\%.$$

(d) The cumulative distribution function is the integral of the probability density; thus,

$$C(h) = \int_0^h p(x)\,dx = \int_0^h 0.4e^{-0.4x}\,dx = 1 - e^{-0.4h}.$$

Solutions for Section 8.7

Exercises

1.

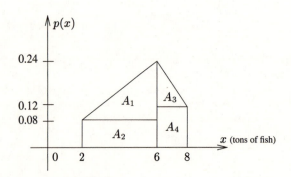

Splitting the figure into four pieces, we see that

$$\text{Area under the curve} = A_1 + A_2 + A_3 + A_4$$
$$= \frac{1}{2}(0.16)4 + 4(0.08) + \frac{1}{2}(0.12)2 + 2(0.12)$$
$$= 1.$$

We expect the area to be 1, since $\int_{-\infty}^\infty p(x)\,dx = 1$ for any probability density function, and $p(x)$ is 0 except when $2 \le x \le 8$.

Problems

5. (a) We can find the proportion of students by integrating the density $p(x)$ between $x = 1.5$ and $x = 2$:

$$P(2) - P(1.5) = \int_{1.5}^2 \frac{x^3}{4}\,dx$$
$$= \left.\frac{x^4}{16}\right|_{1.5}^2$$
$$= \frac{(2)^4}{16} - \frac{(1.5)^4}{16} = 0.684,$$

so that the proportion is 0.684 : 1 or 68.4%.

(b) We find the mean by integrating x times the density over the relevant range:

$$\text{Mean} = \int_0^2 x \left(\frac{x^3}{4}\right) \, dx$$

$$= \int_0^2 \frac{x^4}{4} \, dx$$

$$= \left.\frac{x^5}{20}\right|_0^2$$

$$= \frac{2^5}{20} = 1.6 \text{ hours.}$$

(c) The median will be the time T such that exactly half of the students are finished by time T, or in other words

$$\frac{1}{2} = \int_0^T \frac{x^3}{4} \, dx$$

$$\frac{1}{2} = \left.\frac{x^4}{16}\right|_0^T$$

$$\frac{1}{2} = \frac{T^4}{16}$$

$$T = \sqrt[4]{8} = 1.682 \text{ hours.}$$

9. (a) Since $\mu = 100$ and $\sigma = 15$:

$$p(x) = \frac{1}{15\sqrt{2\pi}} e^{-\frac{1}{2}\left(\frac{x-100}{15}\right)^2}.$$

(b) The fraction of the population with IQ scores between 115 and 120 is (integrating numerically)

$$\int_{115}^{120} p(x) \, dx = \int_{115}^{120} \frac{1}{15\sqrt{2\pi}} e^{-\frac{(x-100)^2}{450}} \, dx$$

$$= \frac{1}{15\sqrt{2\pi}} \int_{115}^{120} e^{-\frac{(x-100)^2}{450}} \, dx$$

$$\approx 0.067 = 6.7\% \text{ of the population.}$$

13. It is not (a) since a probability density must be a non-negative function; not (c) since the total integral of a probability density must be 1; (b) and (d) are probability density functions, but (d) is not a good model. According to (d), the probability that the next customer comes after 4 minutes is 0. In real life there should be a positive probability of not having a customer in the next 4 minutes. So (b) is the best answer.

Solutions for Chapter 8 Review

Exercises

1. The limits of integration are 0 and b, and the rectangle represents the region under the curve $f(x) = h$ between these limits. Thus,

$$\text{Area of rectangle} = \int_0^b h \, dx = \left. hx \right|_0^b = hb.$$

5. Each slice is a circular disk. The radius, r, of the disk increases with h and is given in the problem by $r = \sqrt{h}$. Thus

$$\text{Volume of slice} \approx \pi r^2 \Delta h = \pi h \Delta h.$$

Summing over all slices, we have

$$\text{Total volume} \approx \sum \pi h \Delta h.$$

Taking a limit as $\Delta h \to 0$, we get

$$\text{Total volume} = \lim_{\Delta h \to 0} \sum \pi h \Delta h = \int_0^{12} \pi h\, dh.$$

Evaluating gives

$$\text{Total volume} = \left. \pi \frac{h^2}{2} \right|_0^{12} = 72\pi.$$

Problems

9. **(a)**

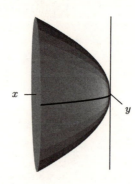

(b) Divide $[0,1]$ into N subintervals of width $\Delta x = \frac{1}{N}$. The volume of the ith disc is $\pi(\sqrt{x_i})^2 \Delta x = \pi x_i \Delta x$. So,
$V \approx \sum_{i=1}^{N} \pi x_i \Delta x$.

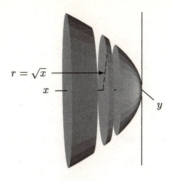

(c)

$$\text{Volume} = \int_0^1 \pi x \, dx = \left. \frac{\pi}{2} x^2 \right|_0^1 = \frac{\pi}{2} \approx 1.57.$$

13. (a) The line $y = ax$ must pass through (l, b). Hence $b = al$, so $a = b/l$.

(b) Cut the cone into N slices, slicing perpendicular to the x–axis. Each piece is almost a cylinder. The radius of the ith cylinder is $r(x_i) = \dfrac{bx_i}{l}$, so the volume

$$V \approx \sum_{i=1}^{N} \pi \left(\frac{bx_i}{l} \right)^2 \Delta x.$$

Therefore, as $N \to \infty$, we get

$$V = \int_0^l \pi b^2 l^{-2} x^2 \, dx$$

$$= \pi \frac{b^2}{l^2} \left[\frac{x^3}{3} \right]_0^l = \left(\pi \frac{b^2}{l^2} \right) \left(\frac{l^3}{3} \right) = \frac{1}{3} \pi b^2 l.$$

17. (a) Since the density is constant, the mass is the product of the area of the plate and its density.

$$\text{Area of the plate} = \int_0^1 (\sqrt{x} - x^2) \, dx = \left. \left(\frac{2}{3} x^{3/2} - \frac{1}{3} x^3 \right) \right|_0^1 = \frac{1}{3} \text{ cm}^2.$$

Thus the mass of the plate is $2 \cdot 1/3 = 2/3$ gm.

(b) See Figure 8.16. Since the region is "fatter" closer to the origin, $\bar{x}$ is less than $1/2$.

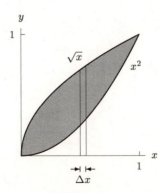

Figure 8.16

(c) To find $\bar{x}$, we slice the region into vertical strips of width Δx. See Figure 8.16.

$$\text{Area of strip } = A_x(x)\Delta x \approx (\sqrt{x} - x^2)\Delta x \text{ cm}^2.$$

Then we have

$$\bar{x} = \frac{\int x\delta A_x(x)\,dx}{\text{Mass}} = \frac{\int_0^1 2x(\sqrt{x} - x^2)\,dx}{2/3} = \frac{3}{2}\int_0^1 2(x^{3/2} - x^3)\,dx = \frac{3}{2}\cdot 2\left(\frac{2}{5}x^{5/2} - \frac{1}{4}x^4\right)\Big|_0^1 = \frac{9}{20}\text{ cm}.$$

This is less than $1/2$, as predicted in part (b). So $\bar{x} = \bar{y} = 9/20$ cm.

21. Let h be height above the bottom of the dam. Then

$$\begin{aligned}
\text{Water force} &= \int_0^{25} (62.4)(25 - h)(60)\,dh \\
&= (62.4)(60)\left(25h - \frac{h^2}{2}\right)\Big|_0^{25} \\
&= (62.4)(60)(625 - 312.5) \\
&= (62.4)(60)(312.5) \\
&= 1{,}170{,}000 \text{ lbs}.
\end{aligned}$$

25. (a) Slice the mountain horizontally into N cylinders of height Δh. The sum of the volumes of the cylinders will be

$$\sum_{i=1}^{N} \pi r^2 \Delta h = \sum_{i=1}^{N} \pi \left(\frac{3.5 \cdot 10^5}{\sqrt{h + 600}}\right)^2 \Delta h.$$

(b)

$$\begin{aligned}
\text{Volume} &= \int_{400}^{14400} \pi \left(\frac{3.5 \cdot 10^5}{\sqrt{h + 600}}\right)^2 dh \\
&= 1.23 \cdot 10^{11} \pi \int_{400}^{14400} \frac{1}{(h + 600)}\,dh \\
&= 1.23 \cdot 10^{11} \pi \ln(h + 600)\Big|_{400}^{14400} dh \\
&= 1.23 \cdot 10^{11} \pi \left[\ln 15000 - \ln 1000\right] \\
&= 1.23 \cdot 10^{11} \pi \ln(15000/1000) \\
&= 1.23 \cdot 10^{11} \pi \ln 15 \approx 1.05 \cdot 10^{12} \text{ cubic feet}.
\end{aligned}$$

29. (a) Divide the cross-section of the blood into rings of radius r, width Δr. See Figure 8.17.

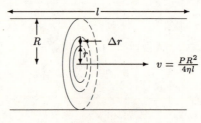

Figure 8.17

Then

$$\text{Area of ring} \approx 2\pi r \Delta r.$$

The velocity of the blood is approximately constant throughout the ring, so

$$\text{Rate blood flows through ring} \approx \text{Velocity} \cdot \text{Area}$$

$$= \frac{P}{4\eta l}(R^2 - r^2) \cdot 2\pi r \Delta r.$$

Thus, summing over all rings, we find the total blood flow:

$$\text{Rate blood flowing through blood vessel} \approx \sum \frac{P}{4\eta l}(R^2 - r^2)2\pi r \Delta r.$$

Taking the limit as $\Delta r \to 0$, we get

$$\text{Rate blood flowing through blood vessel} = \int_0^R \frac{\pi P}{2\eta l}(R^2 r - r^3)dr$$

$$= \frac{\pi P}{2\eta l}\left(\frac{R^2 r^2}{2} - \frac{r^4}{4}\right)\Big|_0^R = \frac{\pi P R^4}{8\eta l}.$$

(b) Since

$$\text{Rate of blood flow} = \frac{\pi P R^4}{8\eta l},$$

if we take $k = \pi P/(8\eta l)$, then we have

$$\text{Rate of blood flow} = kR^4,$$

that is, rate of blood flow is proportional to R^4, in accordance with Poiseuille's Law.

33. Any small piece of mass ΔM on either of the two spheres has kinetic energy $\frac{1}{2}v^2\Delta M$. Since the angular velocity of the two spheres is the same, the actual velocity of the piece ΔM will depend on how far away it is from the axis of revolution. The further away a piece is from the axis, the faster it must be moving and the larger its velocity v. This is because if ΔM is at a distance r from the axis, in one revolution it must trace out a circular path of length $2\pi r$ about the axis. Since every piece in either sphere takes 1 minute to make 1 revolution, pieces farther from the axis must move faster, as they travel a greater distance.

Thus, since the thin spherical shell has more of its mass concentrated farther from the axis of rotation than does the solid sphere, the bulk of it is traveling faster than the bulk of the solid sphere. So, it has the higher kinetic energy.

CAS Challenge Problems

37. (a) The expression for arc length in terms of a definite integral gives

$$A(t) = \int_0^t \sqrt{1 + \left(\frac{1}{2\sqrt{x}}\right)^2}\, dx = \frac{2\sqrt{t}\sqrt{1 + 4t} + \text{arcsinh}\,(2\sqrt{t})}{4}.$$

The integral was evaluated using a computer algebra system; different systems may give the answer in different forms. Some may involve ln instead of arcsinh, which is the inverse function of the hyperbolic sine function.

(b) Figure 8.19 shows that the graphs of $A(t)$ and the graph of $y = t$ look very similar. This suggests that $A(t) \approx t$.

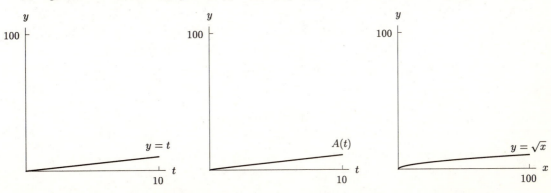

Figure 8.18   **Figure 8.19**

(c) The graph in Figure 8.19 is approximately horizontal and close to the x-axis. Thus, if we measure the arc length up to a certain x-value, the answer is approximately the same as if we had measured the length straight along the x-axis. Hence

$$A(t) \approx x = t.$$

So

$$A(t) \approx t.$$

CHECK YOUR UNDERSTANDING

1. True. Since $y = \pm\sqrt{9 - x^2}$ represent the top and bottom halves of the sphere, slicing disks perpendicular to the x-axis gives

$$\text{Volume of slice} \approx \pi y^2 \Delta x = \pi(9 - x^2)\Delta x$$
$$\text{Volume} = \int_{-3}^{3} \pi(9 - x^2)\, dx.$$

5. False. Volume is always positive, like area.

9. True. One way to look at it is that the center of mass shouldn't change if you change the units by which you measure the masses. If you double the masses, that is no different than using as a new unit of mass half the old unit. Alternatively, let the masses be m_1, m_2, and m_3 located at x_1, x_2, and x_3. Then the center of mass is given by:

$$\bar{x} = \frac{x_1 m_1 + x_2 m_2 + x_3 m_3}{m_1 + m_2 + m_3}.$$

Doubling the masses does not change the center of mass, since it doubles both the numerator and the denominator.

13. False. Work is the product of force and distance moved, so the work done in either case is 200 ft-lb.

17. False. The pressure is positive and when integrated gives a positive force.

21. False. It is true that $p(x) \geq 0$ for all x, but we also need $\int_{-\infty}^{\infty} p(x)dx = 1$. Since $p(x) = 0$ for $x \leq 0$, we need only check the integral from 0 to ∞. We have

$$\int_{0}^{\infty} xe^{-x^2}dx = \lim_{b \to \infty} \left(-\frac{1}{2}e^{-x^2}\right)\bigg|_{0}^{b} = \frac{1}{2}.$$

25. False. Since f is concave down, this means that $f'(x)$ is decreasing, so $f'(x) \leq f'(0) = 3/4$ on the interval $[0, 4]$. However, it could be that $f'(x)$ becomes negative so that $(f'(x))^2$ becomes large, making the integral for the arc length large also. For example, $f(x) = (3/4)x - x^2$ is concave down and $f'(0) = 3/4$, but $f(0) = 0$ and $f(4) = -13$, so the graph of f on the interval $[0, 4]$ has arc length at least 13.

29. False. Note that p is the density function for the population, not the cumulative density function. Thus $p(10) = p(20)$ means that x values near 10 are as likely as x values near 20.

CHAPTER NINE

Solutions for Section 9.1

Exercises

1. Yes, $a = 1$, ratio $= -1/2$.

5. No. Ratio between successive terms is not constant: $\dfrac{2x^2}{x} = 2x$, while $\dfrac{3x^3}{2x^2} = \dfrac{3}{2}x$.

9. No. Ratio between successive terms is not constant: $\dfrac{6z^2}{3z} = 2z$, while $\dfrac{9z^3}{6z^2} = \dfrac{3}{2}z$.

13. Sum $= \dfrac{1}{1 - (-y^2)} = \dfrac{1}{1 + y^2}, |y| < 1$.

17. Using the formula for the sum of an infinite geometric series,

$$\sum_{n=4}^{\infty} \left(\frac{1}{3}\right)^n = \left(\frac{1}{3}\right)^4 + \left(\frac{1}{3}\right)^5 + \cdots = \left(\frac{1}{3}\right)^4 \left(1 + \frac{1}{3} + \left(\frac{1}{3}\right)^2 + \cdots\right) = \frac{\left(\frac{1}{3}\right)^4}{1 - \frac{1}{3}} = \frac{1}{54}$$

Problems

21. (a)

$$P_1 = 0$$
$$P_2 = 250(0.04)$$
$$P_3 = 250(0.04) + 250(0.04)^2$$
$$P_4 = 250(0.04) + 250(0.04)^2 + 250(0.04)^3$$
$$\vdots$$
$$P_n = 250(0.04) + 250(0.04)^2 + 250(0.04)^3 + \cdots + 250(0.04)^{n-1}$$

(b) $P_n = 250(0.04)\left(1 + (0.04) + (0.04)^2 + (0.04)^3 + \cdots + (0.04)^{n-2}\right) = 250\dfrac{0.04(1 - (0.04)^{n-1})}{1 - 0.04}$

(c)

$$P = \lim_{n \to \infty} P_n$$
$$= \lim_{n \to \infty} 250\frac{0.04(1 - (0.04)^{n-1})}{1 - 0.04}$$
$$= \frac{(250)(0.04)}{0.96} = 0.04Q \approx 10.42$$

Thus, $\lim_{n \to \infty} P_n = 10.42$ and $\lim_{n \to \infty} Q_n = 260.42$. We would expect these limits to differ because one is right before taking a tablet, one is right after. We would expect the difference between them to be 250 mg, the amount of ampicillin in one tablet.

25. (a)

$$\text{Total amount of money deposited} = 100 + 92 + 84.64 + \cdots$$
$$= 100 + 100(0.92) + 100(0.92)^2 + \cdots$$
$$= \frac{100}{1 - 0.92} = 1250 \quad \text{dollars}$$

(b) Credit multiplier $= 1250/100 = 12.50$

The 12.50 is the factor by which the bank has increased its deposits, from \$100 to \$1250.

Solutions for Section 9.2

Exercises

1. Since $\lim\limits_{n\to\infty} x^n = 0$ if $|x| < 1$ and $|0.2| < 1$, we have $\lim\limits_{n\to\infty} (0.2)^n = 0$.

5. Since $S_n = \cos(\pi n) = 1$ if n is even and $S_n = \cos(\pi n) = -1$ if n is odd, the values of S_n oscillate between 1 and -1, so the limit does not exist.

9. We use the integral test to determine whether this series converges or diverges. We determine whether the corresponding improper integral $\int_1^\infty \dfrac{1}{x^3} dx$ converges or diverges:

$$\int_1^\infty \frac{1}{x^3} dx = \lim_{b\to\infty} \int_1^b \frac{1}{x^3} dx = \lim_{b\to\infty} \left. \frac{-1}{2x^2} \right|_1^b = \lim_{b\to\infty} \left(\frac{-1}{2b^2} + \frac{1}{2} \right) = \frac{1}{2}.$$

Since the integral $\int_1^\infty \dfrac{1}{x^3} dx$ converges, we conclude from the integral test that the series $\sum\limits_{n=1}^{\infty} \dfrac{1}{n^3}$ converges.

Problems

13. The series $\sum\limits_{n=1}^{\infty} \left(\dfrac{3}{4}\right)^n$ is a convergent geometric series, but $\sum\limits_{n=1}^{\infty} \dfrac{1}{n}$ is the divergent harmonic series.

If $\sum\limits_{n=1}^{\infty} \left(\left(\dfrac{3}{4}\right)^n + \dfrac{1}{n} \right)$ converged, then $\sum\limits_{n=1}^{\infty} \left(\left(\dfrac{3}{4}\right)^n + \dfrac{1}{n} \right) - \sum\limits_{n=1}^{\infty} \left(\dfrac{3}{4}\right)^n = \sum\limits_{n=1}^{\infty} \dfrac{1}{n}$ would converge by Theorem 9.2.

Therefore $\sum\limits_{n=1}^{\infty} \left(\left(\dfrac{3}{4}\right)^n + \dfrac{1}{n} \right)$ diverges.

17. We use the integral test and calculate the corresponding improper integral, $\int_1^\infty 3/(2x-1)^2 dx$:

$$\int_1^\infty \frac{3\, dx}{(2x-1)^2} = \lim_{b\to\infty} \int_1^b \frac{3\, dx}{(2x-1)^2} = \lim_{b\to\infty} \left. \frac{-3/2}{(2x-1)} \right|_1^b = \lim_{b\to\infty} \left(\frac{-3/2}{(2b-1)} + \frac{3}{2} \right) = \frac{3}{2}.$$

Since the integral converges, the series $\sum\limits_{n=1}^{\infty} \dfrac{3}{(2n-1)^2}$ converges.

21. Using left-hand sums for the integral of $f(x) = 1/(4x-3)$ over the interval $1 \le x \le n+1$ with uniform subdivisions of length 1 gives a lower bound on the partial sum:

$$S_n = 1 + \frac{1}{5} + \frac{1}{9} + \cdots + \frac{1}{4n-3} > \int_1^{n+1} \frac{dx}{4x-3} = \left. \frac{1}{4} \ln(4x-3) \right|_1^{n+1} = \frac{1}{4} \ln(4n+1).$$

Since $\ln(4n+1)$ increases without bound as $n \to \infty$, the partial sums of the series are unbounded. Thus, this is not a convergent series.

25. We want to define $\lim\limits_{n\to\infty} S_n = L$ so that S_n is as close to L as we please for all sufficiently large n. Thus, the definition says that for any positive ϵ, there is a value N such that

$$|S_n - L| < \epsilon \quad \text{whenever} \quad n \ge N.$$

29. From Property 1 in Theorem 9.2, we know that if $\sum a_n$ converges, then so does $\sum k a_n$.

Now suppose that $\sum a_n$ diverges and $\sum k a_n$ converges for $k \ne 0$. Thus using Property 1 and replacing $\sum a_n$ by $\sum k a_n$, we know that the following series converges:

$$\sum \frac{1}{k}(k a_n) = \sum a_n.$$

Thus, we have arrived at a contradiction, which means our original assumption, that $\sum\limits_{n=1}^{\infty} k a_n$ converged, must be wrong.

33. **(a)** Let N an integer with $N \geq c$. Consider the series $\displaystyle\sum_{i=N+1}^{\infty} a_i$. The partial sums of this series are increasing because all the terms in the series are positive. We show the partial sums are bounded using the right-hand sum in Figure 9.1. We see that for each positive integer k

$$f(N+1) + f(N+2) + \cdots + f(N+k) \leq \int_N^{N+k} f(x)\,dx.$$

Since $f(n) = a_n$ for all n, and $c \leq N$, we have

$$a_{N+1} + a_{N+2} + \cdots + a_{N+k} \leq \int_c^{N+k} f(x)\,dx.$$

Since $f(x)$ is a positive function, $\int_c^{N+k} f(x)\,dx \leq \int_c^b f(x)\,dx$ for all $b \geq N+k$. Since f is positive and $\int_c^{\infty} f(x)\,dx$ is convergent, $\int_c^{N+k} f(x)\,dx < \int_c^{\infty} f(x)\,dx$, so we have

$$a_{N+1} + a_{N+2} + \cdots + a_{N+k} \leq \int_c^{\infty} f(x)\,dx \quad \text{for all } k.$$

Thus, the partial sums of the series $\displaystyle\sum_{i=N+1}^{\infty} a_i$ are all bounded by the same number, so this series converges. Now use Theorem 9.2, property 2, to conclude that $\displaystyle\sum_{i=1}^{\infty} a_i$ converges.

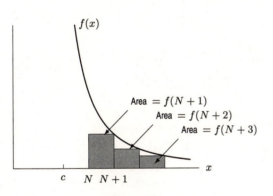

Figure 9.1

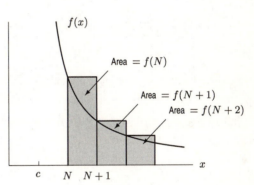

Figure 9.2

(b) We now suppose $\displaystyle\int_c^{\infty} f(x)\,dx$ diverges. In Figure 9.2 we see that for each positive integer k

$$\int_N^{N+k+1} f(x)\,dx \leq f(N) + f(N+1) + \cdots + f(N+k).$$

Since $f(n) = a_n$ for all n, we have

$$\int_N^{N+k+1} f(x)\,dx \leq a_N + a_{N+1} + \cdots + a_{N+k}.$$

Since $f(x)$ is defined for all $x \geq c$, if $\int_c^{\infty} f(x)\,dx$ is divergent, then $\int_N^{\infty} f(x)\,dx$ is divergent. So as $k \to \infty$, the the integral $\int_N^{N+k+1} f(x)\,dx$ diverges, so the partial sums of the series $\displaystyle\sum_{i=N}^{\infty} a_i$ diverge. Thus, the series $\displaystyle\sum_{i=1}^{\infty} a_i$ diverges.

More precisely, suppose the series converged. Then the partial sums would be bounded. (The partial sums would be less than the sum of the series, since all the terms in the series are positive.) But that would imply that the integral converged, by Theorem 9.1 on Convergence of Increasing Bounded Sequences. This contradicts the assumption that $\int_N^{\infty} f(x)\,dx$ is divergent.

Solutions for Section 9.3

Exercises

1. Let $a_n = 1/(n^2 + 2)$. Since $n^2 + 2 > n^2$, we have $1/(n^2 + 2) < 1/n^2$, so

$$0 < a_n < \frac{1}{n^2}.$$

The series $\sum_{n=1}^{\infty} \frac{1}{n^2}$ converges, so the comparison test tells us that the series $\sum_{n=1}^{\infty} \frac{1}{n^2 + 2}$ also converges.

5. Since $\ln n \leq n$ for $n \geq 2$, we have $1/\ln n \geq 1/n$, so the series diverges by comparison with the harmonic series, $\sum 1/n$.

9. Let $a_n = n^2/(n^4 + 1)$. Since $n^4 + 1 > n^4$, we have $\frac{1}{n^4 + 1} < \frac{1}{n^4}$, so

$$a_n = \frac{n^2}{n^4 + 1} < \frac{n^2}{n^4} = \frac{1}{n^2},$$

therefore

$$0 < a_n < \frac{1}{n^2}.$$

Since the series $\sum_{n=1}^{\infty} \frac{1}{n^2}$ converges, the comparison test tells us that the series $\sum_{n=1}^{\infty} \frac{n^2}{n^4 + 1}$ converges also.

13. Since $a_n = (n!)^2/(2n)!$, replacing n by $n + 1$ gives $a_{n+1} = ((n+1)!)^2/(2n+2)!$. Thus,

$$\frac{|a_{n+1}|}{|a_n|} = \frac{\dfrac{((n+1)!)^2}{(2n+2)!}}{\dfrac{(n!)^2}{(2n)!}} = \frac{((n+1)!)^2}{(2n+2)!} \cdot \frac{(2n)!}{(n!)^2}.$$

However, since $(n + 1)! = (n + 1)n!$ and $(2n + 2)! = (2n + 2)(2n + 1)(2n)!$, we have

$$\frac{|a_{n+1}|}{|a_n|} = \frac{(n+1)^2(n!)^2(2n)!}{(2n+2)(2n+1)(2n)!(n!)^2} = \frac{(n+1)^2}{(2n+2)(2n+1)} = \frac{n+1}{4n+2},$$

so

$$L = \lim_{n \to \infty} \frac{|a_{n+1}|}{|a_n|} = \frac{1}{4}.$$

Since $L < 1$, the ratio test tells us that $\sum_{n=1}^{\infty} \frac{(n!)^2}{(2n)!}$ converges.

17. Let $a_n = 1/\sqrt{n}$. Then replacing n by $n+1$ we have $a_{n+1} = 1/\sqrt{n+1}$. Since $\sqrt{n+1} > \sqrt{n}$, we have $\frac{1}{\sqrt{n+1}} < \frac{1}{\sqrt{n}}$, hence $a_{n+1} < a_n$. In addition, $\lim_{n\to\infty} a_n = 0$ so $\sum_{n=0}^{\infty} \frac{(-1)^n}{\sqrt{n}}$ converges by the alternating series test.

Problems

21. The first few terms of the series may be written

$$1 + e^{-1} + e^{-2} + e^{-3} + \cdots;$$

this is a geometric series with $a = 1$ and $x = e^{-1} = 1/e$. Since $|x| < 1$, the geometric series converges to

$$S = \frac{1}{1 - x} = \frac{1}{1 - e^{-1}} = \frac{e}{e - 1}.$$

25. Let $a_n = 1/\sqrt{3n-1}$. Then replacing n by $n+1$ gives $a_{n+1} = 1/\sqrt{3(n+1)-1}$. Since

$$\sqrt{3(n+1)-1} > \sqrt{3n-1},$$

we have

$$a_{n+1} < a_n.$$

In addition, $\lim_{n\to\infty} a_n = 0$ so the alternating series test tells us that the series $\sum_{n=1}^{\infty} \frac{(-1)^{n-1}}{\sqrt{3n-1}}$ converges.

29. **(a)** Assume that n is even. Then

$$1 - \frac{1}{2} + \frac{1}{3} - \frac{1}{4} + \cdots - \frac{1}{n} = \left(1 - \frac{1}{2}\right) + \left(\frac{1}{3} - \frac{1}{4}\right) + \cdots + \left(\frac{1}{n-1} - \frac{1}{n}\right)$$

$$= \frac{1}{1\cdot 2} + \frac{1}{3\cdot 4} + \cdots + \frac{1}{(n-1)\cdot n}.$$

(b) The given series $\frac{1}{1\cdot 2} + \frac{1}{3\cdot 4} + \frac{1}{5\cdot 6} + \cdots$ is term by term less than the series

$$\frac{1}{1\cdot 1} + \frac{1}{2\cdot 2} + \frac{1}{3\cdot 3} + \cdots = \frac{1}{1^2} + \frac{1}{2^2} + \frac{1}{3^2} + \cdots.$$

Since this second series, $\sum 1/n^2$, converges by the integral test, the first series converges.

(c) By parts (a) and (b), the sequence of partial sums for even n converges. The partial sum for odd n equals $1/n$ plus the partial sum for even $n-1$. Thus the partial sums for odd n approach the partial sums for even n, as $n \to \infty$. Therefore the sequence of all partial sums converges, and hence the series converges.

Solutions for Section 9.4

Exercises

1. Yes.

5. The general term can be written as $\frac{1\cdot 3\cdot 5\cdots (2n-1)}{2^n \cdot n!} x^n$ for $n \geq 1$.

9. The general term can be written as $\frac{(x-a)^n}{2^{n-1}\cdot n!}$ for $n \geq 1$.

13. Since $C_n = (n+1)/(2^n + n)$, replacing n by $n+1$ gives $C_{n+1} = (n+2)/(2^{n+1} + n + 1)$. Using the ratio test, we have

$$\frac{|a_{n+1}|}{|a_n|} = |x|\frac{|C_{n+1}|}{|C_n|} = |x|\frac{(n+2)/(2^{n+1}+n+1)}{(n+1)/(2^n+n)} = |x|\frac{n+2}{2^{n+1}+n+1}\cdot\frac{2^n+n}{n+1} = |x|\frac{n+2}{n+1}\cdot\frac{2^n+n}{2^{n+1}+n+1}.$$

Since

$$\lim_{n\to\infty}\frac{n+2}{n+1} = 1$$

and

$$\lim_{n\to\infty}\left(\frac{2^n+n}{2^{n+1}+n+1}\right) = \frac{1}{2}\lim_{n\to\infty}\left(\frac{2^n+n}{2^n+(n+1)/2}\right) = \frac{1}{2},$$

because 2^n dominates n as $n \to \infty$, we have

$$\lim_{n\to\infty}\frac{|a_{n+1}|}{|a_n|} = \frac{1}{2}|x|.$$

Thus the radius of convergence is $R = 2$.

17. Here the coefficient of the nth term is $C_n = (2^n/n!)$. Now we have

$$\left|\frac{a_{n+1}}{a_n}\right| = \left|\frac{(2^{n+1}/(n+1)!)x^{n+1}}{(2^n/n!)x^n}\right| = \frac{2|x|}{n+1} \to 0 \text{ as } n \to \infty.$$

Thus, the radius of convergence is $R = \infty$, and the series converges for all x.

21. We write the series as

$$x - \frac{x^3}{3} + \frac{x^5}{5} - \frac{x^7}{7} + \cdots + (-1)^{n-1} \frac{x^{2n-1}}{2n-1} + \cdots,$$

so

$$a_n = (-1)^{n-1} \frac{x^{2n-1}}{2n-1}.$$

Replacing n by $n+1$, we have

$$a_{n+1} = (-1)^{n+1-1} \frac{x^{2(n+1)-1}}{2(n+1)-1} = (-1)^n \frac{x^{2n+1}}{2n+1}.$$

Thus

$$\frac{|a_{n+1}|}{|a_n|} = \left| \frac{(-1)^n x^{2n+1}}{2n+1} \right| \cdot \left| \frac{2n-1}{(-1)^{n-1} x^{2n-1}} \right| = \frac{2n-1}{2n+1} x^2,$$

so

$$L = \lim_{n \to \infty} \frac{|a_{n+1}|}{|a_n|} = \lim_{n \to \infty} \frac{2n-1}{2n+1} x^2 = x^2.$$

By the ratio test, this series converges if $L < 1$, that is, if $x^2 < 1$, so $R = 1$.

Problems

25. The kth coefficient in the series $\sum k C_k x^k$ is $D_k = k \cdot C_k$. We are given that the series $\sum C_k x^k$ has radius of convergence R by the ratio test, so

$$|x| \lim_{k \to \infty} \frac{|C_{k+1}|}{|C_k|} = \frac{|x|}{R}.$$

Thus, applying the ratio test to the new series, we have

$$\lim_{k \to \infty} \left| \frac{D_{k+1} x^{k+1}}{D_k x^k} \right| = \lim_{k \to \infty} \left| \frac{(k+1)C_{k+1}}{k C_k} \right| |x| = \frac{|x|}{R}.$$

Hence the new series has radius of convergence R.

Solutions for Chapter 9 Review

Exercises

1. Let $a_n = n^2/(3n^2 + 4)$. Since $3n^2 + 4 > 3n^2$, we have $\frac{n^2}{3n^2 + 4} < \frac{1}{3}$, so

$$0 < a_n < \left(\frac{1}{3} \right)^n.$$

The geometric series $\sum_{n=1}^{\infty} \left(\frac{1}{3} \right)^n$ converges, so the comparison test tells us that the series $\sum_{n=1}^{\infty} \left(\frac{n^2}{3n^2 + 4} \right)^n$ also converges.

5. To show that the original series converges, we show that the series $\sum_{n=1}^{\infty} \left(\frac{3}{4} \right)^n$ and $\sum_{n=1}^{\infty} \frac{1}{n^2}$ converge. The first of these is a convergent geometric series, since $|3/4| < 1$. The integral test tells us that series $\sum_{n=1}^{\infty} \frac{1}{n^2}$ converges by comparing it with the convergent integral $\int_0^1 1/x^2 dx$. Theorem 9.2 then tells us that the series $\sum_{n=1}^{\infty} \left(\frac{3^n}{4} + \frac{1}{n^2} \right)$ also converges.

9. Since $a_n = (2n)!/(n!)^2$, replacing n by $n + 1$ gives $a_{n+1} = (2n + 2)!/((n + 1)!)^2$. Thus

$$\frac{a_{n+1}}{a_n} = \frac{\dfrac{(2n + 2)!}{((n + 1)!)^2}}{\dfrac{(2n)!}{(n!)^2}} = \frac{(2n + 2)!}{(n + 1)!(n + 1)!} \cdot \frac{n!n!}{(2n)!}.$$

Since $(2n + 2)! = (2n + 2)(2n + 1)(2n)!$ and $(n + 1)! = (n + 1)n!$, we have

$$\frac{a_{n+1}}{a_n} = \frac{(2n + 2)(2n + 1)}{(n + 1)(n + 1)},$$

therefore

$$L = \lim_{n \to \infty} \frac{a_{n+1}}{a_n} = 4.$$

As $L = 4$ the ratio test tells us that the series $\displaystyle\sum_{n=1}^{\infty} \frac{(2n)!}{(n!)^2}$ diverges.

13. Let $C_n = \dfrac{(2n)!}{(n!)^2}$. Then replacing n by $n + 1$, we have $C_{n+1} = \dfrac{(2n + 2)!}{((n + 1)!)^2}$. Thus, with $a_n = (2n)!x^n/(n!)^2$, we have

$$\frac{|a_{n+1}|}{|a_n|} = |x|\frac{|C_{n+1}|}{|C_n|} = |x|\frac{(2n + 2)!/((n + 1)!)^2}{(2n)!/(n!)^2} = |x|\frac{(2n + 2)!}{(2n)!} \cdot \frac{(n!)^2}{((n + 1)!)^2}.$$

Since $(2n + 2)! = (2n + 2)(2n + 1)(2n)!$ and $(n + 1)! = (n + 1)n!$ we have

$$\frac{|C_{n+1}|}{|C_n|} = \frac{(2n + 2)(2n + 1)}{(n + 1)(n + 1)},$$

so

$$\lim_{n \to \infty} \frac{|a_{n+1}|}{|a_n|} = |x| \lim_{n \to \infty} \frac{|C_{n+1}|}{|C_n|} = |x| \lim_{n \to \infty} \frac{(2n + 2)(2n + 1)}{(n + 1)(n + 1)} = |x| \lim_{n \to \infty} \frac{4n + 2}{n + 1} = 4|x|,$$

so the radius of convergence of this series is $R = 1/4$.

Problems

17. The amount of cephalexin in the body is given by $Q(t) = Q_0 e^{-kt}$, where $Q_0 = Q(0)$ and k is a constant. Since the half-life is 0.9 hours,

$$\frac{1}{2} = e^{-0.9k}, \quad k = -\frac{1}{0.9} \ln \frac{1}{2} \approx 0.8.$$

(a) After 6 hours

$$Q = Q_0 e^{-k(6)} \approx Q_0 e^{-0.8(6)} = Q_0(0.01).$$

Thus, the percentage of the cephalexin that remains after 6 hours $\approx 1\%$.

(b)

$$Q_1 = 250$$
$$Q_2 = 250 + 250(0.01)$$
$$Q_3 = 250 + 250(0.01) + 250(0.01)^2$$
$$Q_4 = 250 + 250(0.01) + 250(0.01)^2 + 250(0.01)^3$$

(c)

$$Q_3 = \frac{250(1 - (0.01)^3)}{1 - 0.01}$$
$$\approx 252.5$$
$$Q_4 = \frac{250(1 - (0.01)^4)}{1 - 0.01}$$
$$\approx 252.5$$

Thus, by the time a patient has taken three cephalexin tablets, the quantity of drug in the body has leveled off to 252.5 mg.

(d) Looking at the answers to part (b) shows that

$$Q_n = 250 + 250(0.01) + 250(0.01)^2 + \cdots + 250(0.01)^{n-1}$$
$$= \frac{250(1 - (0.01)^n)}{1 - 0.01}.$$

(e) In the long run, $n \to \infty$. So,

$$Q = \lim_{n \to \infty} Q_n = \frac{250}{1 - 0.01} = 252.5.$$

21.

$$\text{Present value of first coupon} = \frac{50}{1.06}$$
$$\text{Present value of second coupon} = \frac{50}{(1.06)^2}, \text{ etc.}$$

$$\text{Total present value} = \underbrace{\frac{50}{1.06} + \frac{50}{(1.06)^2} + \cdots + \frac{50}{(1.06)^{10}}}_{\text{coupons}} + \underbrace{\frac{1000}{(1.06)^{10}}}_{\text{principal}}$$

$$= \frac{50}{1.06}\left(1 + \frac{1}{1.06} + \cdots + \frac{1}{(1.06)^9}\right) + \frac{1000}{(1.06)^{10}}$$

$$= \frac{50}{1.06}\left(\frac{1 - \left(\frac{1}{1.06}\right)^{10}}{1 - \frac{1}{1.06}}\right) + \frac{1000}{(1.06)^{10}}$$

$$= 368.004 + 558.395$$

$$= \$926.40$$

25. (a) Since

$$|a_n| = a_n \qquad \text{if } a_n \geq 0$$
$$|a_n| = -a_n \qquad \text{if } a_n < 0,$$

we have

$$a_n + |a_n| = 2|a_n| \qquad \text{if } a_n > 0$$
$$a_n + |a_n| = 0 \qquad \text{if } a_n < 0.$$

Thus, for all n,

$$0 \leq a_n + |a_n| \leq 2|a_n|.$$

(b) If $\sum |a_n|$ converges, then $\sum 2|a_n|$ is convergent, so, by comparison, $\sum (a_n + |a_n|)$ is convergent. Then

$$\sum \left((a_n + |a_n|) - |a_n|\right) = \sum a_n$$

is convergent, as it is the difference of two convergent series.

CHECK YOUR UNDERSTANDING

1. True. A geometric series, $a + ax + ax^2 + \cdots$, is a power series about $x = 0$ with all coefficients equal to a.

5. True. This power series has an interval of convergence centered on $x = 0$. If the power series does not converge for $x = 1$, then the radius of convergence is less than or equal to 1. Thus, $x = 2$ lies outside the interval of convergence, so the series does not converge there.

9. False. It is true that if $\sum |a_n|$ converges, then we know that $\sum a_n$ converges. However, knowing that $\sum a_n$ converges does *not* tell us that $\sum |a_n|$ converges.

For example, if $a_n = (-1)^{n-1}/n$, then $\sum a_n$ converges by the alternating series test. However, $\sum |a_n|$ is the harmonic series which diverges.

13. True. Writing out the terms of this series, we have

$$(1 + (-1)^1) + (1 + (-1)^2) + (1 + (-1)^3) + (1 + (-1)^4) + \cdots$$
$$= (1 - 1) + (1 + 1) + (1 - 1) + (1 + 1) + \cdots$$
$$= 0 + 2 + 0 + 2 + \cdots.$$

17. True. Let $c_n = (-1)^n |a_n|$. Then $|c_n| = |a_n|$ so $\sum |c_n|$ converges, and therefore $\sum c_n = \sum (-1)^n |a_n|$ converges.

21. False. Consider the power series

$$(x - 1) - \frac{(x - 1)^2}{2} + \frac{(x - 1)^3}{3} + \cdots + (-1)^{n-1} \frac{(x - 1)^n}{n} + \cdots,$$

whose interval of convergence is $0 < x \le 2$. This series converges at one endpoint, $x = 2$, but not at the other, $x = 0$.

25. False. If $a_n b_n = 1/n^2$ and $a_n = b_n = 1/n$, then $\sum a_n b_n$ converges, but $\sum a_n$ and $\sum b_n$ do not converge.

CHAPTER TEN

Solutions for Section 10.1

Exercises

1. Let $\dfrac{1}{1+x} = (1+x)^{-1}$. Then $f(0) = 1$.

$$
\begin{array}{ll}
f'(x) = -1!(1+x)^{-2} & f'(0) = -1, \\
f''(x) = 2!(1+x)^{-3} & f''(0) = 2!, \\
f'''(x) = -3!(1+x)^{-4} & f'''(0) = -3!, \\
f^{(4)}(x) = 4!(1+x)^{-5} & f^{(4)}(0) = 4!, \\
f^{(5)}(x) = -5!(1+x)^{-6} & f^{(5)}(0) = -5!, \\
f^{(6)}(x) = 6!(1+x)^{-7} & f^{(6)}(0) = 6!, \\
f^{(7)}(x) = -7!(1+x)^{-8} & f^{(7)}(0) = -7!, \\
f^{(8)}(x) = 8!(1+x)^{-9} & f^{(8)}(0) = 8!.
\end{array}
$$

$$
\begin{aligned}
P_4(x) &= 1 - x + x^2 - x^3 + x^4, \\
P_6(x) &= 1 - x + x^2 - x^3 + x^4 - x^5 + x^6, \\
P_8(x) &= 1 - x + x^2 - x^3 + x^4 - x^5 + x^6 - x^7 + x^8.
\end{aligned}
$$

5. Let $f(x) = \arctan x$. Then $f(0) = \arctan 0 = 0$, and

$$
\begin{array}{ll}
f'(x) = 1/(1+x^2) = (1+x^2)^{-1} & f'(0) = 1, \\
f''(x) = (-1)(1+x^2)^{-2}2x & f''(0) = 0, \\
f'''(x) = 2!(1+x^2)^{-3}2^2x^2 + (-1)(1+x^2)^{-2}2 & f'''(0) = -2, \\
f^{(4)}(x) = -3!(1+x^2)^{-4}2^3x^3 + 2!(1+x^2)^{-3}2^3x \\
\qquad + 2!(1+x^2)^{-3}2^2x & f^{(4)}(0) = 0.
\end{array}
$$

Therefore,

$$
P_3(x) = P_4(x) = x - \frac{1}{3}x^3.
$$

9. Let $f(x) = \dfrac{1}{\sqrt{1+x}} = (1+x)^{-1/2}$. Then $f(0) = 1$.

$$
\begin{array}{ll}
f'(x) = -\frac{1}{2}(1+x)^{-3/2} & f'(0) = -\frac{1}{2}, \\
f''(x) = \frac{3}{2^2}(1+x)^{-5/2} & f''(0) = \frac{3}{2^2}, \\
f'''(x) = -\frac{3\cdot5}{2^3}(1+x)^{-7/2} & f'''(0) = -\frac{3\cdot5}{2^3}, \\
f^{(4)}(x) = \frac{3\cdot5\cdot7}{2^4}(1+x)^{-9/2} & f^{(4)}(0) = \frac{3\cdot5\cdot7}{2^4}
\end{array}
$$

Then,

$$
\begin{aligned}
P_2(x) &= 1 - \frac{1}{2}x + \frac{1}{2!}\frac{3}{2^2}x^2 = 1 - \frac{1}{2}x + \frac{3}{8}x^2, \\
P_3(x) &= P_2(x) - \frac{1}{3!}\frac{3\cdot5}{2^3}x^3 = 1 - \frac{1}{2}x + \frac{3}{8}x^2 - \frac{5}{16}x^3, \\
P_4(x) &= P_3(x) + \frac{1}{4!}\frac{3\cdot5\cdot7}{2^4}x^4 = 1 - \frac{1}{2}x + \frac{3}{8}x^2 - \frac{5}{16}x^3 + \frac{35}{128}x^4.
\end{aligned}
$$

13. Let $f(x) = e^x$. Since $f^{(k)}(x) = e^x = f(x)$ for all $k \geq 1$, the Taylor polynomial of degree 4 for $f(x) = e^x$ about $x = 1$ is

$$
\begin{aligned}
P_4(x) &= e^1 + e^1(x-1) + \frac{e^1}{2!}(x-1)^2 + \frac{e^1}{3!}(x-1)^3 + \frac{e^1}{4!}(x-1)^4 \\
&= e\left[1 + (x-1) + \frac{1}{2}(x-1)^2 + \frac{1}{6}(x-1)^3 + \frac{1}{24}(x-1)^4\right].
\end{aligned}
$$

Problems

17. As we can see from Problem 15, a is the y-intercept of $f(x)$, b is the slope of the tangent line to $f(x)$ at $x = 0$ and c tells us the concavity of $f(x)$ near $x = 0$.
So $a < 0$, $b > 0$ and $c > 0$.

21.

$$f(x) = 4x^2 - 7x + 2 \quad f(0) = 2$$
$$f'(x) = 8x - 7 \quad\quad f'(0) = -7$$
$$f''(x) = 8 \quad\quad\quad f''(0) = 8,$$

so $P_2(x) = 2 + (-7)x + \frac{8}{2}x^2 = 4x^2 - 7x + 2$. We notice that $f(x) = P_2(x)$ in this case.

25.

$$\lim_{x \to 0} \frac{1 - \cos x}{x^2} = \lim_{x \to 0} \frac{1 - (1 - \frac{x^2}{2!} + \frac{x^4}{4!})}{x^2} = \lim_{x \to 0} \left(\frac{1}{2} - \frac{x^2}{4!} \right) = \frac{1}{2}.$$

29. (a) $f(x) = e^{x^2}$.
$f'(x) = 2xe^{x^2}$, $f''(x) = 2(1 + 2x^2)e^{x^2}$, $f'''(x) = 4(3x + 2x^3)e^{x^2}$,
$f^{(4)}(x) = 4(3 + 6x^2)e^{x^2} + 4(3x + 2x^3)2xe^{x^2}$.
The Taylor polynomial about $x = 0$ is

$$P_4(x) = 1 + \frac{0}{1!}x + \frac{2}{2!}x^2 + \frac{0}{3!}x^3 + \frac{12}{4!}x^4$$

$$= 1 + x^2 + \frac{1}{2}x^4.$$

(b) $f(x) = e^x$. The Taylor polynomial of degree 2 is

$$Q_2(x) = 1 + \frac{x}{1!} + \frac{x^2}{2!} = 1 + x + \frac{1}{2}x^2.$$

If we substitute x^2 for x in the Taylor polynomial for e^x of degree 2, we will get $P_4(x)$, the Taylor polynomial for e^{x^2} of degree 4:

$$Q_2(x^2) = 1 + x^2 + \frac{1}{2}(x^2)^2$$

$$= 1 + x^2 + \frac{1}{2}x^4$$

$$= P_4(x).$$

(c) Let $Q_{10}(x) = 1 + \frac{x}{1!} + \frac{x^2}{2!} + \cdots + \frac{x^{10}}{10!}$ be the Taylor polynomial of degree 10 for e^x about $x = 0$. Then

$$P_{20}(x) = Q_{10}(x^2)$$

$$= 1 + \frac{x^2}{1!} + \frac{(x^2)^2}{2!} + \cdots + \frac{(x^2)^{10}}{10!}$$

$$= 1 + \frac{x^2}{1!} + \frac{x^4}{2!} + \cdots + \frac{x^{20}}{10!}.$$

(d) Let $e^x \approx Q_5(x) = 1 + \frac{x}{1!} + \cdots + \frac{x^5}{5!}$. Then

$$e^{-2x} \approx Q_5(-2x)$$

$$= 1 + \frac{-2x}{1!} + \frac{(-2x)^2}{2!} + \frac{(-2x)^3}{3!} + \frac{(-2x)^4}{4!} + \frac{(-2x)^5}{5!}$$

$$= 1 - 2x + 2x^2 - \frac{4}{3}x^3 + \frac{2}{3}x^4 - \frac{4}{15}x^5.$$

Solutions for Section 10.2

Exercises

1.

$$f(x) = \tfrac{1}{1-x} = (1-x)^{-1} \qquad\qquad f(0) = 1,$$
$$f'(x) = -(1-x)^{-2}(-1) = (1-x)^{-2} \qquad f'(0) = 1,$$
$$f''(x) = -2(1-x)^{-3}(-1) = 2(1-x)^{-3} \qquad f''(0) = 2,$$
$$f'''(x) = -6(1-x)^{-4}(-1) = 6(1-x)^{-4} \qquad f'''(0) = 6.$$

$$f(x) = \frac{1}{1-x} = 1 + 1 \cdot x + \frac{2x^2}{2!} + \frac{6x^3}{3!} + \cdots$$
$$= 1 + x + x^2 + x^3 + \cdots$$

5.

$$f(x) = \sin x \qquad f(\tfrac{\pi}{4}) = \tfrac{\sqrt{2}}{2},$$
$$f'(x) = \cos x \qquad f'(\tfrac{\pi}{4}) = \tfrac{\sqrt{2}}{2},$$
$$f''(x) = -\sin x \qquad f''(\tfrac{\pi}{4}) = -\tfrac{\sqrt{2}}{2},$$
$$f'''(x) = -\cos x \qquad f'''(\tfrac{\pi}{4}) = -\tfrac{\sqrt{2}}{2}.$$

$$\sin x = \frac{\sqrt{2}}{2} + \frac{\sqrt{2}}{2}\left(x - \frac{\pi}{4}\right) - \frac{\sqrt{2}}{2}\frac{(x-\frac{\pi}{4})^2}{2!} - \frac{\sqrt{2}}{2}\frac{(x-\frac{\pi}{4})^3}{3!} - \cdots$$
$$= \frac{\sqrt{2}}{2} + \frac{\sqrt{2}}{2}\left(x - \frac{\pi}{4}\right) - \frac{\sqrt{2}}{4}\left(x - \frac{\pi}{4}\right)^2 - \frac{\sqrt{2}}{12}\left(x - \frac{\pi}{4}\right)^3 - \cdots$$

9.

$$f(x) = \tfrac{1}{x} \qquad f(1) = 1$$
$$f'(x) = -\tfrac{1}{x^2} \qquad f'(1) = -1$$
$$f''(x) = \tfrac{2}{x^3} \qquad f''(1) = 2$$
$$f'''(x) = -\tfrac{6}{x^4} \qquad f'''(1) = -6$$

$$\frac{1}{x} = 1 - (x-1) + \frac{2(x-1)^2}{2!} - \frac{6(x-1)^3}{3!} + \cdots$$
$$= 1 - (x-1) + (x-1)^2 - (x-1)^3 + \cdots.$$

13. The general term can be written as $(-1)^n x^n$ for $n \geq 0$.

17. The general term can be written as $(-1)^k x^{2k+1}/(2k+1)$ for $k \geq 0$.

Problems

21. (a)

$$f(x) = \ln(1+2x) \qquad f(0) = 0$$
$$f'(x) = \tfrac{2}{1+2x} \qquad f'(0) = 2$$
$$f''(x) = -\tfrac{4}{(1+2x)^2} \qquad f''(0) = -4$$
$$f'''(x) = \tfrac{16}{(1+2x)^3} \qquad f'''(0) = 16$$

$$\ln(1+2x) = 2x - 2x^2 + \frac{8}{3}x^3 + \cdots$$

(b) To get the expression for $\ln(1+2x)$ from the series for $\ln(1+x)$, substitute $2x$ for x in the series

$$\ln(1+x) = x - \frac{x^2}{2} + \frac{x^3}{3} - \frac{x^4}{4} + \cdots$$

to get

$$\ln(1 + 2x) = 2x - \frac{(2x)^2}{2} + \frac{(2x)^3}{3} - \frac{(2x)^4}{4} + \cdots$$

$$= 2x - 2x^2 + \frac{8x^3}{3} - 4x^4 + \cdots$$

(c) Since the interval of convergence for $\ln(1 + x)$ is $-1 < x < 1$, substituting $2x$ for x suggests the interval of convergence of $\ln(1 + 2x)$ is $-1 < 2x < 1$, or $-\frac{1}{2} < x < \frac{1}{2}$.

25. The Taylor series for $\ln(1 - x)$ is

$$\ln(1 - x) = -x - \frac{x^2}{2} - \frac{x^3}{3} - \cdots - \frac{x^n}{n} - \cdots,$$

so

$$\lim_{n \to \infty} \frac{|a_{n+1}|}{|a_n|} = |x| \lim_{n \to \infty} \frac{1/(n + 1)}{1/n} = |x| \lim_{n \to \infty} \left| \frac{n}{n + 1} \right| = |x|.$$

Thus the series converges for $|x| < 1$, and the radius of convergence is 1. Note: This series can be obtained from the series for $\ln(1 + x)$ by replacing x by $-x$ and has the same radius of convergence as the series for $\ln(1 + x)$.

29. This is the series for $1/(1 - x)$ with x replaced by $1/4$, so the series converges to $1/(1 - (1/4)) = 4/3$.

33. This is the series for e^x with $x = 3$ substituted. Thus

$$1 + 3 + \frac{9}{2!} + \frac{27}{3!} + \frac{81}{4!} + \cdots = 1 + 3 + \frac{3^2}{2!} + \frac{3^3}{3!} + \frac{3^4}{4!} + \cdots = e^3.$$

37. Since $x - \frac{1}{2}x^2 + \frac{1}{3}x^3 + \cdots = \ln(1 + x)$, we solve $\ln(1 + x) = 0.2$, giving $1 + x = e^{0.2}$, so $x = e^{0.2} - 1$.

41. We define $e^{i\theta}$ to be

$$e^{i\theta} = 1 + i\theta + \frac{(i\theta)^2}{2!} + \frac{(i\theta)^3}{3!} + \frac{(i\theta)^4}{4!} + \frac{(i\theta)^5}{5!} + \frac{(i\theta)^6}{6!} + \cdots$$

Suppose we consider the expression $\cos\theta + i\sin\theta$, with $\cos\theta$ and $\sin\theta$ replaced by their Taylor series:

$$\cos\theta + i\sin\theta = \left(1 - \frac{\theta^2}{2!} + \frac{\theta^4}{4!} - \frac{\theta^6}{6!} + \cdots \right) + i \left(\theta - \frac{\theta^3}{3!} + \frac{\theta^5}{5!} - \cdots \right)$$

Reordering terms, we have

$$\cos\theta + i\sin\theta = 1 + i\theta - \frac{\theta^2}{2!} - \frac{i\theta^3}{3!} + \frac{\theta^4}{4!} + \frac{i\theta^5}{5!} - \frac{\theta^6}{6!} - \cdots$$

Using the fact that $i^2 = -1$, $i^3 = -i$, $i^4 = 1$, $i^5 = i, \cdots$, we can rewrite the series as

$$\cos\theta + i\sin\theta = 1 + i\theta + \frac{(i\theta)^2}{2!} + \frac{(i\theta)^3}{3!} + \frac{(i\theta)^4}{4!} + \frac{(i\theta)^5}{5!} + \frac{(i\theta)^6}{6!} + \cdots$$

Amazingly enough, this series is the Taylor series for e^x with $i\theta$ substituted for x. Therefore, we have shown that

$$\cos\theta + i\sin\theta = e^{i\theta}.$$

Solutions for Section 10.3

Exercises

1. We'll use

$$\sqrt{1+y} = (1+y)^{\frac{1}{2}} = 1 + \left(\frac{1}{2}\right)y + \left(\frac{1}{2}\right)\left(\frac{-1}{2}\right)\frac{y^2}{2!}$$
$$+ \left(\frac{1}{2}\right)\left(\frac{-1}{2}\right)\left(\frac{-3}{2}\right)\frac{y^3}{3!} + \cdots$$
$$= 1 + \frac{y}{2} - \frac{y^2}{8} + \frac{y^3}{16} - \cdots.$$

Substitute $y = -2x$.

$$\sqrt{1-2x} = 1 + \frac{(-2x)}{2} - \frac{(-2x)^2}{8} + \frac{(-2x)^3}{16} - \cdots$$
$$= 1 - x - \frac{x^2}{2} - \frac{x^3}{2} - \cdots$$

5. Substituting $x = -2y$ into $\ln(1+x) = x - \frac{x^2}{2} + \frac{x^3}{3} - \frac{x^4}{4} + \cdots$ gives

$$\ln(1-2y) = (-2y) - \frac{(-2y)^2}{2} + \frac{(-2y)^3}{3} - \frac{(-2y)^4}{4} + \cdots$$
$$= -2y - 2y^2 - \frac{8}{3}y^3 - 4y^4 - \cdots.$$

9.

$$\frac{z}{e^{z^2}} = ze^{-z^2} = z\left(1 + (-z^2) + \frac{(-z^2)^2}{2!} + \frac{(-z^2)^3}{3!} + \cdots\right)$$
$$= z - z^3 + \frac{z^5}{2!} - \frac{z^7}{3!} + \cdots$$

13. Multiplying out gives $(1+x)^3 = 1 + 3x + 3x^2 + x^3$. Since this polynomial equals the original function for all x, it must be the Taylor series. The general term is $0 \cdot x^n$ for $n \geq 4$.

17.

$$\frac{a}{\sqrt{a^2+x^2}} = \frac{a}{a(1+\frac{x^2}{a^2})^{\frac{1}{2}}} = \left(1 + \frac{x^2}{a^2}\right)^{-\frac{1}{2}}$$
$$= 1 + \left(-\frac{1}{2}\right)\frac{x^2}{a^2} + \frac{1}{2!}\left(-\frac{1}{2}\right)\left(-\frac{3}{2}\right)\left(\frac{x^2}{a^2}\right)^2$$
$$+ \frac{1}{3!}\left(-\frac{1}{2}\right)\left(-\frac{3}{2}\right)\left(-\frac{5}{2}\right)\left(\frac{x^2}{a^2}\right)^3 + \cdots$$
$$= 1 - \frac{1}{2}\left(\frac{x}{a}\right)^2 + \frac{3}{8}\left(\frac{x}{a}\right)^4 - \frac{5}{16}\left(\frac{x}{a}\right)^6 + \cdots$$

Problems

21. The Taylor series about 0 for $y = \dfrac{1}{1-x^2}$ is

$$y = 1 + x^2 + x^4 + x^6 + \cdots.$$

The series for $y = (1 + x)^{1/4}$ is, using the binomial expansion,

$$y = 1 + \frac{1}{4}x + \frac{1}{4}\left(-\frac{3}{4}\right)\frac{x^2}{2!} + \frac{1}{4}\left(-\frac{3}{4}\right)\left(-\frac{7}{4}\right)\frac{x^3}{3!} + \cdots.$$

The series for $y = \sqrt{1 + \frac{x}{2}} = (1 + \frac{x}{2})^{1/2}$ is, again using the binomial expansion,

$$y = 1 + \frac{1}{2}\cdot\frac{x}{2} + \frac{1}{2}\left(-\frac{1}{2}\right)\cdot\frac{x^2}{8} + \frac{1}{2}\left(-\frac{1}{2}\right)\left(-\frac{3}{2}\right)\cdot\frac{x^3}{48} + \cdots.$$

Similarly for $y = \dfrac{1}{\sqrt{1-x}} = (1-x)^{-(1/2)}$,

$$y = 1 + \left(-\frac{1}{2}\right)(-x) + \left(-\frac{1}{2}\right)\left(-\frac{3}{2}\right)\cdot\frac{x^2}{2!} + \left(-\frac{1}{2}\right)\left(-\frac{3}{2}\right)\left(-\frac{5}{2}\right)\cdot\frac{-x^3}{3!} + \cdots.$$

Near 0, let's truncate these series after their x^2 terms:

$$\frac{1}{1-x^2} \approx 1 + x^2,$$

$$(1+x)^{1/4} \approx 1 + \frac{1}{4}x - \frac{3}{32}x^2,$$

$$\sqrt{1 + \frac{x}{2}} \approx 1 + \frac{1}{4}x - \frac{1}{32}x^2,$$

$$\frac{1}{\sqrt{1-x}} \approx 1 + \frac{1}{2}x + \frac{3}{8}x^2.$$

Thus $\frac{1}{1-x^2}$ looks like a parabola opening upward near the origin, with y-axis as the axis of symmetry, so (a) = I. Now $\frac{1}{\sqrt{1-x}}$ has the largest positive slope ($\frac{1}{2}$), and is concave up (because the coefficient of x^2 is positive). So (d) = II.

The last two both have positive slope ($\frac{1}{4}$) and are concave down. Since $(1+x)^{\frac{1}{4}}$ has the smallest second derivative (i.e., the most negative coefficient of x^2), (b) = IV and therefore (c) = III.

25.

$$E = kQ\left(\frac{1}{(R-1)^2} - \frac{1}{(R+1)^2}\right)$$

$$= \frac{kQ}{R^2}\left(\frac{1}{(1-\frac{1}{R})^2} - \frac{1}{(1+\frac{1}{R})^2}\right)$$

Since $|\frac{1}{R}| < 1$, we can expand the two terms using the binomial expansion:

$$\frac{1}{(1-\frac{1}{R})^2} = \left(1 - \frac{1}{R}\right)^{-2}$$

$$= 1 - 2\left(-\frac{1}{R}\right) + (-2)(-3)\frac{(-\frac{1}{R})^2}{2!} + (-2)(-3)(-4)\frac{(-\frac{1}{R})^3}{3!} + \cdots$$

$$\frac{1}{(1+\frac{1}{R})^2} = \left(1 + \frac{1}{R}\right)^{-2}$$

$$= 1 - 2\left(\frac{1}{R}\right) + (-2)(-3)\frac{(\frac{1}{R})^2}{2!} + (-2)(-3)(-4)\frac{(\frac{1}{R})^3}{3!} + \cdots$$

Substituting, we get:

$$E = \frac{kQ}{R^2}\left[1 + \frac{2}{R} + \frac{3}{R^2} + \frac{4}{R^3} + \cdots - \left(1 - \frac{2}{R} + \frac{3}{R^2} - \frac{4}{R^3} + \cdots\right)\right] \approx \frac{kQ}{R^2}\left(\frac{4}{R} + \frac{8}{R^3}\right),$$

using only the first two non-zero terms.

29. (a) Factoring the expression for $t_1 - t_2$, we get

$$\Delta t = t_1 - t_2 = \frac{2l_2}{c(1 - v^2/c^2)} - \frac{2l_1}{c\sqrt{1 - v^2/c^2}} - \frac{2l_2}{c\sqrt{1 - v^2/c^2}} + \frac{2l_1}{c(1 - v^2/c^2)}$$

$$= \frac{2(l_1 + l_2)}{c(1 - v^2/c^2)} - \frac{2(l_1 + l_2)}{c\sqrt{1 - v^2/c^2}}$$

$$= \frac{2(l_1 + l_2)}{c}\left(\frac{1}{1 - v^2/c^2} - \frac{1}{\sqrt{1 - v^2/c^2}}\right).$$

Expanding the two terms within the parentheses in terms of v^2/c^2 gives

$$\left(1 - \frac{v^2}{c^2}\right)^{-1} = 1 + \frac{v^2}{c^2} + \frac{(-1)(-2)}{2!}\left(\frac{-v^2}{c^2}\right)^2 + \frac{(-1)(-2)(-3)}{3!}\left(\frac{-v^2}{c^2}\right)^3 + \cdots$$

$$= 1 + \frac{v^2}{c^2} + \frac{v^4}{c^4} + \frac{v^6}{c^6} + \cdots$$

$$\left(1 - \frac{v^2}{c^2}\right)^{-1/2} = 1 + \frac{1}{2}\frac{v^2}{c^2} + \frac{\left(\frac{-1}{2}\right)\left(\frac{-3}{2}\right)}{2!}\left(\frac{-v^2}{c^2}\right)^2 + \frac{\left(\frac{-1}{2}\right)\left(\frac{-3}{2}\right)\left(\frac{-5}{2}\right)}{3!}\left(\frac{-v^2}{c^2}\right)^3 + \cdots$$

$$= 1 + \frac{1}{2}\frac{v^2}{c^2} + \frac{3}{8}\frac{v^4}{c^4} + \frac{5}{16}\frac{v^6}{c^6} + \cdots$$

Thus, we have

$$\Delta t = \frac{2(l_1 + l_2)}{c}\left(1 + \frac{v^2}{c^2} + \frac{v^4}{c^4} + \frac{v^6}{c^6} + \cdots - 1 - \frac{1}{2}\frac{v^2}{c^2} - \frac{3}{8}\frac{v^4}{c^4} - \frac{5}{16}\frac{v^6}{c^6} - \cdots\right)$$

$$= \frac{2(l_1 + l_2)}{c}\left(\frac{1}{2}\frac{v^2}{c^2} + \frac{5}{8}\frac{v^4}{c^4} + \frac{11}{16}\frac{v^6}{c^6} + \cdots\right)$$

$$\Delta t \approx \frac{(l_1 + l_2)}{c}\left(\frac{v^2}{c^2} + \frac{5}{4}\frac{v^4}{c^4}\right).$$

(b) For small v. we can neglect all but the first nonzero term, so

$$\Delta t \approx \frac{(l_1 + l_2)}{c} \cdot \frac{v^2}{c^2} = \frac{(l_1 + l_2)}{c^3}v^2.$$

Thus, Δt is proportional to v^2 with constant of proportionality $(l_1 + l_2)/c^3$.

33. (a) We take the left-hand Riemann sum with the formula

Left-hand sum $= (1 + 0.9608 + 0.8521 + 0.6977 + 0.5273)(0.2) = 0.8076.$

Similarly,

Right-hand sum $= (0.9608 + 0.8521 + 0.6977 + 0.5273 + 0.3679)(0.2) = 0.6812.$

(b) Since

$$e^x = 1 + x + \frac{x^2}{2!} + \frac{x^3}{3!} + \dots,$$

$$e^{-x^2} \approx 1 + (-x^2) + \frac{(-x^2)^2}{2!} + \frac{(-x^2)^3}{3!}$$

$$= 1 - x^2 + \frac{x^4}{2} - \frac{x^6}{6}.$$

(c)

$$\int_0^1 e^{-x^2}\,dx \approx \int_0^1 \left(1 - x^2 + \frac{x^4}{2} - \frac{x^6}{6}\right)dx$$

$$= \left(x - \frac{x^3}{3} + \frac{x^5}{10} - \frac{x^7}{42}\right)\Bigg|_0^1 = 0.74286.$$

(d) We can improve the left and right sum values by averaging them to get 0.74439 or by increasing the number of subdivisions. We can improve on the estimate using the Taylor approximation by taking more terms.

Solutions for Section 10.4

Exercises

1. Let $f(x) = (1-x)^{1/3}$, so $f(0.5) = (0.5)^{1/3}$. The error bound in the Taylor approximation of degree 3 for $f(0.5) = 0.5^{\frac{1}{3}}$ about $x = 0$ is:

$$|E_3| = |f(0.5) - P_3(0.5)| \leq \frac{M \cdot |0.5 - 0|^4}{4!} = \frac{M(0.5)^4}{24},$$

where $|f^{(4)}(x)| \leq M$ for $0 \leq x \leq 0.5$. Now, $f^{(4)}(x) = -\frac{80}{81}(1-x)^{-(11/3)}$. By looking at the graph of $(1-x)^{-(11/3)}$, we see that $|f^{(4)}(x)|$ is maximized for x between 0 and 0.5 when $x = 0.5$. Thus,

$$|f^{(4)}| \leq \frac{80}{81}\left(\frac{1}{2}\right)^{-(11/3)} = \frac{80}{81} \cdot 2^{11/3},$$

so

$$|E_3| \leq \frac{80 \cdot 2^{11/3} \cdot (0.5)^4}{81 \cdot 24} \approx 0.033.$$

Problems

5. (a) The Taylor polynomial of degree 0 about $t = 0$ for $f(t) = e^t$ is simply $P_0(x) = 1$. Since $e^t \geq 1$ on $[0, 0.5]$, the approximation is an underestimate.

(b) Using the zero degree error bound, if $|f'(t)| \leq M$ for $0 \leq t \leq 0.5$, then

$$|E_0| \leq M \cdot |t| \leq M(0.5).$$

Since $|f'(t)| = |e^t| = e^t$ is increasing on $[0, 0.5]$,

$$|f'(t)| \leq e^{0.5} < \sqrt{4} = 2.$$

Therefore

$$|E_0| \leq (2)(0.5) = 1.$$

(Note: By looking at a graph of $f(t)$ and its 0th degree approximation, it is easy to see that the greatest error occurs when $t = 0.5$, and the error is $e^{0.5} - 1 \approx 0.65 < 1$. So our error bound works.)

9. (a) The vertical distance between the graph of $y = \cos x$ and $y = P_{10}(x)$ at $x = 6$ is no more than 4, so

$$|\text{Error in } P_{10}(6)| \leq 4.$$

Since at $x = 6$ the $\cos x$ and $P_{20}(x)$ graphs are indistinguishable in this figure, the error must be less than the smallest division we can see, which is about 0.2 so,

$$|\text{Error in } P_{20}(6)| \leq 0.2.$$

(b) The maximum error occurs at the ends of the interval, that is, at $x = -9$, $x = 9$. At $x = 9$, the graphs of $y = \cos x$ and $y = P_{20}(x)$ are no more than 1 apart, so

$$\left|\begin{array}{l}\text{Maximum error in } P_{20}(x) \\ \text{for } -9 \leq x \leq 9\end{array}\right| \leq 1.$$

(c) We are looking for the largest x-interval on which the graphs of $y = \cos x$ and $y = P_{10}(x)$ are indistinguishable. This is hard to estimate accurately from the figure, though $-4 \leq x \leq 4$ certainly satisfies this condition.

13. (a)

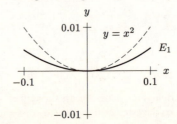

The graph of E_1 looks like a parabola. Since the graph of E_1 is sandwiched between the graph of $y = x^2$ and the x axis, we have

$$|E_1| \le x^2 \quad \text{for} \quad |x| \le 0.1.$$

(b)

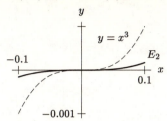

The graph of E_2 looks like a cubic, sandwiched between the graph of $y = x^3$ and the x axis, so

$$|E_2| \le x^3 \quad \text{for} \quad |x| \le 0.1.$$

(c) Using the Taylor expansion

$$e^x = 1 + x + \frac{x^2}{2!} + \frac{x^3}{3!} + \cdots$$

we see that

$$E_1 = e^x - (1 + x) = \frac{x^2}{2!} + \frac{x^3}{3!} + \frac{x^4}{4!} + \cdots.$$

Thus for small x, the $x^2/2!$ term dominates, so

$$E_1 \approx \frac{x^2}{2!},$$

and so E_1 is approximately a quadratic.

Similarly

$$E_2 = e^x - (1 + x + \frac{x^2}{2}) = \frac{x^3}{3!} + \frac{x^4}{4!} + \cdots.$$

Thus for small x, the $x^3/3!$ term dominates, so

$$E_2 \approx \frac{x^3}{3!}$$

and so E_2 is approximately a cubic.

Solutions for Section 10.5

Exercises

1. No, a Fourier series has terms of the form $\cos nx$, not $\cos^n x$.

5.

$$a_0 = \frac{1}{2\pi} \int_{-\pi}^{\pi} f(x)\, dx = \frac{1}{2\pi}\left[\int_{-\pi}^{0} -1\, dx + \int_{0}^{\pi} 1\, dx\right] = 0$$

$$a_1 = \frac{1}{\pi} \int_{-\pi}^{\pi} f(x) \cos x\, dx = \frac{1}{\pi}\left[\int_{-\pi}^{0} -\cos x\, dx + \int_{0}^{\pi} \cos x\, dx\right]$$

$$= \frac{1}{\pi}\left[-\sin x\Big|_{-\pi}^{0} + \sin x\Big|_{0}^{\pi}\right] = 0.$$

Similarly, a_2 and a_3 are both 0.

(In fact, notice $f(x) \cos nx$ is an odd function, so $\int_{-\pi}^{\pi} f(x) \cos nx = 0$.)

$$b_1 = \frac{1}{\pi} \int_{-\pi}^{\pi} f(x) \sin x\, dx = \frac{1}{\pi}\left[\int_{-\pi}^{0} -\sin x\, dx + \int_{0}^{\pi} \sin x\, dx\right]$$

$$= \frac{1}{\pi}\left[\cos x\Big|_{-\pi}^{0} + (-\cos x)\Big|_{0}^{\pi}\right] = \frac{4}{\pi}.$$

$$b_2 = \frac{1}{\pi} \int_{-\pi}^{\pi} f(x) \sin 2x \, dx = \frac{1}{\pi} \left[\int_{-\pi}^{0} -\sin 2x \, dx + \int_{0}^{\pi} \sin 2x \, dx \right]$$

$$= \frac{1}{\pi} \left[\frac{1}{2} \cos 2x \Big|_{-\pi}^{0} + \left(-\frac{1}{2} \cos 2x \right) \Big|_{0}^{\pi} \right] = 0.$$

$$b_3 = \frac{1}{\pi} \int_{-\pi}^{\pi} f(x) \sin 3x \, dx = \frac{1}{\pi} \left[\int_{-\pi}^{0} -\sin 3x \, dx + \int_{0}^{\pi} \sin 3x \, dx \right]$$

$$= \frac{1}{\pi} \left[\frac{1}{3} \cos 3x \Big|_{-\pi}^{0} + \left(-\frac{1}{3} \cos 3x \right) \Big|_{0}^{\pi} \right] = \frac{4}{3\pi}.$$

Thus, $F_1(x) = F_2(x) = \frac{4}{\pi} \sin x$ and $F_3(x) = \frac{4}{\pi} \sin x + \frac{4}{3\pi} \sin 3x$.

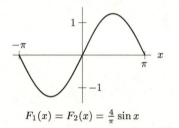

$F_1(x) = F_2(x) = \frac{4}{\pi} \sin x$

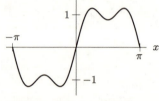

$F_3(x) = \frac{4}{\pi} \sin x + \frac{4}{3\pi} \sin 3x$

9.

$$a_0 = \frac{1}{2\pi} \int_{-\pi}^{\pi} h(x) \, dx = \frac{1}{2\pi} \int_{0}^{\pi} x \, dx = \frac{\pi}{4}$$

As in Problem 10, we use the integral table (III-15 and III-16) to find formulas for a_n and b_n.

$$a_n = \frac{1}{\pi} \int_{-\pi}^{\pi} h(x) \cos(nx) \, dx = \frac{1}{\pi} \int_{0}^{\pi} x \cos nx \, dx = \frac{1}{\pi} \left(\frac{x}{n} \sin(nx) + \frac{1}{n^2} \cos(nx) \right) \Big|_{0}^{\pi}$$

$$= \frac{1}{\pi} \left(\frac{1}{n^2} \cos(n\pi) - \frac{1}{n^2} \right)$$

$$= \frac{1}{n^2 \pi} \left(\cos(n\pi) - 1 \right).$$

Note that since $\cos(n\pi) = (-1)^n$, $a_n = 0$ if n is even and $a_n = -\frac{2}{n^2 \pi}$ if n is odd.

$$b_n = \frac{1}{\pi} \int_{-\pi}^{\pi} h(x) \cos(nx) \, dx = \frac{1}{\pi} \int_{0}^{\pi} x \sin x \, dx$$

$$= \frac{1}{\pi} \left(-\frac{x}{n} \cos(nx) + \frac{1}{n^2} \sin(nx) \right) \Big|_{0}^{\pi}$$

$$= \frac{1}{\pi} \left(-\frac{\pi}{n} \cos(n\pi) \right)$$

$$= -\frac{1}{n} \cos(n\pi)$$

$$= \frac{1}{n} (-1)^{n+1} \quad \text{if } n \geq 1$$

We have that the nth Fourier polynomial for h (for $n \geq 1$) is

$$H_n(x) = \frac{\pi}{4} + \sum_{i=1}^{n} \left(\frac{1}{i^2 \pi} \left(\cos(i\pi) - 1 \right) \cdot \cos(ix) + \frac{(-1)^{i+1} \sin(ix)}{i} \right).$$

This can also be written as

$$H_n(x) = \frac{\pi}{4} + \sum_{i=1}^{n} \frac{(-1)^{i+1}\sin(ix)}{i} + \sum_{i=1}^{\left[\frac{n}{2}\right]} \frac{-2}{(2i-1)^2\pi} \cos((2i-1)x)$$

where $\left[\frac{n}{2}\right]$ denotes the biggest integer smaller than or equal to $\frac{n}{2}$. In particular, we have the graphs in Figure 10.1.

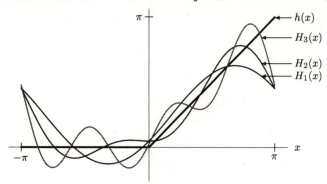

Figure 10.1

Problems

13. Since the period is 2, we make the substitution $t = \pi x - \pi$. Thus, $x = \frac{t+\pi}{\pi}$. We find the Fourier coefficients. Notice that all of the integrals are the same as in Problem 12 except for an extra factor of 2. Thus, $a_0 = 1$, $a_n = 0$, and $b_n = \frac{4}{\pi n}(-1)^{n+1}$, so:

$$G_4(t) = 1 + \frac{4}{\pi}\sin t - \frac{2}{\pi}\sin 2t + \frac{4}{3\pi}\sin 3t - \frac{1}{\pi}\sin 4t.$$

Again, we substitute back in to get a Fourier polynomial in terms of x:

$$F_4(x) = 1 + \frac{4}{\pi}\sin(\pi x - \pi) - \frac{2}{\pi}\sin(2\pi x - 2\pi)$$

$$+ \frac{4}{3\pi}\sin(3\pi x - 3\pi) - \frac{1}{\pi}\sin(4\pi x - 4\pi)$$

$$= 1 - \frac{4}{\pi}\sin(\pi x) - \frac{2}{\pi}\sin(2\pi x) - \frac{4}{3\pi}\sin(3\pi x) - \frac{1}{\pi}\sin(4\pi x).$$

Notice in this case, the terms in our series are $\sin(n\pi x)$, not $\sin(2\pi n x)$, as in Problem 12. In general, the terms will be $\sin(n\frac{2\pi}{b}x)$, where b is the period.

17. Since each square in the graph has area $\left(\frac{\pi}{4}\right)\cdot(0.2)$,

$$a_0 = \frac{1}{2\pi}\int_{-\pi}^{\pi} f(x)\,dx$$

$$= \frac{1}{2\pi}\cdot\left(\frac{\pi}{4}\right)\cdot(0.2)\ [\text{Number of squares under graph above }x\text{-axis}$$

$$- \text{Number of squares above graph below }x\text{ axis}]$$

$$\approx \frac{1}{2\pi}\cdot\left(\frac{\pi}{4}\right)\cdot(0.2)\cdot[13 + 11 - 14] = 0.25.$$

Approximate the Fourier coefficients using Riemann sums.

$$a_1 = \frac{1}{\pi} \int_{-\pi}^{\pi} f(x) \cos x \, dx$$

$$\approx \frac{1}{\pi} \left[f(-\pi) \cos(-\pi) + f\left(-\frac{\pi}{2}\right) \cos\left(-\frac{\pi}{2}\right) + f(0) \cos(0) + f\left(\frac{\pi}{2}\right) \cos\left(\frac{\pi}{2}\right) \right] \cdot \frac{\pi}{2}$$

$$= \frac{1}{\pi} \left[(0.92)(-1) + (1)(0) + (-1.7)(1) + (0.7)(0) \right] \cdot \frac{\pi}{2}$$

$$= -1.31$$

Similarly for b_1:

$$b_1 = \frac{1}{\pi} \int_{-\pi}^{\pi} f(x) \sin x \, dx$$

$$\approx \frac{1}{\pi} \left[f(-\pi) \sin(-\pi) + f\left(-\frac{\pi}{2}\right) \sin\left(-\frac{\pi}{2}\right) + f(0) \sin(0) + f\left(\frac{\pi}{2}\right) \sin\left(\frac{\pi}{2}\right) \right] \cdot \frac{\pi}{2}$$

$$= \frac{1}{\pi} \left[(0.92)(0) + (1)(-1) + (-1.7)(0) + (0.7)(1) \right] \cdot \frac{\pi}{2}$$

$$= -0.15.$$

So our first Fourier approximation is

$$F_1(x) = 0.25 - 1.31 \cos x - 0.15 \sin x.$$

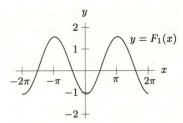

Similarly for a_2:

$$a_2 = \frac{1}{\pi} \int_{-\pi}^{\pi} f(x) \cos 2x \, dx$$

$$\approx \frac{1}{\pi} \left[f(-\pi) \cos(-2\pi) + f\left(-\frac{\pi}{2}\right) \cos(-\pi) + f(0) \cos(0) + f\left(\frac{\pi}{2}\right) \cos(-\pi) \right] \cdot \frac{\pi}{2}$$

$$= \frac{1}{\pi} \left[(0.92)(1) + (1)(-1) + (-1.7)(1) + (0.7)(-1) \right] \cdot \frac{\pi}{2}$$

$$= -1.24$$

Similarly for b_2:

$$b_2 = \frac{1}{\pi} \int_{-\pi}^{\pi} f(x) \sin 2x \, dx$$

$$\approx \frac{1}{\pi} \left[f(-\pi) \sin(-2\pi) + f\left(-\frac{\pi}{2}\right) \sin(-\pi) + f(0) \sin(0) + f\left(\frac{\pi}{2}\right) \sin(-\pi) \right] \cdot \frac{\pi}{2}$$

$$= \frac{1}{\pi} \left[(0.92)(0) + (1)(0) + (-1.7)(0) + (0.7)(0) \right] \cdot \frac{\pi}{2}$$

$$= 0.$$

So our second Fourier approximation is

$$F_2(x) = 0.25 - 1.31 \cos x - 0.15 \sin x - 1.24 \cos 2x.$$

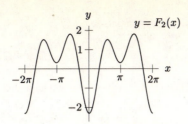

As you can see from comparing our graphs of F_1 and F_2 to the original, our estimates of the Fourier coefficients are not very accurate.

There are other methods of estimating the Fourier coefficients such as taking other Riemann sums, using Simpson's rule, and using the trapezoid rule. With each method, the greater the number of subdivisions, the more accurate the estimates of the Fourier coefficients.

The actual function graphed in the problem was

$$y = \frac{1}{4} - 1.3 \cos x - \frac{\sin(\frac{3}{5})}{\pi} \sin x - \frac{2}{\pi} \cos 2x - \frac{\cos 1}{3\pi} \sin 2x$$

$$= 0.25 - 1.3 \cos x - 0.18 \sin x - 0.63 \cos 2x - 0.057 \sin 2x.$$

21. (a)

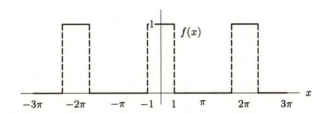

The energy of the pulse train f is

$$E = \frac{1}{\pi} \int_{-\pi}^{\pi} (f(x))^2 \, dx = \frac{1}{\pi} \int_{-1}^{1} 1^2 = \frac{1}{\pi}(1 - (-1)) = \frac{2}{\pi}.$$

Next, find the Fourier coefficients:

$$a_0 = \text{average value of } f \text{ on } [-\pi, \pi] = \frac{1}{2\pi}(\text{Area}) = \frac{1}{2\pi}(2) = \frac{1}{\pi},$$

$$a_k = \frac{1}{\pi} \int_{-\pi}^{\pi} f(x) \cos kx \, dx = \frac{1}{\pi} \int_{-1}^{1} \cos kx \, dx = \frac{1}{k\pi} \sin kx \Big|_{-1}^{1}$$

$$= \frac{1}{k\pi}(\sin k - \sin(-k)) = \frac{1}{k\pi}(2 \sin k),$$

$$b_k = \frac{1}{\pi} \int_{-\pi}^{\pi} f(x) \sin kx \, dx = \frac{1}{\pi} \int_{-1}^{1} \sin kx \, dx = -\frac{1}{k\pi} \cos kx \Big|_{-1}^{1}$$

$$= -\frac{1}{k\pi}(\cos k - \cos(-k)) = \frac{1}{k\pi}(0) = 0.$$

The energy of f contained in the constant term is

$$A_0^2 = 2a_0^2 = 2\left(\frac{1}{\pi}\right)^2 = \frac{2}{\pi^2}$$

which is

$$\frac{A_0^2}{E} = \frac{2/\pi^2}{2/\pi} = \frac{1}{\pi} \approx 0.3183 = 31.83\% \quad \text{of the total.}$$

The fraction of energy contained in the first harmonic is

$$\frac{A_1^2}{E} = \frac{a_1^2}{E} = \frac{\left(\frac{2\sin 1}{\pi}\right)^2}{\frac{2}{\pi}} \approx 0.4508 = 45.08\%.$$

The fraction of energy contained in both the constant term and the first harmonic together is

$$\frac{A_0^2}{E} + \frac{A_1^2}{E} \approx 0.7691 = 76.91\%.$$

(b) The fraction of energy contained in the second harmonic is

$$\frac{A_2^2}{E} = \frac{a_2^2}{E} = \frac{\left(\frac{\sin 2}{\pi}\right)^2}{\frac{2}{\pi}} \approx 0.1316 = 13.16\%$$

so the fraction of energy contained in the constant term and first two harmonics is

$$\frac{A_0^2}{E} + \frac{A_1^2}{E} + \frac{A_2^2}{E} \approx 0.7691 + 0.1316 = 0.9007 = 90.07\%.$$

Therefore, the constant term and the first two harmonics are needed to capture 90% of the energy of f.

(c)

$$F_3(x) = \frac{1}{\pi} + \frac{2\sin 1}{\pi}\cos x + \frac{\sin 2}{\pi}\cos 2x + \frac{2\sin 3}{3\pi}\cos 3x$$

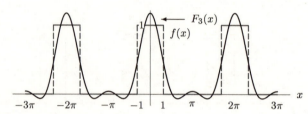

25. The easiest way to do this is to use Problem 24.

$$\int_{-\pi}^{\pi} \sin^2 mx\, dx = \int_{-\pi}^{\pi}(1 - \cos^2 mx)\, dx = \int_{-\pi}^{\pi} dx - \int_{-\pi}^{\pi}\cos^2 mx\, dx$$
$$= 2\pi - \pi \quad \text{using Problem 24}$$
$$= \pi.$$

Solutions for Chapter 10 Review

Exercises

1. $e^x \approx 1 + e(x - 1) + \dfrac{e}{2}(x - 1)^2$

5. $f'(x) = 3x^2 + 14x - 5$, $f''(x) = 6x + 14$, $f'''(x) = 6$. The Taylor polynomial about $x = 1$ is

$$P_3(x) = 4 + \frac{12}{1!}(x - 1) + \frac{20}{2!}(x - 1)^2 + \frac{6}{3!}(x - 1)^3$$
$$= 4 + 12(x - 1) + 10(x - 1)^2 + (x - 1)^3.$$

Notice that if you multiply out and collect terms in $P_3(x)$, you will get $f(x)$ back.

9. Substituting $y = -4z^2$ into $\dfrac{1}{1 + y} = 1 - y + y^2 - y^3 + \cdots$ gives

$$\frac{1}{1 - 4z^2} = 1 + 4z^2 + 16z^4 + 64z^6 + \cdots.$$

Problems

13. Infinite geometric series with $a = 1$, $x = -1/3$, so

$$\text{Sum} = \frac{1}{1 - (-1/3)} = \frac{3}{4}.$$

17. Factoring out a 3, we see

$$3\left(1 + 1 + \frac{1}{2!} + \frac{1}{3!} + \frac{1}{4!} + \frac{1}{5!} + \cdots\right) = 3e^1 = 3e.$$

21. The graph in Figure 10.2 suggests that the Taylor polynomials converge to $f(x) = \dfrac{1}{1+x}$ on the interval $(-1, 1)$. The Taylor expansion is

$$f(x) = \frac{1}{1+x} = 1 - x + x^2 - x^3 + x^4 - \cdots,$$

so the ratio test gives

$$\lim_{n \to \infty} \frac{|a_{n+1}|}{|a_n|} = \lim_{n \to \infty} \frac{|(-1)^{n+1} x^{n+1}|}{|(-1)^n x^n|} = |x|.$$

Thus, the series converges if $|x| < 1$; that is $-1 < x < 1$.

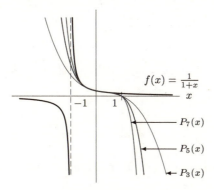

Figure 10.2

25. (a) Since $\sqrt{4 - x^2} = 2\sqrt{1 - x^2/4}$, we use the Binomial expansion

$$\sqrt{4 - x^2} \approx 2\left(1 + \frac{1}{2}\left(-\frac{x^2}{4}\right) + \frac{1}{2!}\left(\frac{1}{2}\right)\left(-\frac{1}{2}\right)\left(-\frac{x^2}{4}\right)^2\right)$$

$$= 2\left(1 - \frac{x^2}{8} - \frac{x^4}{128}\right) = 2 - \frac{x^2}{4} - \frac{x^4}{64}.$$

(b) Substituting the Taylor series in the integral gives

$$\int_0^1 \sqrt{4 - x^2}\, dx \approx \int_0^1 \left(2 - \frac{x^2}{4} - \frac{x^4}{64}\right) dx = 2x - \frac{x^3}{12} - \frac{x^5}{320}\Big|_0^1 = 1.9135.$$

(c) Since $x = 2\sin t$, we have $dx = 2\cos t\, dt$; in addition $t = 0$ when $x = 0$ and $t = \pi/6$ when $x = 1$. Thus

$$\int_0^1 \sqrt{4 - x^2}\, dx = \int_0^{\pi/6} \sqrt{4 - 4\sin^2 t} \cdot 2\cos t\, dt$$

$$= \int_0^{\pi/6} 2 \cdot 2\sqrt{1 - \sin^2 t}\cos t\, dt = 4\int_0^{\pi/6} \cos^2 t\, dt.$$

Using the table of integrals, we find

$$4\int_0^{\pi/6} \cos^2 t\, dt = 4 \cdot \frac{1}{2}(\cos t \sin t + t)\Big|_0^{\pi/6} = 2\left(\cos\frac{\pi}{6}\sin\frac{\pi}{6} + \frac{\pi}{6}\right) = \frac{\sqrt{3}}{2} + \frac{\pi}{3}.$$

(d) Using a calculator, $(\sqrt{3}/3) + (\pi/3) = 1.9132$, so the answers to parts (b) and (c) agree to three decimal places.

29. (a) To find when V takes on its minimum values, set $\frac{dV}{dr} = 0$. So

$$-V_0 \frac{d}{dr}\left(2\left(\frac{r_0}{r}\right)^6 - \left(\frac{r_0}{r}\right)^{12}\right) = 0$$

$$-V_0\left(-12r_0^6 r^{-7} + 12r_0^{12} r^{-13}\right) = 0$$

$$12r_0^6 r^{-7} = 12r_0^{12} r^{-13}$$

$$r_0^6 = r^6$$

$$r = r_0.$$

Rewriting $V'(r)$ as $\dfrac{12r_0^6 V_0}{r^7}\left(1 - \left(\dfrac{r_0}{r}\right)^6\right)$, we see that $V'(r) > 0$ for $r > r_0$ and $V'(r) < 0$ for $r < r_0$. Thus, $V = -V_0(2(1)^6 - (1)^{12}) = -V_0$ is a minimum.

(Note: We discard the negative root $-r_0$ since the distance r must be positive.)

(b)

$$V(r) = -V_0\left(2\left(\frac{r_0}{r}\right)^6 - \left(\frac{r_0}{r}\right)^{12}\right) \qquad V(r_0) = -V_0$$

$$V'(r) = -V_0(-12r_0^6 r^{-7} + 12r_0^{12} r^{-13}) \qquad V'(r_0) = 0$$

$$V''(r) = -V_0(84r_0^6 r^{-8} - 156r_0^{12} r^{-14}) \qquad V''(r_0) = 72V_0 r_0^{-2}$$

The Taylor series is thus:

$$V(r) = -V_0 + 72V_0 r_0^{-2} \cdot (r - r_0)^2 \cdot \frac{1}{2} + \cdots$$

(c) The difference between V and its minimum value $-V_0$ is

$$V - (-V_0) = 36V_0 \frac{(r - r_0)^2}{r_0^2} + \cdots$$

which is approximately proportional to $(r - r_0)^2$ since terms containing higher powers of $(r - r_0)$ have relatively small values for r near r_0.

(d) From part (a) we know that $dV/dr = 0$ when $r = r_0$, hence $F = 0$ when $r = r_0$. Since, if we discard powers of $(r - r_0)$ higher than the second,

$$V(r) \approx -V_0\left(1 - 36\frac{(r - r_0)^2}{r_0^2}\right)$$

giving

$$F = -\frac{dV}{dr} \approx 72 \cdot \frac{r - r_0}{r_0^2}(-V_0) = -72V_0 \frac{r - r_0}{r_0^2}.$$

So F is approximately proportional to $(r - r_0)$.

33. (a) Notice $g'(0) = 0$ because g has a critical point at $x = 0$. So, for $n \geq 2$,

$$g(x) \approx P_n(x) = g(0) + \frac{g''(0)}{2!}x^2 + \frac{g'''(0)}{3!}x^3 + \cdots + \frac{g^{(n)}(0)}{n!}x^n.$$

(b) The Second Derivative test says that if $g''(0) > 0$, then 0 is a local minimum and if $g''(0) < 0$, 0 is a local maximum.

(c) Let $n = 2$. Then $P_2(x) = g(0) + \dfrac{g''(0)}{2!}x^2$. So, for x near 0,

$$g(x) - g(0) \approx \frac{g''(0)}{2!}x^2.$$

If $g''(0) > 0$, then $g(x) - g(0) \geq 0$, as long as x stays near 0. In other words, there exists a small interval around $x = 0$ such that for any x in this interval $g(x) \geq g(0)$. So $g(0)$ is a local minimum.

The case when $g''(0) < 0$ is treated similarly; then $g(0)$ is a local maximum.

37. (a) Expand $f(x)$ into its Fourier series:

$$f(x) = a_0 + a_1 \cos x + a_2 \cos 2x + a_3 \cos 3x + \cdots + a_k \cos kx + \cdots$$
$$+ b_1 \sin x + b_2 \sin 2x + b_3 \sin 3x + \cdots + b_k \sin kx + \cdots$$

Then differentiate term-by-term:

$$f'(x) = -a_1 \sin x - 2a_2 \sin 2x - 3a_3 \sin 3x - \cdots - ka_k \sin kx - \cdots$$
$$+ b_1 \cos x + 2b_2 \cos 2x + 3b_3 \cos 3x + \cdots + kb_k \cos kx + \cdots$$

Regroup terms:

$$f'(x) = +b_1 \cos x + 2b_2 \cos 2x + 3b_3 \cos 3x + \cdots + kb_k \cos kx + \cdots$$
$$- a_1 \sin x - 2a_2 \sin 2x - 3a_3 \sin 3x - \cdots - ka_k \sin kx - \cdots$$

which forms a Fourier series for the derivative $f'(x)$. The Fourier coefficient of $\cos kx$ is kb_k and the Fourier coefficient of $\sin kx$ is $-ka_k$. Note that there is no constant term as you would expect from the formula ka_k with $k = 0$. Note also that if the kth harmonic f is absent, so is that of f'.

(b) If the amplitude of the kth harmonic of f is

$$A_k = \sqrt{a_k^2 + b_k^2}, \quad k \geq 1,$$

then the amplitude of the kth harmonic of f' is

$$\sqrt{(kb_k)^2 + (-ka_k)^2} = \sqrt{k^2(b_k^2 + a_k^2)} = k\sqrt{a_k^2 + b_k^2} = kA_k.$$

(c) The energy of the kth harmonic of f' is k^2 times the energy of the kth harmonic of f.

CAS Challenge Problems

41. (a) The Taylor polynomials of degree 7 are

$$\text{For } \sin x, \quad P_7(x) = x - \frac{x^3}{6} + \frac{x^5}{120} - \frac{x^7}{5040}$$
$$\text{For } \sin x \cos x, \quad Q_7(x) = x - \frac{2x^3}{3} + \frac{2x^5}{15} - \frac{4x^7}{315}$$

(b) The coefficient of x^3 in $Q_7(x)$ is $-2/3$, and the coefficient of x^3 in $P_7(x)$ is $-1/6$, so the ratio is

$$\frac{-2/3}{-1/6} = 4.$$

The corresponding ratios for x^5 and x^7 are

$$\frac{2/15}{1/120} = 16 \quad \text{and} \quad \frac{-4/315}{-1/5040} = 64.$$

(c) It appears that the ratio is always a power of 2. For x^3, it is $4 = 2^2$; for x^5, it is $16 = 2^4$; for x^7, it is $64 = 2^6$. This suggests that in general, for the coefficient of x^n, it is 2^{n-1}.

(d) From the identity $\sin(2x) = 2\sin x \cos x$, we expect that $P_7(2x) = 2Q_7(x)$. So, if a_n is the coefficient of x^n in $P_7(x)$, and if b_n is the coefficient of x^n in $Q_7(x)$, then, since the x^n terms $P_7(2x)$ and $2Q_7(x)$ must be equal, we have

$$a_n(2x)^n = 2b_n x^n.$$

Dividing both sides by x^n and combining the powers of 2, this gives the pattern we observed. For $a_n \neq 0$,

$$\frac{b_n}{a_n} = 2^{n-1}.$$

CHECK YOUR UNDERSTANDING

1. False. For example, both $f(x) = x^2$ and $g(x) = x^2 + x^3$ have $P_2(x) = x^2$.

5. False. The Taylor series for $\sin x$ about $x = \pi$ is calculated by taking derivatives and using the formula

$$f(a) + f'(a)(x - a) + \frac{f''(a)}{2!}(x - a)^2 + \cdots.$$

The series for $\sin x$ about $x = \pi$ turns out to be

$$-(x - \pi) + \frac{(x - \pi)^3}{3!} - \frac{(x - \pi)^5}{5!} + \cdots.$$

9. False. The derivative of $f(x)g(x)$ is not $f'(x)g'(x)$. If this statement were true, the Taylor series for $(\cos x)(\sin x)$ would have all zero terms.

13. True. For large x, the graph of $P_{10}(x)$ looks like the graph of its highest powered term, $x^{10}/10!$. But e^x grows faster than any power, so e^x gets further and further away from $x^{10}/10! \approx P_{10}(x)$.

17. True. Since f is even, $f(x)\sin(mx)$ is odd for any m, so

$$b_m = \frac{1}{\pi} \int_{-\pi}^{\pi} f(x)\sin x(mx)\, dx = 0.$$

21. False. The quadratic approximation to $f_1(x)f_2(x)$ near $x = 0$ is

$$f_1(0)f_2(0) + (f_1'(0)f_2(0) + f_1(0)f_2'(0))x + \frac{f_1''(0)f_2(0) + 2f_1'(0)f_2'(0) + f_1(0)f_2''(0)}{2}x^2.$$

On the other hand, we have
$$L_1(x) = f_1(0) + f_1'(0)x, \quad L_2(x) = f_2(0) + f_2'(0)x,$$

so

$$L_1(x)L_2(x) = (f_1(0) + f_1'(0)x)(f_2(0) + f_2'(0)x) = f_1(0)f_2(0) + (f_1'(0)f_2(0) + f_2'(0)f_1(0))x + f_1'(0)f_2'(0)x^2.$$

The first two terms of the right side agree with the quadratic approximation to $f_1(x)f_2(x)$ near $x = 0$, but the term of degree 2 does not.

For example, the linear approximation to e^x is $1+x$, but the quadratic approximation to $(e^x)^2 = e^{2x}$ is $1+2x+2x^2$, not $(1 + x)^2 = 1 + 2x + x^2$.

CHAPTER ELEVEN

Solutions for Section 11.1

Exercises

1. **(a)** (III) An island can only sustain the population up to a certain size. The population will grow until it reaches this limiting value.
   (b) (V) The ingot will get hot and then cool off, so the temperature will increase and then decrease.
   (c) (I) The speed of the car is constant, and then decreases linearly when the breaks are applied uniformly.
   (d) (II) Carbon-14 decays exponentially.
   (e) (IV) Tree pollen is seasonal, and therefore cyclical.

5. If y satisfies the differential equation, then we must have

$$\frac{d\left(5 + 3e^{kx}\right)}{dx} = 10 - 2(5 + 3e^{kx})$$

$$3ke^{kx} = 10 - 10 - 6e^{kx}$$

$$3ke^{kx} = -6e^{kx}$$

$$k = -2.$$

So, if $k = -2$ the formula for y solves the differential equation.

9. If $y = \sin 2t$, then $\frac{dy}{dt} = 2\cos 2t$, and $\frac{d^2y}{dt^2} = -4\sin 2t$.
   Thus $\frac{d^2y}{dt^2} + 4y = -4\sin 2t + 4\sin 2t = 0$.

Problems

13. **(a)** $P = \frac{1}{1+e^{-t}} = (1 + e^{-t})^{-1}$
   $\frac{dP}{dt} = -(1 + e^{-t})^{-2}(-e^{-t}) = \frac{e^{-t}}{(1+e^{-t})^2}.$
   Then $P(1 - P) = \frac{1}{1+e^{-t}}\left(1 - \frac{1}{1+e^{-t}}\right) = \left(\frac{1}{1+e^{-t}}\right)\left(\frac{e^{-t}}{1+e^{-t}}\right) = \frac{e^{-t}}{(1+e^{-t})^2} = \frac{dP}{dt}.$
   (b) As t tends to ∞, e^{-t} goes to 0. Thus $\displaystyle\lim_{t\to\infty} \frac{1}{1+e^{-t}} = 1.$

Solutions for Section 11.2

Exercises

1. There are many possible answers. One possibility is shown in Figures 11.1 and 11.2.

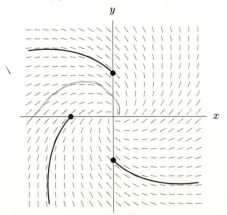

Figure 11.1

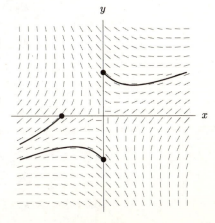

Figure 11.2

Problems

5. **(a)**

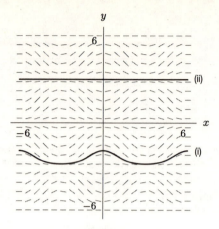

Figure 11.3

(b) We can see that the slope lines are horizontal when y is an integer multiple of π. We conclude from Figure 11.3 that the solution is $y = n\pi$ in this case.

To check this, we note that if $y = n\pi$, then $(\sin x)(\sin y) = (\sin x)(\sin n\pi) = 0 = y'$. Thus $y = n\pi$ is a solution to $y' = (\sin x)(\sin y)$, and it passes through $(0, n\pi)$.

9. (a) II   (b) VI   (c) IV   (d) I   (e) III   (f) V

Solutions for Section 11.3

Exercises

1. **(a)**

Table 11.1 *Euler's method for*
$y' = x + y$ *with* $y(0) = 1$

x	y	$\Delta y =$(slope)Δx
0	1	$0.1 = (1)(0.1)$
0.1	1.1	$0.12 = (1.2)(0.1)$
0.2	1.22	$0.142 = (1.42)(0.1)$
0.3	1.362	$0.1662 = (1.662)(0.1)$
0.4	1.5282	

So $y(0.4) \approx 1.5282$.

(b)

Table 11.2 *Euler's method for*
$y' = x + y$ *with* $y(-1) = 0$

x	y	$\Delta y =$(slope)Δx
-1	0	$-0.1 = (-1)(0.1)$
-0.9	-0.1	$-0.1 = (-1)(0.1)$
-0.8	-0.2	$-0.1 = (-1)(0.1)$
-0.7	-0.3	
$\vdots$	$\vdots$	Notice that y
0	-1	decreases by 0.1
$\vdots$	$\vdots$	for every step
0.4	-1.4	

So $y(0.4) = -1.4$. (This answer is exact.)

Problems

5. (a) (i)

Table 11.3 *Euler's method for*
$y' = (\sin x)(\sin y)$, *starting at* $(0, 2)$

x	y	$\Delta y =$(slope)Δx
0	2	$0 = (\sin 0)(\sin 2)(0.1)$
0.1	2	$0.009 = (\sin 0.1)(\sin 2)(0.1)$
0.2	2.009	$0.018 = (\sin 0.2)(\sin 2.009)(0.1)$
0.3	2.027	

(ii)

Table 11.4 *Euler's method for*
$y' = (\sin x)(\sin y)$, *starting at*
$(0, \pi)$

x	y	$\Delta y =$(slope)Δx
0	π	$0 = (\sin 0)(\sin \pi)(0.1)$
0.1	π	$0 = (\sin 0.1)(\sin \pi)(0.1)$
0.2	π	$0 = (\sin 0.2)(\sin \pi)(0.1)$
0.3	π	

(b) The slope field shows that the slope of the solution curve through $(0, \pi)$ is always 0. Thus the solution curve is the horizontal line with equation $y = \pi$.

9. (a) Using one step, $\frac{\Delta B}{\Delta t} = 0.05$, so $\Delta B = \left(\frac{\Delta B}{\Delta t}\right) \Delta t = 50$. Therefore we get an approximation of $B \approx 1050$ after one year.

(b) With two steps, $\Delta t = 0.5$ and we have

Table 11.5

t	B	$\Delta B = (0.05B)\Delta t$
0	1000	25
0.5	1025	25.63
1.0	1050.63	

(c) Keeping track to the nearest hundredth with $\Delta t = 0.25$, we have

Table 11.6

t	B	$\Delta B = (0.05B)\Delta t$
0	1000	12.5
0.25	1012.5	12.66
0.5	1025.16	12.81
0.75	1037.97	12.97
1	1050.94	

(d) In part (a), we get our approximation by making a single increment, ΔB, where ΔB is just $0.05B$. If we think in terms of interest, ΔB is just like getting one end of the year interest payment. Since ΔB is 0.05 times the balance B, it is like getting 5% interest at the end of the year.

(e) Part (b) is equivalent to computing the final amount in an account that begins with $1000 and earns 5% interest compounded twice annually. Each step is like computing the interest after 6 months. When $t = 0.5$, for example, the interest is $\Delta B = (0.05B) \cdot \frac{1}{2}$, and we add this to $1000 to get the new balance.

Similarly, part (c) is equivalent to the final amount in an account that has an initial balance of $1000 and earns 5% interest compounded quarterly.

Solutions for Section 11.4

Exercises

1. $\frac{dP}{dt} = 0.02P$ implies that $\frac{dP}{P} = 0.02\,dt$.

   $\int \frac{dP}{P} = \int 0.02\,dt$ implies that $\ln|P| = 0.02t + C$.

   $|P| = e^{0.02t+C}$ implies that $P = Ae^{0.02t}$, where $A = \pm e^C$.
   We are given $P(0) = 20$. Therefore, $P(0) = Ae^{(0.02)\cdot 0} = A = 20$. So the solution is $P = 20e^{0.02t}$.

5. Separating variables and integrating both sides gives

   $$\int \frac{1}{L}dL = \frac{1}{2}\int dp$$

   or

   $$\ln|L| = \frac{1}{2}p + C.$$

   This can be written

   $$L(p) = \pm e^{(1/2)p+C} = Ae^{p/2}.$$

   The initial condition $L(0) = 100$ gives $100 = A$, so

   $$L(p) = 100e^{p/2}.$$

9. $\frac{1}{z}\frac{dz}{dt} = 5$ implies $\frac{dz}{z} = 5\,dt$.
   Integrating and moving terms, we have $z = Ae^{5t}$. Using the fact that $z(1) = 5$, we have $z(1) = Ae^5 = 5$, so $A = \frac{5}{e^5}$.
   Therefore, $z = \frac{5}{e^5}e^{5t} = 5e^{5t-5}$.

13. $\frac{dP}{dt} = P + 4$ implies that $\frac{dP}{P+4} = dt$.

    $\int \frac{dP}{P+4} = \int dt$ implies that $\ln|P+4| = t + C$.

    $P + 4 = Ae^t$ implies that $P = Ae^t - 4$. $P = 100$ when $t = 0$, so $P(0) = Ae^0 - 4 = 100$, and $A = 104$. Therefore $P = 104e^t - 4$.

17. We know that the general solution to a differential equation of the form

    $$\frac{dy}{dt} = k(y - A)$$

    is

    $$y = Ce^{kt} + A.$$

    Thus, in our case, we get

    $$y = Ce^{t/2} + 200.$$

    We know that at $t = 0$ we have $y = 50$, so solving for C we get

    $$y = Ce^{t/2} + 200$$
    $$50 = Ce^{0/2} + 200$$
    $$-150 = Ce^0$$
    $$C = -150.$$

    Thus we get

    $$y = 200 - 150e^{t/2}.$$

21. $\frac{dz}{dt} = te^z$ implies $e^{-z}dz = tdt$ implies $\int e^{-z}\,dz = \int t\,dt$ implies $-e^{-z} = \frac{t^2}{2} + C$.
    Since the solution passes through the origin, $z = 0$ when $t = 0$, we must have $-e^{-0} = \frac{0}{2} + C$, so $C = -1$. Thus $-e^{-z} = \frac{t^2}{2} - 1$, or $z = -\ln(1 - \frac{t^2}{2})$.

25. $\frac{dw}{d\theta} = \theta w^2 \sin\theta^2$ implies that $\int \frac{dw}{w^2} = \int \theta \sin\theta^2\,d\theta$ implies that $-\frac{1}{w} = -\frac{1}{2}\cos\theta^2 + C$. According to the initial conditions, $w(0) = 1$, so $-1 = -\frac{1}{2} + C$ and $C = -\frac{1}{2}$. Thus $-\frac{1}{w} = -\frac{1}{2}\cos\theta^2 - \frac{1}{2}$ implies that $\frac{1}{w} = \frac{\cos\theta^2 + 1}{2}$ implies that $w = \frac{2}{\cos\theta^2 + 1}$.

Problems

29. $\frac{dQ}{dt} - \frac{Q}{k} = 0$ so $\frac{dQ}{dt} = \frac{Q}{k}$. This is now the same problem as Problem 30, except the constant factor on the right is $\frac{1}{k}$ instead of k. Thus the solution is $Q = Ae^{\frac{1}{k}t}$ for any constant A.

33. Separating variables and integrating gives

$$\int \frac{1}{aP+b}\,dP = \int dt.$$

This gives

$$\frac{1}{a}\ln|aP+b| = t + C$$
$$\ln|aP+b| = at + D$$
$$aP+b = \pm e^{at+D} = Ae^{at}$$

or

$$P(t) = \frac{1}{a}(Ae^{at} - b).$$

37. Separating variables and integrating gives

$$\int \frac{1}{L-b}\,dL = \int k(x+a)\,dx$$

or

$$\ln|L-b| = k\left(\frac{1}{2}x^2 + ax\right) + C.$$

Solving for L gives

$$L(x) = b + Ae^{k\left(\frac{1}{2}x^2 + ax\right)}.$$

41. Since $\frac{dy}{dt} = -y\ln(\frac{y}{2})$, we have $\frac{dy}{y\ln(\frac{y}{2})} = -\,dt$, so that $\int \frac{dy}{y\ln(\frac{y}{2})} = \int(-\,dt)$.
Substituting $w = \ln(\frac{y}{2})$, $dw = \frac{1}{y}\,dy$ gives:

$$\int \frac{dw}{w} = \int(-\,dt)$$

so

$$\ln|w| = \ln\left|\ln\left(\frac{y}{2}\right)\right| = -t + C.$$

Since $y(0) = 1$, we have $C = \ln|\ln\frac{1}{2}| = \ln|-\ln 2| = \ln(\ln 2)$. Thus $\ln|\ln(\frac{y}{2})| = -t + \ln(\ln 2)$, or

$$\left|\ln\left(\frac{y}{2}\right)\right| = e^{-t+\ln(\ln 2)} = (\ln 2)e^{-t}$$

Again, since $y(0) = 1$, we see that $-\ln(y/2) = (\ln 2)e^{-t}$ and thus $y = 2(2^{-e^{-t}})$. (Note that $\ln(y/2) = (\ln 2)e^{-t}$ does not satisfy $y(0) = 1$.)

45. (a), (b)

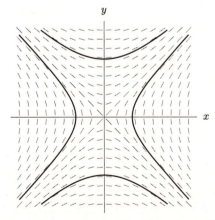

(c) Since $\frac{dy}{dx} = \frac{x}{y}$, we have $\int y\,dy = \int x\,dx$ and thus $\frac{y^2}{2} = \frac{x^2}{2} + C$, or $y^2 - x^2 = 2C$. This is the equation of the hyperbolas in part (b).

Solutions for Section 11.5

Exercises

1. (a) = (I), (b) = (IV), (c) = (III). Graph (II) represents an egg originally at $0°$ C which is moved to the kitchen table ($20°$ C) two minutes after the egg in part (a) is moved.

5. The equilibrium solutions of a differential equation are those functions satisfying the differential equation whose derivative is everywhere 0. Graphically, this means that a function is an equilibrium solution if it is a horizontal line that lies on the slope field. Looking at the figure in the problem, it appears that the equilibrium solutions for this problem are at $y = 1$ and $y = 3$. An equilibrium solution is stable if a small change in the initial value conditions gives a solution which tends toward equilibrium as $t \to \infty$. we see that $y = 3$ is a stable solution, while $y = 1$ is an unstable solution. See Figure 11.4.

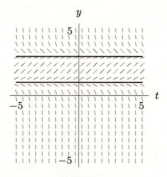

Figure 11.4

Problems

9. (a) Suppose $Y(t)$ is the quantity of oil in the well at time t. We know that the oil in the well decreases at a rate proportional to $Y(t)$, so

$$\frac{dY}{dt} = -kY.$$

Integrating, and using the fact that initially $Y = Y_0 = 10^6$, we have

$$Y = Y_0 e^{-kt} = 10^6 e^{-kt}.$$

In six years, $Y = 500,000 = 5 \cdot 10^5$, so

$$5 \cdot 10^5 = 10^6 e^{-k \cdot 6}$$

so

$$0.5 = e^{-6k}$$
$$k = -\frac{\ln 0.5}{6} = 0.1155.$$

When $Y = 600,000 = 6 \cdot 10^5$,

$$\text{Rate at which oil decreasing} = \left|\frac{dY}{dt}\right| = kY = 0.1155(6 \cdot 10^5) = 69{,}300 \text{ barrels/year.}$$

(b) We solve the equation

$$5 \cdot 10^4 = 10^6 e^{-0.1155t}$$
$$0.05 = e^{-0.1155t}$$
$$t = \frac{\ln 0.05}{-0.1155} = 25.9 \text{ years.}$$

13. **(a)** $\frac{dB}{dt} = \frac{r}{100}B$. The constant of proportionality is $\frac{r}{100}$.

(b) Solving, we have

$$\frac{dB}{B} = \frac{r\,dt}{100}$$

$$\int \frac{dB}{B} = \int \frac{r}{100}\,dt$$

$$\ln |B| = \frac{r}{100}t + C$$

$$B = e^{(r/100)t+C} = Ae^{(r/100)t}, \qquad A = e^C.$$

A is the initial amount in the account, since A is the amount at time $t = 0$.

(c)

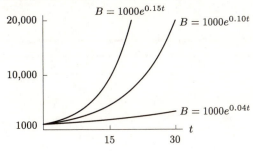

17. **(a)**

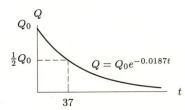

(b) $\dfrac{dQ}{dt} = -kQ$

(c) Since $25\% = 1/4$, it takes two half-lives $= 74$ hours for the drug level to be reduced to 25%. Alternatively, $Q = Q_0 e^{-kt}$ and $\frac{1}{2} = e^{-k(37)}$, we have

$$k = -\frac{\ln(1/2)}{37} \approx 0.0187.$$

Therefore $Q = Q_0 e^{-0.0187t}$. We know that when the drug level is 25% of the original level that $Q = 0.25Q_0$. Setting these equal, we get

$$0.25 = e^{-0.0187t}.$$

giving

$$t = -\frac{\ln(0.25)}{0.0187} \approx 74 \text{ hours} \approx 3 \text{ days}.$$

21. The rate of disintegration is proportional to the quantity of carbon-14 present. Let Q be the quantity of carbon-14 present at time t, with $t = 0$ in 1977. Then

$$Q = Q_0 e^{-kt},$$

where Q_0 is the quantity of carbon-14 present in 1977 when $t = 0$. Then we know that

$$\frac{Q_0}{2} = Q_0 e^{-k(5730)}$$

so that

$$k = -\frac{\ln(1/2)}{5730} = 0.000121.$$

Thus

$$Q = Q_0 e^{-0.000121t}.$$

The quantity present at any time is proportional to the rate of disintegration at that time so

$$Q_0 = c8.2 \qquad \text{and} \qquad Q = c13.5$$

where c is a constant of proportionality. Thus substituting for Q and Q_0 in

$$Q = Q_0 e^{-0.000121t}$$

gives

$$c13.5 = c8.2e^{-0.000121t}$$

so

$$t = -\frac{\ln(13.5/8.2)}{0.000121} \approx -4120.$$

Thus Stonehenge was built about 4120 years before 1977, in about 2150 B.C.

Solutions for Section 11.6

Exercises

1. Since mg is constant and $a = dv/dt$, differentiating $ma = mg - kv$ gives

$$m\frac{da}{dt} = -k\frac{dv}{dt} = -ma.$$

Thus, the differential equation is

$$\frac{da}{dt} = -\frac{k}{m}a.$$

Solving for a gives

$$a = a_0 e^{-kt/m}.$$

At $t = 0$, we have $a = g$, the acceleration due to gravity. Thus, $a_0 = g$, so

$$a = ge^{-kt/m}.$$

Problems

5. Let $D(t)$ be the quantity of dead leaves, in grams per square centimeter. Then $\frac{dD}{dt} = 3 - 0.75D$, where t is in years. We factor out -0.75 and then separate variables.

$$\frac{dD}{dt} = -0.75(D - 4)$$

$$\int \frac{dD}{D - 4} = \int -0.75 \, dt$$

$$\ln|D - 4| = -0.75t + C$$

$$|D - 4| = e^{-0.75t+C} = e^{-0.75t}e^{C}$$

$$D = 4 + Ae^{-0.75t}, \text{ where } A = \pm e^{C}.$$

If initially the ground is clear, the solution looks like the following graph:

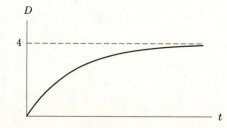

The equilibrium level is 4 grams per square centimeter, regardless of the initial condition.

9. We are given that

$$BC = 2OC.$$

If the point A has coordinates (x, y) then $OC = x$ and $AC = y$. The slope of the tangent line, y', is given by

$$y' = \frac{AC}{BC} = \frac{y}{BC},$$

so

$$BC = \frac{y}{y'}.$$

Substitution into $BC = 2OC$ gives

$$\frac{y}{y'} = 2x,$$

so

$$\frac{y'}{y} = \frac{1}{2x}.$$

Separating variables to integrate this differential equation gives

$$\int \frac{dy}{y} = \int \frac{dx}{2x}$$

$$\ln|y| = \frac{1}{2}\ln|x| + C = \ln\sqrt{|x|} + \ln A$$

$$|y| = A\sqrt{|x|}$$

$$y = \pm(A\sqrt{x}).$$

Thus, in the first quadrant, the curve has equation $y = A\sqrt{x}$.

13. (a) If I is intensity and l is the distance traveled through the water, then for some $k > 0$,

$$\frac{dI}{dl} = -kI.$$

(The proportionality constant is negative because intensity decreases with distance). Thus $I = Ae^{-kl}$. Since $I = A$ when $l = 0$, A represents the initial intensity of the light.

(b) If 50% of the light is absorbed in 10 feet, then $0.50A = Ae^{-10k}$, so $e^{-10k} = \frac{1}{2}$, giving

$$k = \frac{-\ln\frac{1}{2}}{10} = \frac{\ln 2}{10}.$$

In 20 feet, the percentage of light left is

$$e^{-\frac{\ln 2}{10}\cdot 20} = e^{-2\ln 2} = (e^{\ln 2})^{-2} = 2^{-2} = \frac{1}{4},$$

so $\frac{3}{4}$ or 75% of the light has been absorbed. Similarly, after 25 feet,

$$e^{-\frac{\ln 2}{10}\cdot 25} = e^{-2.5\ln 2} = (e^{\ln 2})^{-\frac{5}{2}} = 2^{-\frac{5}{2}} \approx 0.177.$$

Approximately 17.7% of the light is left, so 82.3% of the light has been absorbed.

17. (a) The quantity and the concentration both increase with time. As the concentration increases, the rate at which the drug is excreted also increases, and so the rate at which the drug builds up in the blood decreases; thus the graph of concentration against time is concave down. The concentration rises until the rate of excretion exactly balances the rate at which the drug is entering; at this concentration there is a horizontal asymptote. (See Figure 11.5.)

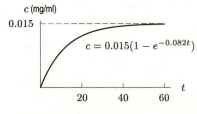

Figure 11.5

(b) Let's start by writing a differential equation for the quantity, $Q(t)$.

$$\text{Rate at which quantity of drug changes} = \text{Rate in} - \text{Rate out}$$

$$\frac{dQ}{dt} = 43.2 - 0.082Q$$

where Q is measured in mg. We want an equation for concentration $c(t) = Q(t)/v$, where $c(t)$ is measured in mg/ml and v is volume, so $v = 35,000$ ml.

$$\frac{1}{v}\frac{dQ}{dt} = \frac{43.2}{v} - 0.082\frac{Q}{v},$$

giving

$$\frac{dc}{dt} = \frac{43.2}{35,000} - 0.082c.$$

(c) Factor out -0.082 and separate variables to solve.

$$\frac{dc}{dt} = -0.082(c - 0.015)$$

$$\int \frac{dc}{c - 0.015} = -0.082 \int dt$$

$$\ln|c - 0.015| = -0.082t + B$$

$$c - 0.015 = Ae^{-0.082t} \quad \text{where} \quad A = \pm e^B$$

Since $c = 0$ when $t = 0$, we have $A = -0.015$, so

$$c = 0.015 - 0.015e^{-0.082t} = 0.015(1 - e^{-0.082t}).$$

Thus $c \to 0.015$ mg/ml as $t \to \infty$.

21. (a) Concentration of carbon monoxide $= \dfrac{\text{Quantity in room}}{\text{Volume}}$.

If $Q(t)$ represents the quantity of carbon monoxide in the room at time t, $c(t) = Q(t)/60$.

$$\begin{array}{c}\text{Rate quantity of} \\ \text{carbon monoxide in room} \\ \text{changes}\end{array} = \text{rate in} - \text{rate out}$$

Now

$$\text{Rate in} = 5\%(0.002\text{m}^3/\text{min}) = 0.05(0.002) = 0.0001\text{m}^3/\text{min}.$$

Since smoky air is leaving at $0.002\text{m}^3/\text{min}$, containing a concentration $c(t) = Q(t)/60$ of carbon monoxide

$$\text{Rate out} = 0.002\frac{Q(t)}{60}$$

Thus

$$\frac{dQ}{dt} = 0.0001 - \frac{0.002}{60}Q$$

Since $c = Q/60$, we can substitute $Q = 60c$, giving

$$\frac{d(60c)}{dt} = 0.0001 - \frac{0.002}{60}(60c)$$

$$\frac{dc}{dt} = \frac{0.0001}{60} - \frac{0.002}{60}c$$

(b) Factoring the right side of the differential equation and separating gives

$$\frac{dc}{dt} = -\frac{0.0001}{3}(c - 0.05) \approx 3 \times 10^{-5}(c - 0.05)$$

$$\int \frac{dc}{c - 0.05} = -\int 3 \times 10^{-5} dt$$

$$\ln|c - 0.05| = -3 \times 10^{-5}t + K$$

$$c - 0.05 = Ae^{-3\times 10^{-5}t} \quad \text{where} A = \pm e^K.$$

Since $c = 0$ when $t = 0$, we have $A = -0.05$, so

$$c = 0.05 - 0.05e^{-3\times 10^{-5}t}$$

(c) As $t \to \infty$, $e^{-3\times 10^{-5}t} \to 0$ so $c \to 0.05$.

Thus in the long run, the concentration of carbon monoxide tends to 5%, the concentration of the incoming air.

Solutions for Section 11.7

Exercises

1. A continuous growth rate of 0.2% means that

$$\frac{1}{P}\frac{dP}{dt} = 0.2\% = 0.002.$$

Separating variables and integrating gives

$$\int \frac{dP}{P} = \int 0.002\, dt$$
$$P = P_0 e^{0.002t} = (6.6 \times 10^6)e^{0.002t}.$$

Problems

5.

Table 11.7

Year	P	$\frac{dP}{dt} \approx \frac{P(t+10)-P(t-10)}{20}$
1790	3.9	
1800	5.3	$(7.2 - 3.9)/20 = 0.165$
1810	7.2	$(9.6 - 5.3)/20 = 0.215$
1820	9.6	$(12.9 - 7.2)/20 = 0.285$
1830	12.9	$(17.1 - 9.6)/20 = 0.375$
1840	17.1	$(23.2 - 12.9)/20 = 0.515$
1850	23.2	$(31.4 - 17.1)/20 = 0.715$
1860	31.4	$(38.6 - 23.2)/20 = 0.770$
1870	38.6	$(50.2 - 31.4)/20 = 0.940$
1880	50.2	$(62.9 - 38.6)/20 = 1.215$
1890	62.9	$(76.0 - 50.2)/20 = 1.290$
1900	76.0	$(92.0 - 62.9)/20 = 1.455$
1910	92.0	$(105.7 - 76.0)/20 = 1.485$
1920	105.7	$(122.8 - 92.0)/20 = 1.540$
1930	122.8	$(131.7 - 105.7)/20 = 1.300$
1940	131.7	$(150.7 - 122.8)/20 = 1.395$
1950	150.7	

According to these calculations, the largest value of dP/dt occurs in 1920 when the rate of change is $\frac{dP}{dt} = 1.540$ million people/year. The population in 1920 was 105.7 million. If we assume that the limiting value, L, is twice the population when it is changing most quickly, then $L = 2 \times 105.7 = 211.4$ million. This is greater than the estimate of 187 million computed in the text and closer to the actual 1990 population of 248.7 million.

9. (a) Let I be the number of informed people at time t, and I_0 the number who know initially. Then this model predicts that $\frac{dI}{dt} = k(M - I)$ for some positive constant k. Solving this, we find the solution is

$$I = M - (M - I_0)e^{-kt}.$$

We sketch the solution with $I_0 = 0$. Notice that $\frac{dI}{dt}$ is largest when I is smallest, so the information spreads fastest in the beginning, at $t = 0$. In addition, the graph below shows that $I \to M$ as $t \to \infty$, meaning that everyone gets the information eventually.

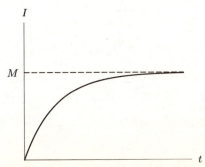

(b) In this case, the model suggests that $\frac{dI}{dt} = kI(M - I)$ for some positive constant k. This is a logistic model with carrying capacity M. We sketch the solutions for three different values of I_0 below.

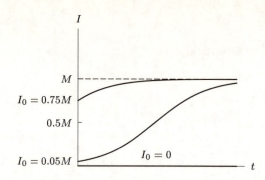

(i) If $I_0 = 0$ then $I = 0$ for all t. In other words, if nobody knows something, it doesn't spread by word of mouth!

(ii) If $I_0 = 0.05M$, then $\frac{dI}{dt}$ is increasing up to $I = \frac{M}{2}$. Thus, the information is spreading fastest at $I = \frac{M}{2}$.

(iii) If $I_0 = 0.75M$, then $\frac{dI}{dt}$ is always decreasing for $I > \frac{M}{2}$, so $\frac{dI}{dt}$ is largest when $t = 0$.

13.

(a) 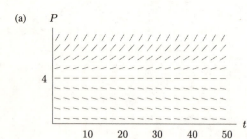   (b)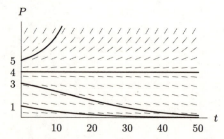

(c) There are two equilibrium values, $P = 0$, and $P = 4$. The first, representing extinction, is stable. The equilibrium value $P = 4$ is unstable because the populations increase if greater than 4, and decrease if less than 4. Notice that the equilibrium values can be obtained by setting $dP/dt = 0$:

$$\frac{dP}{dt} = 0.02P^2 - 0.08P = 0.02P(P - 4) = 0$$

so

$$P = 0 \text{ or } P = 4.$$

Solutions for Section 11.8

Exercises

1. Since

$$\frac{dS}{dt} = -aSI,$$
$$\frac{dI}{dt} = aSI - bI,$$
$$\frac{dR}{dt} = bI$$

we have

$$\frac{dS}{dt} + \frac{dI}{dt} + \frac{dR}{dt} = -aSI + aSI - bI + bI = 0.$$

Thus $\frac{d}{dt}(S + I + R) = 0$, so $S + I + R = $ constant.

5. If $w = 2$ and $r = 2$, then $\frac{dw}{dt} = -2$ and $\frac{dr}{dt} = 2$, so initially the number of worms decreases and the number of robins increases. In the long run, however, the populations will oscillate; they will even go back to $w = 2$ and $r = 2$.

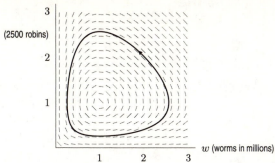

9. The numbers of robins begins to increase while the number of worms remains approximately constant. See Figure 11.6. The numbers of robins and worms oscillate periodically between 0.2 and 3, with the robin population lagging behind the worm population.

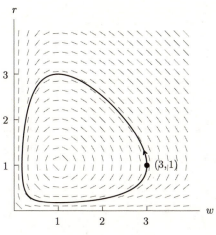

Figure 11.6

13. x decreases quickly while y increases more slowly.

Problems

17. (a) Predator-prey, because x decreases while alone, but is helped by y, whereas y increases logistically when alone, and is harmed by x. Thus x is predator, y is prey.

(b)

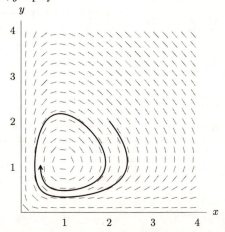

Provided neither initial population is zero, both populations tend to about 1. If x is initially zero, but y is not, then $y \to \infty$. If y is initially zero, but x is not, then $x \to 0$.

21. (a) Since the guerrillas are hard to find, the rate at which they are put out of action is proportional to the number of chance encounters between a guerrilla and a conventional soldier, which is in turn proportional to the number of guerrillas and to the number of conventional soldiers. Thus the rate at which guerrillas are put out of action is proportional to the product of the strengths of the two armies.

(b)

$$\frac{dx}{dt} = -xy$$

$$\frac{dy}{dt} = -x$$

(c) Thinking of y as a function of x and x a function of of t, then by the chain rule: $\frac{dy}{dt} = \frac{dy}{dx}\frac{dx}{dt}$ so:

$$\frac{dy}{dx} = \frac{dy/dt}{dx/dt} = \frac{-x}{-xy} = \frac{1}{y}$$

Separating variables:

$$\int y\,dy = \int dx$$

$$\frac{y^2}{2} = x + C$$

The value of C is determined by the initial strengths of the two armies.

(d) The sign of C determines which side wins the battle. Looking at the general solution $\frac{y^2}{2} = x + C$, we see that if $C > 0$ the y-intercept is at $\sqrt{2C}$, so y wins the battle by virtue of the fact that it still has troops when $x = 0$. If $C < 0$ then the curve intersects the axes at $x = -C$, so x wins the battle because it has troops when $y = 0$. If $C = 0$, then the solution goes to the point $(0, 0)$, which represents the case of mutual annihilation.

(e) We assume that an army wins if the opposing force goes to 0 first. Figure 11.7 shows that the conventional force wins if $C > 0$ and the guerrillas win if $C < 0$. Neither side wins if $C = 0$ (all soldiers on both sides are killed in this case).

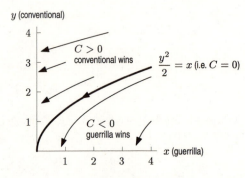

Figure 11.7

Solutions for Section 11.9

Exercises

1. (a) $dS/dt = 0$ where $S = 0$ or $I = 0$ (both axes).

$dI/dt = 0.0026I(S - 192)$, so $dI/dt = 0$ where $I = 0$ or $S = 192$.

Thus every point on the S axis is an equilibrium point (corresponding to no one being sick).

(b) In region I, where $S > 192$, $\dfrac{dS}{dt} < 0$ and $\dfrac{dI}{dt} > 0$.

In region II, where $S < 192$, $\dfrac{dS}{dt} < 0$ and $\dfrac{dI}{dt} < 0$. See Figure 11.8.

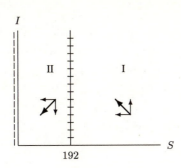

Figure 11.8

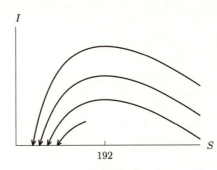

Figure 11.9

(c) If the trajectory starts with $S_0 > 192$, then I increases to a maximum when $S = 192$. If $S_0 < 192$, then I always decreases. See Figure 11.8. Regardless of the initial conditions, the trajectory always goes to a point on the S-axis (where $I = 0$). The S-intercept represents the number of students who never get the disease. See Figure 11.9.

Problems

5. We first find the nullclines. Vertical nullclines occur where $\frac{dx}{dt} = 0$, which happens when $x = 0$ or $y = \frac{1}{3}(2 - x)$. Horizontal nullclines occur where $\frac{dy}{dt} = y(1 - 2x) = 0$, which happens when $y = 0$ or $x = \frac{1}{2}$. These nullclines are shown in Figure 11.10.

Equilbrium points (also shown in Figure 11.10) occur at the intersections of vertical and horizontal nullclines. There are three such points for this system of equations; $(0, 0)$, $(\frac{1}{2}, \frac{1}{2})$ and $(2, 0)$.

The nullclines divide the positive quadrant into four regions as shown in Figure 11.10. Trajectory directions for these regions are shown in Figure 11.11.

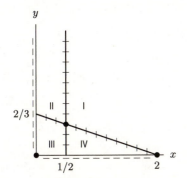

Figure 11.10: Nullclines and equilibrium points (dots)

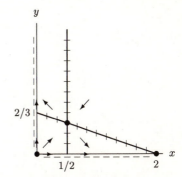

Figure 11.11: General directions of trajectories and equilibrium points (dots)

9. We assume that x, $y \geq 0$ and then find the nullclines. $\frac{dx}{dt} = x(1 - \frac{x}{2} - y) = 0$ when $x = 0$ or $y + \frac{x}{2} = 1$. $\frac{dy}{dt} = y(1 - \frac{y}{3} - x) = 0$ when $y = 0$ or $x + \frac{y}{3} = 1$. We find the equilibrium points. They are $(2, 0)$, $(0, 3)$, $(0, 0)$, and $(\frac{4}{5}, \frac{3}{5})$. The nullclines and equilibrium points are shown in Figure 11.12.

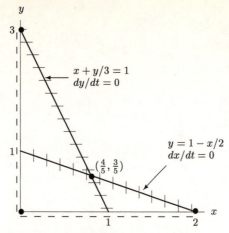

Figure 11.12: Nullclines and equilibrium points (dots)

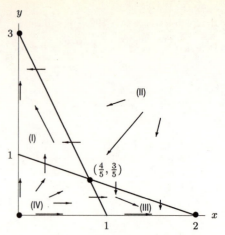

Figure 11.13: General directions of trajectories and equilibrium points (dots)

Figure 11.13 shows that if the initial point is in sector (I), the trajectory heads towards the equilibrium point $(0, 3)$. Similarly, if the trajectory begins in sector (III), then it heads towards the equilibrium $(2, 0)$ over time. If the trajectory begins in sector (II) or (IV), it can go to any of the three equilibrium points $(2, 0)$, $(0, 3)$, or $\left(\frac{4}{5}, \frac{3}{5}\right)$.

13. (a)

$$\frac{dx}{dt} = 0 \text{ when } x = \frac{10.5}{0.45} = 23.3$$

$$\frac{dy}{dt} = 0 \text{ when } 8.2x - 0.8y - 142 = 0$$

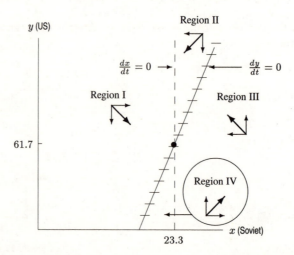

Figure 11.14: Nullclines and equilibrium point (dot) for US-Soviet arms race

There is an equilibrium point where the trajectories cross at $x = 23.3$, $y = 61.7$

In region I, $\dfrac{dx}{dt} > 0$, $\dfrac{dy}{dt} < 0$.

In region II, $\dfrac{dx}{dt} < 0$, $\dfrac{dy}{dt} < 0$.

In region III, $\dfrac{dx}{dt} < 0$, $\dfrac{dy}{dt} > 0$.

In region IV, $\dfrac{dx}{dt} > 0$, $\dfrac{dy}{dt} > 0$.

(b)

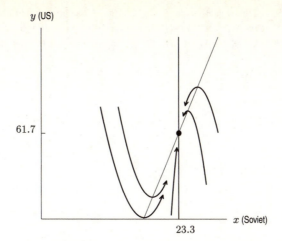

Figure 11.15: Trajectories for US-Soviet arms race.

(c) All the trajectories tend towards the equilibrium point $x = 23.3$, $y = 61.7$. Thus the model predicts that in the long run the arms race will level off with the Soviet Union spending 23.3 billion dollars a year on arms and the US 61.7 billion dollars.

(d) As the model predicts, yearly arms expenditure did tend towards 23 billion for the Soviet Union and 62 billion for the US.

Solutions for Section 11.10

Exercises

1. If $y = 2\cos t + 3\sin t$, then $y' = -2\sin t + 3\cos t$ and $y'' = -2\cos t - 3\sin t$. Thus, $y'' + y = 0$.

5. If $y(t) = A\sin(\omega t) + B\cos(\omega t)$ then
$$y' = \omega A\cos(\omega t) - \omega B\sin(\omega t)$$
$$y'' = -\omega^2 A\sin(\omega t) - \omega^2 B\cos(\omega t)$$
therefore
$$y'' + \omega^2 y = -\omega^2 A\sin(\omega t) - \omega^2 B\cos(2t) + \omega^2(A\sin(\omega t) + B\cos(\omega t)) = 0$$
for all values of A and B, so the given function is a solution.

9. The amplitude is $\sqrt{3^2 + 7^2} = \sqrt{58}$.

Problems

13. At $t = 0$, we find that $y = 2$, which is clearly the highest point since $-1 \leq \cos 3t \leq 1$. Thus, at $t = 0$ the mass is at its highest point. Since $y' = -6\sin 3t$, we see $y' = 0$ when $t = 0$. Thus, at $t = 0$ the object is at rest, although it will move down after $t = 0$.

17. First, we note that the solutions of:
(a) $x'' + x = 0$ are $x = A\cos t + B\sin t$;
(b) $x'' + 4x = 0$ are $x = A\cos 2t + B\sin 2t$;
(c) $x'' + 16x = 0$ are $x = A\cos 4t + B\sin 4t$.
This follows from what we know about the general solution to $x'' + \omega^2 x = 0$.
The period of the solutions to (a) is 2π, the period of the solutions to (b) is π, and the period of the solutions of (c) is $\frac{\pi}{2}$. Since the t-scales are the same on all of the graphs, we see that graphs (I) and (IV) have the same period, which is twice the period of graph (III). Graph (II) has twice the period of graphs (I) and (IV). Since each graph represents a solution, we have the following:

- equation (a) goes with graph (II)
  equation (b) goes with graphs (I) and (IV)
  equation (c) goes with graph (III)

- The graph of (I) passes through $(0, 0)$, so $0 = A\cos 0 + B\sin 0 = A$. Thus, the equation is $x = B\sin 2t$. Since the amplitude is 2, we see that $x = 2\sin 2t$ is the equation of the graph. Similarly, the equation for (IV) is $x = -3\sin 2t$. The graph of (II) also passes through $(0, 0)$, so, similarly, the equation must be $x = B\sin t$. In this case, we see that $B = -1$, so $x = -\sin t$.

  Finally, the graph of (III) passes through $(0, 1)$, and 1 is the maximum value. Thus, $1 = A\cos 0 + B\sin 0$, so $A = 1$. Since it reaches a local maximum at $(0, 1)$, $x'(0) = 0 = -4A\sin 0 + 4B\cos 0$, so $B = 0$. Thus, the solution is $x = \cos 4t$.

21. (a) Since a mass of 3 kg stretches the spring by 2 cm, the spring constant k is given by

$$3g = 2k \quad \text{so} \quad k = \frac{3g}{2}.$$

See Figure 11.16.

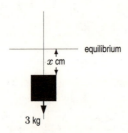

equilibrium

x cm

3 kg

Figure 11.16

Suppose we measure the displacement x from the equilibrium; then, using

$$\text{Mass} \cdot \text{Acceleration} = \text{Force}$$

gives

$$3x'' = -kx = -\frac{3gx}{2}$$

$$x'' + \frac{g}{2}x = 0$$

Since at time $t = 0$, the brick is 5 cm below the equilibrium and not moving, the initial conditions are $x(0) = 5$ and $x'(0) = 0$.

(b) The solution to the differential equation is

$$x = A\cos\left(\sqrt{\frac{g}{2}}t\right) + B\sin\left(\sqrt{\frac{g}{2}}t\right).$$

Since $x(0) = 5$, we have

$$x = A\cos(0) + B\sin(0) = 5 \quad \text{so} \quad A = 5.$$

In addition,

$$x'(t) = -5\sqrt{\frac{g}{2}}\sin\left(\sqrt{\frac{g}{2}}t\right) + B\sqrt{\frac{g}{2}}\cos\left(\sqrt{\frac{g}{2}}t\right)$$

so

$$x'(0) = -5\sqrt{\frac{g}{2}}\sin(0) + B\sqrt{\frac{g}{2}}\cos(0) = 0 \quad \text{so} \quad B = 0.$$

Thus,

$$x = 5\cos\sqrt{\frac{g}{2}}t.$$

25. The equation we have for the charge tells us that:

$$\frac{d^2Q}{dt^2} = -\frac{Q}{LC},$$

where L and C are positive.

If we let $\omega = \sqrt{\frac{1}{LC}}$, we know the solution is of the form:

$$Q = C_1\cos\omega t + C_2\sin\omega t.$$

Since $Q(0) = 0$, we find that $C_1 = 0$, so $Q = C_2 \sin \omega t$.

Since $Q'(0) = 4$, and $Q' = \omega C_2 \cos \omega t$, we have $C_2 = \dfrac{4}{\omega}$, so $Q = \dfrac{4}{\omega} \sin \omega t$.

But we want the maximum charge, meaning the amplitude of Q, to be $2\sqrt{2}$ coulombs. Thus, we have $\dfrac{4}{\omega} = 2\sqrt{2}$, which gives us $\omega = \sqrt{2}$.

So we now have: $\sqrt{2} = \frac{1}{\sqrt{LC}} = \frac{1}{\sqrt{10C}}$. Thus, $C = \frac{1}{20}$ farads.

Solutions for Section 11.11

Exercises

1. The characteristic equation is $r^2 + 4r + 3 = 0$, so $r = -1$ or -3.
   Therefore $y(t) = C_1 e^{-t} + C_2 e^{-3t}$.

5. The characteristic equation is $r^2 + 7 = 0$, so $r = \pm\sqrt{7}i$.
   Therefore $s(t) = C_1 \cos \sqrt{7}t + C_2 \sin \sqrt{7}t$.

9. The characteristic equation is $r^2 + r + 1 = 0$, so $r = -\frac{1}{2} \pm \frac{\sqrt{3}}{2}i$.
   Therefore $p(t) = C_1 e^{-t/2} \cos \frac{\sqrt{3}}{2}t + C_2 e^{-t/2} \sin \frac{\sqrt{3}}{2}t$.

13. The characteristic equation is
$$r^2 + 5r + 6 = 0$$
which has the solutions $r = -2$ and $r = -3$ so that
$$y(t) = Ae^{-3t} + Be^{-2t}$$

The initial condition $y(0) = 1$ gives
$$A + B = 1$$

and $y'(0) = 0$ gives
$$-3A - 2B = 0$$

so that $A = -2$ and $B = 3$ and
$$y(t) = -2e^{-3t} + 3e^{-2t}$$

17. The characteristic equation is $r^2 + 6r + 5 = 0$, so $r = -1$ or -5.
    Therefore $y(t) = C_1 e^{-t} + C_2 e^{-5t}$.
    $y'(t) = -C_1 e^{-t} - 5C_2 e^{-5t}$
    $y'(0) = 0 = -C_1 - 5C_2$
    $y(0) = 1 = C_1 + C_2$
        Therefore $C_2 = -1/4$, $C_1 = 5/4$ and $y(t) = \frac{5}{4}e^{-t} - \frac{1}{4}e^{-5t}$.

21. The characteristic equation is
$$r^2 + 5r + 6 = 0$$
which has the solutions $r = -2$ and $r = -3$ so that
$$y(t) = Ae^{-2t} + Be^{-3t}$$

The initial condition $y(0) = 1$ gives
$$A + B = 1$$

and $y(1) = 0$ gives
$$Ae^{-2} + Be^{-3} = 0$$

so that $A = \dfrac{1}{1-e}$ and $B = -\dfrac{e}{1-e}$ and
$$y(t) = \frac{1}{1-e}e^{-2t} + \frac{-e}{1-e}e^{-3t}$$

Problems

25. (a) $x'' + 4x = 0$ represents an undamped oscillator, and so goes with (IV).

(b) $x'' - 4x = 0$ has characteristic equation $r^2 - 4 = 0$ and so $r = \pm 2$. The solution is $C_1 e^{-2t} + C_2 e^{2t}$. This represents non-oscillating motion, so it goes with (II).

(c) $x'' - 0.2x' + 1.01x = 0$ has characteristic equation $r^2 - 0.2 + 1.01 = 0$ so $b^2 - 4ac = 0.04 - 4.04 = -4$, and $r = 0.1 \pm i$. So the solution is

$$C_1 e^{(0.1+i)t} + C_2 e^{(0.1-i)t} = e^{0.1t}(A \sin t + B \cos t).$$

The negative coefficient in the x' term represents an amplifying force. This is reflected in the solution by $e^{0.1t}$, which increases as t increases, so this goes with (I).

(d) $x'' + 0.2x' + 1.01x$ has characteristic equation $r^2 + 0.2r + 1.01 = 0$ so $b^2 - 4ac = -4$. This represents a damped oscillator. We have $r = -0.1 \pm i$ and so the solution is $x = e^{-0.1t}(A \sin t + B \cos t)$, which goes with (III).

29. Recall that $F_{\text{drag}} = -c\frac{ds}{dt}$, so to find the largest coefficient of damping we look at the coefficient of s'. Thus spring (iii) has the largest coefficient of damping.

33. The stiffest spring exerts the greatest restoring force for a small displacement. Recall that by Hooke's Law $F_{\text{spring}} = -ks$, so we look for the differential equation with the greatest coefficient of s. This is spring (ii).

37. The characteristic equation is $r^2 + r - 2 = 0$, so $r = 1$ or -2. Therefore $z(t) = C_1 e^t + C_2 e^{-2t}$. Since $e^t \to \infty$ as $t \to \infty$, we must have $C_1 = 0$. Therefore $z(t) = C_2 e^{-2t}$. Furthermore, $z(0) = 3 = C_2$, so $z(t) = 3e^{-2t}$.

41. In this case, the differential equation describing charge is $8Q'' + 2Q' + \frac{1}{4}Q = 0$, so the characteristic equation is $8r^2 + 2r + \frac{1}{4} = 0$. This quadratic equation has solutions

$$r = \frac{-2 \pm \sqrt{4 - 4 \cdot 8 \cdot \frac{1}{4}}}{16} = -\frac{1}{8} \pm \frac{1}{8}i.$$

Thus, the equation for charge is

$$Q(t) = e^{-\frac{1}{8}t}\left(A \sin \frac{t}{8} + B \cos \frac{t}{8}\right).$$

$$Q'(t) = -\frac{1}{8}e^{-\frac{1}{8}t}\left(A \sin \frac{t}{8} + B \cos \frac{t}{8}\right) + e^{-\frac{1}{8}t}\left(\frac{1}{8}A \cos \frac{t}{8} - \frac{1}{8}B \sin \frac{t}{8}\right)$$

$$= \frac{1}{8}e^{-\frac{1}{8}t}\left((A - B)\cos \frac{t}{8} + (-A - B)\sin \frac{t}{8}\right).$$

(a) We have

$$Q(0) = B = 0,$$
$$Q'(0) = \frac{1}{8}(A - B) = 2.$$

Thus, $B = 0$, $A = 16$, and

$$Q(t) = 16e^{-\frac{1}{8}t}\sin \frac{t}{8}.$$

(b) We have

$$Q(0) = B = 2,$$
$$Q'(0) = \frac{1}{8}(A - B) = 0.$$

Thus, $B = 2$, $A = 2$, and

$$Q(t) = 2e^{-\frac{1}{8}t}\left(\sin \frac{t}{8} + \cos \frac{t}{8}\right).$$

(c) By increasing the inductance, we have gone from the overdamped case to the underdamped case. We find that while the charge still tends to 0 as $t \to \infty$, the charge in the underdamped case oscillates between positive and negative values. In the over-damped case of Problem 39, the charge starts nonnegative and remains positive.

Solutions for Chapter 11 Review

Exercises

1. **(a)** Yes   **(b)** No   **(c)** Yes

     (d) No   **(e)** Yes   **(f)** Yes

     (g) No   **(h)** Yes   **(i)** No

     (j) Yes   **(k)** Yes   **(l)** No

5. This equation is separable, so we integrate, giving

$$\int \frac{1}{10 + 0.5H} \, dH = \int dt$$

so

$$\frac{1}{0.5} \ln |10 + 0.5H| = t + C.$$

Thus

$$H = Ae^{0.5t} - 20.$$

9. $\frac{dP}{dt} = 0.03P + 400$ so $\int \frac{dP}{P + \frac{40000}{3}} = \int 0.03 dt$.

$\ln |P + \frac{40000}{3}| = 0.03t + C$ giving $P = Ae^{0.03t} - \frac{40000}{3}$. Since $P(0) = 0$, $A = \frac{40000}{3}$, therefore $P = \frac{40000}{3}(e^{0.03t} - 1)$.

13. $\frac{dy}{dx} = \frac{y(3-x)}{x(\frac{1}{2}y - 4)}$ gives $\int \frac{(\frac{1}{2}y - 4)}{y} dy = \int \frac{(3-x)}{x} dx$ so $\int (\frac{1}{2} - \frac{4}{y}) dy = \int (\frac{3}{x} - 1) dx$. Thus $\frac{1}{2}y - 4\ln|y| = 3\ln|x| - x + C$.

Since $y(1) = 5$, we have $\frac{5}{2} - 4\ln 5 = \ln|1| - 1 + C$ so $C = \frac{7}{2} - 4\ln 5$. Thus,

$$\frac{1}{2}y - 4\ln|y| = 3\ln|x| - x + \frac{7}{2} - 4\ln 5.$$

We cannot solve for y in terms of x, so we leave the equation in this form.

17. $\frac{dy}{dx} = \frac{y(100-x)}{x(20-y)}$ gives $\int (\frac{20-y}{y}) dy = \int (\frac{100-x}{x}) dx$. Thus, $20\ln|y| - y = 100\ln|x| - x + C$. The curve passes through $(1, 20)$, so $20\ln 20 - 20 = -1 + C$ giving $C = 20\ln 20 - 19$. Therefore, $20\ln|y| - y = 100\ln|x| - x + 20\ln 20 - 19$. We cannot solve for y in terms of x, so we leave the equation in this form.

21. $e^{-\cos\theta} \frac{dz}{d\theta} = \sqrt{1 - z^2} \sin\theta$ implies $\int \frac{dz}{\sqrt{1-z^2}} = \int e^{\cos\theta} \sin\theta \, d\theta$ implies $\arcsin z = -e^{\cos\theta} + C$. According to the initial conditions: $z(0) = \frac{1}{2}$, so $\arcsin \frac{1}{2} = -e^{\cos 0} + C$, therefore $\frac{\pi}{6} = -e + C$, and $C = \frac{\pi}{6} + e$. Thus $z = \sin(-e^{\cos\theta} + \frac{\pi}{6} + e)$.

25. The characteristic equation of $9z'' - z = 0$ is

$$9r^2 - 1 = 0.$$

If this is written in the form $r^2 + br + c = 0$, we have that $r^2 - 1/9 = 0$ and

$$b^2 - 4c = 0 - (4)(-1/9) = 4/9 > 0$$

This indicates overdamped motion and since the roots of the characteristic equation are $r = \pm 1/3$, the general solution is

$$y(t) = C_1 e^{\frac{1}{3}t} + C_2 e^{-\frac{1}{3}t}.$$

29. The characteristic equation of $x'' + 2x' + 10x = 0$ is

$$r^2 + 2r + 10 = 0$$

We have that

$$b^2 - 4c = 2^2 - 4(10) = -36 < 0$$

This indicates underdamped motion and since the roots of the characteristic equation are $r = -1 \pm 3i$, the general solution is

$$y(t) = C_1 e^{-t} \cos 3t + C_2 e^{-t} \sin 3t$$

Problems

33. Recall that $s'' + bs' + cs = 0$ is overdamped if the discriminant $b^2 - 4c > 0$, critically damped if $b^2 - 4c = 0$, and underdamped if $b^2 - 4c < 0$. This has discriminant $b^2 - 4c = b^2 + 64$. Since $b^2 + 64$ is always positive, the solution is always overdamped.

37. Let $V(t)$ be the volume of water in the tank at time t, then

$$\frac{dV}{dt} = k\sqrt{V}$$

This is a separable equation which has the solution

$$V(t) = \left(\frac{kt}{2} + C\right)^2$$

Since $V(0) = 200$ this gives $200 = C^2$ so

$$V(t) = \left(\frac{kt}{2} + \sqrt{200}\right)^2.$$

However, $V(1) = 180$ therefore

$$180 = \left(\frac{k}{2} + \sqrt{200}\right)^2,$$

so that $k = 2\left(\sqrt{180} - \sqrt{200}\right) = -1.45146$. Therefore,

$$V(t) = \left(-0.726t + \sqrt{200}\right)^2.$$

The tank will be half-empty when $V(t) = 100$, so we solve

$$100 = \left(-0.726t + \sqrt{200}\right)^2$$

to obtain $t = 5.7$ days. The tank will be half empty in 5.7 days.

The volume after 4 days is $V(4)$ which is approximately 126.32 liters.

41. Let I be the number of infected people. Then, the number of healthy people in the population is $M - I$. The rate of infection is

$$\text{Infection rate} = \frac{0.01}{M}(M - I)I.$$

and the rate of recovery is

$$\text{Recovery rate} = 0.009I.$$

Therefore,

$$\frac{dI}{dt} = \frac{0.01}{M}(M - I)I - 0.009I$$

or

$$\frac{dI}{dt} = 0.001I\left(1 - 10\frac{I}{M}\right).$$

This is a logistic differential equation, and so the solution will look like the following graph:

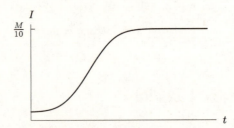

The limiting value for I is $\frac{1}{10}M$, so 1/10 of the population is infected in the long run.

CAS Challenge Problems

45. (a) We find the equilibrium solutions by setting $dP/dt = 0$, that is, $P(P - 1)(2 - P) = 0$, which gives three solutions, $P = 0$, $P = 1$, and $P = 2$.

(b) To get your computer algebra system to check that P_1 and P_2 are solutions, substitute one of them into the equation and form an expression consisting of the difference between the right and left hand sides, then ask the CAS to simplify that expression. Do the same for the other function. In order to avoid too much typing, define P_1 and P_2 as functions in your system.

(c) Substituting $t = 0$ gives

$$P_1(0) = 1 - \frac{1}{\sqrt{4}} = 1/2$$

$$P_2(0) = 1 + \frac{1}{\sqrt{4}} = 3/2.$$

We can find the limits using a computer algebra system. Alternatively, setting $u = e^t$, we can use the limit laws to calculate

$$\lim_{t \to \infty} \frac{e^t}{\sqrt{3 + e^{2t}}} = \lim_{u \to \infty} \frac{u}{\sqrt{3 + u^2}} = \lim_{u \to \infty} \sqrt{\frac{u^2}{3 + u^2}}$$

$$= \sqrt{\lim_{u \to \infty} \frac{u^2}{3 + u^2}} = \sqrt{\lim_{u \to \infty} \frac{1}{\frac{3}{u^2} + 1}}$$

$$= \sqrt{\frac{1}{\lim\limits_{u \to \infty} \frac{3}{u^2} + 1}} = \sqrt{\frac{1}{0 + 1}} = 1.$$

Therefore, we have

$$\lim_{t \to \infty} P_1(t) = 1 - 1 = 0$$

$$\lim_{t \to \infty} P_2(t) = 1 + 1 = 2.$$

To predict these limits without having a formula for P, looking at the original differential equation. We see if $0 < P < 1$, then $P(P-1)(2-P) < 0$, so $P' < 0$. Thus, if $0 < P(0) < 1$, then $P'(0) < 0$, so P is initially decreasing, and tends toward the equilibrium solution $P = 0$. On the other hand, if $1 < P < 2$, then $P(P-1)(2-P) > 0$, so $P' > 0$. So, if $1 < P(0) < 2$, then $P'(0) > 0$, so P is initially increasing and tends towards the equilibrium solution $P = 2$.

CHECK YOUR UNDERSTANDING

1. False. Suppose $k = -1$. The equation $y'' - y = 0$ or $y'' = y$ has solutions $y = e^t$ and $y = e^{-t}$ and general solution $y = C_1 e^t + C_2 e^{-t}$.

5. False. This is a logistic equation with equilibrium values $P = 0$ and $P = 2$. Solution curves do not cross the line $P = 2$ and do not go from $(0, 1)$ to $(1, 3)$.

9. True. No matter what initial value you pick, the solution curve has the x-axis as an asymptote.

13. True. Rewrite the equation as $dy/dx = xy + x = x(y + 1)$. Since the equation now has the form $dy/dx = f(x)g(y)$, it can be solved by separation of variables.

17. True. Since $f'(x) = g(x)$, we have $f''(x) = g'(x)$. Since $g(x)$ is increasing, $g'(x) > 0$ for all x, so $f''(x) > 0$ for all x. Thus the graph of f is concave up for all x.

21. False. Let $g(x) = 0$ for all x and let $f(x) = 17$. Then $f'(x) = g(x)$ and $\lim_{x \to \infty} g(x) = 0$, but $\lim_{x \to \infty} f(x) = 17$.

25. True. The slope of the graph of f is $dy/dx = 2x - y$. Thus when $x = a$ and $y = b$, the slope is $2a - b$.

29. False. Since $f'(1) = 2(1) - 5 = -3$, the point $(1, 5)$ could not be a critical point of f.

33. True. We will use the hint. Let $w = g(x) - f(x)$. Then:

$$\frac{dw}{dx} = g'(x) - f'(x) = (2x - g(x)) - (2x - f(x)) = f(x) - g(x) = -w.$$

Thus $dw/dx = -w$. This equation is the equation for exponential decay and has the general solution $w = Ce^{-x}$. Thus,

$$\lim_{x \to \infty} (g(x) - f(x)) = \lim_{x \to \infty} Ce^{-x} = 0.$$

37. If we differentiate implicitly the equation for the family, we get $2x - 2y\,dy/dx = 0$. When we solve, we get the differential equation we want $dy/dx = x/y$.

CHAPTER TWELVE

Solutions for Section 12.1

Exercises

1. The gravitational force on a 100 kg object which is $7,000,000$ meters from the center of the earth (or about 600 km above the earth's surface) is about 820 newtons.

5. The amount of money spent on beef equals the product of the unit price p and the quantity C of beef consumed:

$$M = pC = pf(I, p).$$

Thus, we multiply each entry in Table 12.1 on page 561 of the text by the price at the top of the column. This yields Table 12.1.

Table 12.1 *Amount of money spent on beef ($/household/week)*

		Price			
		3.00	3.50	4.00	4.50
	20	7.95	9.07	10.04	10.94
	40	12.42	14.18	15.76	17.46
Income	60	15.33	17.50	19.88	21.78
	80	16.05	18.52	20.76	22.82
	100	17.37	20.20	22.40	24.89

9. The distance of a point $P = (x, y, z)$ from the yz-plane is $|x|$, from the xz-plane is $|y|$, and from the xy-plane is $|z|$. So A is closest to the yz-plane, since it has the smallest x-coordinate in absolute value. B lies on the xz-plane, since its y-coordinate is 0. C is farthest from the xy-plane, since it has the largest z-coordinate in absolute value.

13. The distance formula: $d = \sqrt{(x_2 - x_1)^2 + (y_2 - y_1)^2 + (z_2 - z_1)^2}$ gives us the distance between any pair of points (x_1, y_1, z_1) and (x_2, y_2, z_2). Thus, we find

$$\text{Distance from } P_1 \text{ to } P_2 = 2\sqrt{2}$$
$$\text{Distance from } P_2 \text{ to } P_3 = \sqrt{6}$$
$$\text{Distance from } P_1 \text{ to } P_3 = \sqrt{10}$$

So P_2 and P_3 are closest to each other.

17. The graph is a plane parallel to the xz-plane, and passing through the point $(0, 1, 0)$. See Figure 12.1.

Figure 12.1

Problems

21. Each entry is the square of the y coordinate, so a possible formula is

$$f(x, y) = y^2.$$

25. (a) Holding x fixed at 4 means that we are considering an injection of 4 mg of the drug; letting t vary means we are watching the effect of this dose as time passes. Thus the function $f(4, t)$ describes the concentration of the drug in the blood resulting from a 4 mg injection as a function of time. Figure 12.2 shows the graph of $f(4, t) = te^{-t}$. Notice that the concentration in the blood from this dose is at a maximum at 1 hour after injection, and that the concentration in the blood eventually approaches zero.

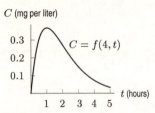

Figure 12.2: The function $f(4, t)$ shows the concentration in the blood resulting from a 4 mg injection

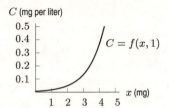

Figure 12.3: The function $f(x, 1)$ shows the concentration in the blood 1 hour after the injection

(b) Holding t fixed at 1 means that we are focusing on the blood 1 hour after the injection; letting x vary means we are considering the effect of different doses at that instant. Thus, the function $f(x, 1)$ gives the concentration of the drug in the blood 1 hour after injection as a function of the amount injected. Figure 12.3 shows the graph of $f(x, 1) = e^{-(5-x)} = e^{x-5}$. Notice that $f(x, 1)$ is an increasing function of x. This makes sense: If we administer more of the drug, the concentration in the bloodstream is higher.

29. By drawing the top four corners, we find that the length of the edge of the cube is 5. See Figure 12.4. We also notice that the edges of the cube are parallel to the coordinate axis. So the x-coordinate of the the center equals

$$-1 + \frac{5}{2} = 1.5.$$

The y-coordinate of the center equals

$$-2 + \frac{5}{2} = 0.5.$$

The z-coordinate of the center equals

$$2 - \frac{5}{2} = -0.5.$$

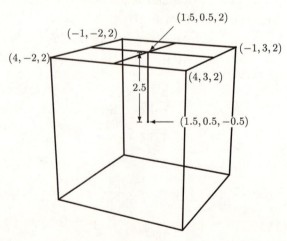

Figure 12.4

33. The distance from the y-axis is $d = \sqrt{x^2 + z^2}$, so f is ruled out because it depends on y. Also, g takes different values at points the same distance from the y axis, for example, $g(1, 0, 0) = 0$ but $g(1/\sqrt{2}, 0, 1/\sqrt{2}) = 1/4$. So g is ruled out. On the other hand, $h(x, y, z) = 1/\sqrt{d^2 + b^2}$, so (since b is a constant), h is a function of d alone.

Solutions for Section 12.2

Exercises

1. The graph is a horizontal plane 3 units above the xy-plane. See Figure 12.5.

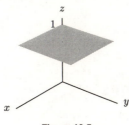

Figure 12.5

5. The graph is a plane with x-intercept 6, and y-intercept 3, and z-intercept 4. See Figure 12.6.

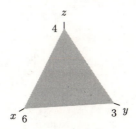

Figure 12.6

9. (a) The value of z only depends on the distance from the point (x, y) to the origin. Therefore the graph has a circular symmetry around the z-axis. There are two such graphs among those depicted in the figure in the text: I and V. The one corresponding to $z = \frac{1}{x^2+y^2}$ is I since the function blows up as (x, y) gets close to $(0, 0)$.

(b) For similar reasons as in part (a), the graph is circularly symmetric about the z-axis, hence the corresponding one must be V.

(c) The graph has to be a plane, hence IV.

(d) The function is independent of x, hence the corresponding graph can only be II. Notice that the cross-sections of this graph parallel to the yz-plane are parabolas, which is a confirmation of the result.

(e) The graph of this function is depicted in III. The picture shows the cross-sections parallel to the zx-plane, which have the shape of the cubic curves $z = x^3 -$ constant.

Problems

13. (a) This is a bowl; z increases as the distance from the origin increases, from a minimum of 0 at $x = y = 0$.

(b) Neither. This is an upside-down bowl. This function will decrease from 1, at $x = y = 0$, to arbitrarily large negative values as x and y increase due to the negative squared terms of x and y. It will look like the bowl in part (a) except flipped over and raised up slightly.

(c) This is a plate. Solving the equation for z gives $z = 1 - x - y$ which describes a plane whose x and y slopes are -1. It is perfectly flat, but not horizontal.

(d) Within its domain, this function is a bowl. It is undefined at points at which $x^2 + y^2 > 5$, but within those limits it describes the bottom half of a sphere of radius $\sqrt{5}$ centered at the origin.

(e) This function is a plate. It is perfectly flat and horizontal.

17. (a) Cross-sections with x fixed at $x = b$ are in Figure 12.7.

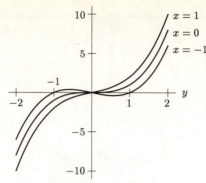

Figure 12.7: Cross-section
$f(a, y) = y^3 + ay$, with $a = -1, 0, 1$

(b) Cross-section with y fixed at $y = 6$ are in Figure 12.8.

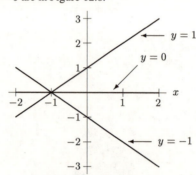

Figure 12.8: Cross-section
$f(x, b) = b^3 + bx$, with $b = -1, 0, 1$

21. (a) The plane $y = 1$ intersects the graph in the parabola $z = (x^2 + 1)\sin(1) + x = x^2 \sin(1) + x + \sin(1)$. Since $\sin(1)$ is a constant, $z = x^2 \sin(1) + x + \sin(1)$ is a quadratic function whose graph is a parabola.

Any plane of the form $y = a$ will do as long as a is not a multiple of π.

(b) The plane $y = \pi$ intersects the graph in the straight line $z = \pi^2 x$. (Since $\sin \pi = 0$, the equation becomes linear, $z = \pi^2 x$ if $y = \pi$.)

(c) The plane $x = 0$ intersects the graph in the curve $z = \sin y$.

Solutions for Section 12.3

Exercises

1. The contour where $f(x, y) = x + y = c$, or $y = -x + c$, is the graph of the straight line with slope -1 as shown in Figure 12.9. Note that we have plotted the contours for $c = -3, -2, -1, 0, 1, 2, 3$.

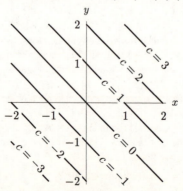

Figure 12.9

5. The contour where $f(x, y) = -x^2 - y^2 + 1 = c$, where $c \leq 1$, is the graph of the circle centered at $(0, 0)$, with radius $\sqrt{1 - c}$ as shown in Figure 12.10. Note that we have plotted the contours for $c = -3, -2, -1, 0, 1$. The contours become more closely packed as we move further from the origin.

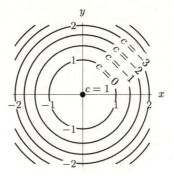

Figure 12.10

9. The contour where $f(x, y) = \cos(\sqrt{x^2 + y^2}) = c$, where $-1 \leq c \leq 1$, is a set of circles centered at $(0, 0)$, with radius $\cos^{-1} c + 2k\pi$ with $k = 0, 1, 2, ..$ and $-\cos^{-1} c + 2k\pi$, with $k = 1, 2, 3, ...$ as shown in Figure 12.11. Note that we have plotted contours for $c = 0, 0.2, 0.4, 0.6, 0.8, 1$.

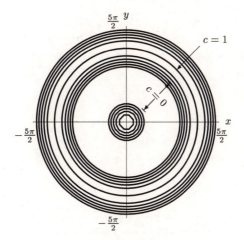

Figure 12.11

13. See Figure 12.12.

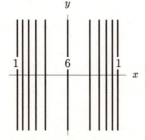

Figure 12.12

Problems

17. **(a)** The point representing 13% and $6000 on the graph lies between the 120 and 140 contours. We estimate the monthly payment to be about $137.

    (b) Since the interest rate has dropped, we will be able to borrow more money and still make a monthly payment of $137. To find out how much we can afford to borrow, we find where the interest rate of 11% intersects the $137 contour and read off the loan amount to which these values correspond. Since the $137 contour is not shown, we estimate its position from the $120 and $140 contours. We find that we can borrow an amount of money that is more than $6000 but less than $6500. So we can borrow about $250 more without increasing the monthly payment.

    (c) The entries in the table will be the amount of loan at which each interest rate intersects the 137 contour. Using the $137 contour from (b) we make table 12.2.

Table 12.2   *Amount borrowed at a monthly payment of $137.*

Interest Rate (%)	0	1	2	3	4	5	6	7
Loan Amount ($)	8200	8000	7800	7600	7400	7200	7000	6800
Interest rate (%)	8	9	10	11	12	13	14	15
Loan Amount ($)	6650	6500	6350	6250	6100	6000	5900	5800

21. **(a)** The profit is given by the following:

$$\pi = \text{Revenue from } q_1 + \text{Revenue from } q_2 - \text{Cost}.$$

Measuring π in thousands, we obtain:

$$\pi = 3q_1 + 12q_2 - 4.$$

(b) A contour diagram of π follows. Note that the units of π are in thousands.

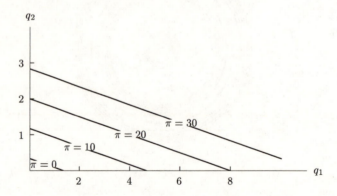

25. Suppose P_0 is the production given by L_0 and K_0, so that

$$P_0 = f(L_0, K_0) = cL_0^\alpha K_0^\beta.$$

We want to know what happens to production if L_0 is increased to $2L_0$ and K_0 is increased to $2K_0$:

$$\begin{aligned}
P &= f(2L_0, 2K_0) \\
&= c(2L_0)^\alpha (2K_0)^\beta \\
&= c2^\alpha L_0^\alpha 2^\beta K_0^\beta \\
&= 2^{\alpha+\beta} cL_0^\alpha K_0^\beta \\
&= 2^{\alpha+\beta} P_0.
\end{aligned}$$

Thus, doubling L and K has the effect of multiplying P by $2^{\alpha+\beta}$. Notice that if $\alpha+\beta > 1$, then $2^{\alpha+\beta} > 2$, if $\alpha+\beta = 1$, then $2^{\alpha+\beta} = 2$, and if $\alpha+\beta < 1$, then $2^{\alpha+\beta} < 2$. Thus, $\alpha+\beta > 1$ gives increasing returns to scale, $\alpha+\beta = 1$ gives constant returns to scale, and $\alpha+\beta < 1$ gives decreasing returns to scale.

29. (a) See Figure 12.13.

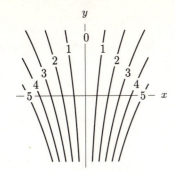

Figure 12.13

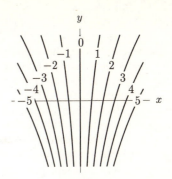

Figure 12.14

(b) See Figure 12.14.

33. (a)

(i)

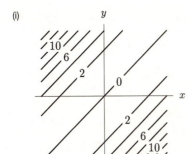

(ii)

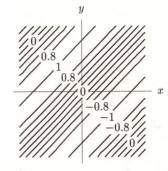

(b) The function $f(x, y) = g(y - x)$ is constant on lines $y - x = k$. Thus all lines parallel to $y = x$ are level curves of f.

Solutions for Section 12.4

Exercises

1. A table of values is linear if the rows are all linear and have the same slope and the columns are all linear and have the same slope. We see that the table might represent a linear function since the slope in each row is 3 and the slope in each column is -4.

5. (a) Yes.
  (b) The coefficient of m is 15 dollars per month. It represents the monthly charge to use this service. The coefficient of t is 0.05 dollars per minute. Each minute the customer is on-line costs 5 cents.
  (c) The intercept represents the base charge. It costs \$35 just to get hooked up to this service.
  (d) We have $f(3, 800) = 120$. A customer who uses this service for three months and is on-line for a total of 800 minutes is charged \$120.

9. A contour diagram is linear if the contours are parallel straight lines, equally spaced for equally spaced values of z. This contour diagram could represent a linear function.

13. In the diagram the contours correspond to values of the function that are 15 units apart, i.e., there are contours for $-90, -75, -60$, etc. An increase of 3 units in the y direction moves you from one contour to the next and changes the function by -15, so the y slope is $-15/3 = -5$. Similarly, an increase of 6 in the x direction crosses two contour lines and changes the function by 30; so the x slope is $30/6 = 5$. Hence $f(x, y) = c + 5x - 5y$. We see from the diagram that $f(8, 4) = -75$. Solving for c gives $c = -95$. Therefore the function is $f(x, y) = -95 + 5x - 5y$

Problems

17. Let the equation of the plane be

$$z = c + mx + ny$$

Since we know the points: $(4, 0, 0)$, $(0, 3, 0)$, and $(0, 0, 2)$ are all on the plane, we know that they satisfy the same equation. We can use these values of (x, y, z) to find c, m, and n. Putting these points into the equation we get:

$$0 = c + m \cdot 4 + n \cdot 0 \quad \text{so } c = -4m$$

$$0 = c + m \cdot 0 + n \cdot 3 \quad \text{so } c = -3n$$

$$2 = c + m \cdot 0 + n \cdot 0 \quad \text{so } c = 2$$

Because we have a value for c, we can solve for m and n to get

$$c = 2, m = -\frac{1}{2}, n = -\frac{2}{3}.$$

So the linear function is

$$f(x, y) = 2 - \frac{1}{2}x - \frac{2}{3}y.$$

21. The time in minutes to go 10 miles at a speed of s mph is $(10/s)(60) = 600/s$. Thus the 120 lb person going 10 mph uses $(7.4)(600/10) = 444$ calories, and the 180 lb person going 8 mph uses $(7.0)(600/8) = 525$ calories. The 120 lb person burns $444/120 = 3.7$ calories per pound for the trip, while the 180 lb person burns $525/180 = 2.9$ calories per pound for the trip.

25.

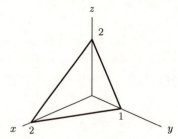

29. **(a)** Expenditure, E, is given by the equation:

$$E = (\text{price of raw material 1})m_1 + (\text{price of raw material 2})m_2 + C$$

where C denotes all the other expenses (assumed to be constant). Since the prices of the raw materials are constant, but m_1 and m_2 are variables, we have a linear function.
(b) Revenue, R, is given by the equation:

$$R = (p_1)q_1 + (p_2)q_2.$$

Since p_1 and p_2 are constant, while q_1 and q_2 are variables, we again have a linear function.
(c) Revenue is again given by the equation,

$$R = (p_1)q_1 + (p_2)q_2.$$

Since p_2 and q_2 are now constant, the term $(p_2)q_2$ is also constant. However, since p_1 and q_1 are variables, the $(p_1)q_1$ term means that the function is not linear.

Solutions for Section 12.5

Exercises

1. **(a)** Observe that setting $f(x, y, z) = c$ gives a cylinder about the x-axis, with radius $\sqrt{c}$. These surfaces are in graph (I).
(b) By the same reasoning the level curves for $h(x, y, z)$ are cylinders about the y-axis, so they are represented in graph (II).

5. The bottom half of the ellipsoid is represented by

$$z = f(x, y) = -\sqrt{2(1 - x^2 - y^2)}$$

$$g(x, y, z) = x^2 + y^2 + \frac{z^2}{2} = 1.$$

Other answers are possible

9. No, because $z = \sqrt{x^2 + 3y^2}$ and $z = -\sqrt{x^2 + 3y^2}$, so some z-values correspond to two points on the surface.

13. If we solve for z, we get $z = (1 - x^2 - y)^2$, so the level surface is the graph of $f(x, y) = (1 - x^2 - y)^2$.

17. An elliptic paraboloid.

Problems

21. In the xz-plane, the equation $x^2/4 + z^2 = 1$ is an ellipse, with widest points at $x = \pm 2$ on the x-axis and crossing the z-axis at $z = \pm 1$. Since the equation has no y term, the level surface is a cylinder of elliptical cross-section, centered along the y-axis.

25. The level surfaces are the graphs of $\sin(x + y + z) = k$ for constant k (with $-1 \leq k \leq 1$). This means $x + y + z = \sin^{-1}(k) + 2\pi n$, or $\pi - \sin^{-1}(k) + 2n\pi$ for all integers n. Therefore for each value of k, with $-1 \leq k \leq 1$, we get an infinite family of parallel planes. So the level surfaces are families of parallel planes.

29. Starting with the equation $z = \sqrt{x^2 + y^2}$, we flip the cone and shift it up one, yielding $z = 1 - \sqrt{x^2 + y^2}$. This is a cone with vertex at $(0, 0, 1)$ that intersects the xy-plane in a circle of radius 1. Interchanging the variables, we see that $y = 1 - \sqrt{x^2 + z^2}$ is an equation whose graph includes the desired cone C. Finally, we express this equation as a level surface $g(x, y, z) = 1 - \sqrt{x^2 + z^2} - y = 0$.

Solutions for Section 12.6

Exercises

1. No, $1/(x^2 + y^2)$ is not defined at the origin, so is not continuous at all points in the square $-1 \leq x \leq 1, -1 \leq y \leq 1$.

5. The function $\tan(\theta)$ is undefined when $\theta = \pi/2 \approx 1.57$. Since there are points in the square $-2 \leq x \leq 2, -2 \leq y \leq 2$ with $x \cdot y = \pi/2$ (e.g. $x = 1, y = \pi/2$) the function $\tan(xy)$ is not defined inside the square, hence not continuous.

9. Since f doesn't depend on y we have:

$$\lim_{(x,y) \to (0,0)} f(x, y) = \lim_{x \to 0} \frac{x}{x^2 + 1} = \frac{0}{0 + 1} = 0.$$

Problems

13. Points along the positive x-axis are of the form $(x, 0)$; at these points the function looks like $x/2x = 1/2$ everywhere (except at the origin, where it is undefined). On the other hand, along the y-axis, the function looks like $y^2/y = y$, which approaches 0 as we get closer to the origin. Since approaching the origin along two different paths yields numbers that aren't the same, the limit doesn't exist.

17. (a) We have $f(x, 0) = 0$ for all x and $f(0, y) = 0$ for all y, so these are both continuous (constant) functions of one variable.

  (b) The contour diagram suggests that the contours of f are lines through the origin. Providing it is not vertical, the equation of such a line is

$$y = mx.$$

To confirm that such lines are contours of f, we must show that f is constant along these lines. Substituting into the function, we get

$$f(x, y) = f(x, mx) = \frac{x(mx)}{x^2 + (mx)^2} = \frac{mx^2}{x^2 + m^2 x^2} = \frac{m}{1 + m^2} = \text{constant}.$$

Since $f(x, y)$ is constant along the line $y = mx$, such lines are contained in contours of f.

(c) We consider the limit of $f(x, y)$ as $(x, y) \to (0, 0)$ along the line $y = mx$. We can see that

$$\lim_{x \to 0} f(x, mx) = \frac{m}{1 + m^2}.$$

Therefore, if $m = 1$ we have

$$\lim_{\substack{(x,y) \to (0,0) \\ y=x}} f(x, y) = \frac{1}{2}$$

whereas if $m = 0$ we have

$$\lim_{\substack{(x,y) \to (0,0) \\ y=0}} f(x, y) = 0.$$

Thus, no matter how close we are to the origin, we can find points (x, y) where the value $f(x, y)$ is 1/2 and points (x, y) where the value $f(x, y)$ is 0. So the limit $\lim_{(x,y) \to (0,0)} f(x, y)$ does not exist. Thus, f is not continuous at $(0, 0)$, even though the one-variable functions $f(x, 0)$ and $f(0, y)$ are continuous at $(0, 0)$. See Figures 17 and 17

21. The function, f is continuous at all points (x, y) with $x \neq 3$. We analyze the continuity of f at the point $(3, a)$. We have:

$$\lim_{(x,y) \to (3,a), x<3} f(x, y) = \lim_{y \to a}(c + y) = c + a$$

$$\lim_{(x,y) \to (3,a), x>3} f(x, y) = \lim_{x>3, x\to 3}(5 - x) = 2.$$

We want to see if we can find one value of c such that $c + a = 2$ for all a. This would mean that $c = 2 - a$, but then c would be dependent on a. Therefore, we cannot make the function continuous everywhere.

Solutions for Chapter 12 Review

Exercises

1. These conditions describe a line parallel to the z-axis which passes through the xy-plane at $(2, 1, 0)$.

5. Contours are lines of the form $3x - 5y + 1 = c$ as shown in Figure 12.15. Note that for the regions of x and y given, the c values range from $-12 < c < 12$ and are equally spaced by 4.

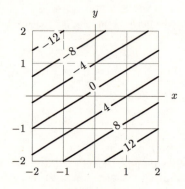

Figure 12.15

9. The function h decreases as y increases: each increase of y by 2 takes you down one contour and hence changes the function by 2, so the slope in the y direction is -1. The slope in the x direction is 2, so the formula is $h(x, y) = c + 2x - y$. From the diagram we see that $h(0, 0) = 4$, so $c = 4$. Therefore, the formula for this linear function is $h(x, y) = 4 + 2x - y$.

13. The paraboloid is $z = x^2 + y^2 + 5$, so it is represented by

$$z = f(x, y) = x^2 + y^2 + 5$$

and

$$g(x, y, z) = x^2 + y^2 + 5 - z = 0.$$

Other answers are possible.

Problems

17. Might be true. The function $z = x^2 - y^2 + 1$ has this property. The level curve $z = 1$ is the lines $y = x$ and $y = -x$.

21. We complete the square

$$x^2 + 4x + y^2 - 6y + z^2 + 12z = 0$$
$$x^2 + 4x + 4 + y^2 - 6y + 9 + z^2 + 12z + 36 = 4 + 9 + 36$$
$$(x + 2)^2 + (y - 3)^2 + (z + 6)^2 = 49$$

The center is $(-2, 3, -6)$ and the radius is 7.

25. The level surfaces have equation $\cos(x + y + z) = c$. For each value of c between -1 and 1, the level surface is an infinite family of planes parallel to $x + y + z = \arccos(c)$. For example, the level surface $\cos(x + y + z) = 0$ is the family of planes

$$x + y + z = \frac{\pi}{2} \pm 2n\pi, \quad n = 0, 1, 2, \ldots.$$

29. One possible equation: $z = (x - y)^2$. See Figure 12.16.

Figure 12.16

33. (a) For $g(x, t) = \cos 2t \sin x$, our snapshots for fixed values of t are still one arch of the sine curve. The amplitudes, which are governed by the $\cos 2t$ factor, now change twice as fast as before. That is, the string is vibrating twice as fast.

(b) For $y = h(x, t) = \cos t \sin 2x$, the vibration of the string is more complicated. If we hold t fixed at any value, the snapshot now shows one full period, i.e. one crest and one trough, of the sine curve. The magnitude of the sine curve is time dependent, given by $\cos t$. Now the center of the string, $x = \pi/2$, remains stationary just like the end points. This is a vibrating string with the center held fixed, as shown in Figure 12.17.

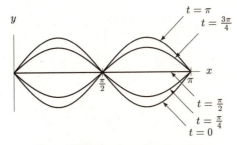

Figure 12.17: Another vibrating string:
$$y = h(x, t) = \cos t \sin 2x$$

CHECK YOUR UNDERSTANDING

1. True. Since each choice of x and y determines a unique value for $f(x, y)$, choosing $x = 10$ yields a unique value of $f(10, y)$ for any choice of y.

5. False. Fixing $w = k$ gives the one-variable function $g(v) = e^v/k$, which is an increasing exponential function if $k > 0$, but is decreasing if $k < 0$.

9. True. For example, consider the weekly beef consumption C of a household as a function of total income I and the cost of beef per pound p. It is possible that consumption increases as income increases (for fixed p) and consumption decreases as the price of beef increases (for fixed I).

13. False. The plane $z = 2$ is parallel to the xy-plane.

17. True. The origin is the closest point in the yz-plane to the point $(3, 0, 0)$, and its distance to $(3, 0, 0)$ is 3.

21. True. The cross-section with $y = 1$ is the line $z = x + 1$.

25. True. The intersection, where $f(x, y) = g(x, y)$, is given by $x^2 + y^2 = 1 - x^2 - y^2$, or $x^2 + y^2 = 1/2$. This is a circle of radius $1/\sqrt{2}$ parallel to the xy-plane at height $z = 1/2$.

29. False. For example, the y-axis intersects the graph of $f(x, y) = 1 - x^2 - y^2$ twice, at $y = \pm 1$.

33. False. The graph could be a hemisphere, a bowl-shape, or any surface formed by rotating a curve about a vertical line.

37. False. The fact that the $f = 10$ and $g = 10$ contours are identical only says that one horizontal slice through each graph is the same, but does not imply that the entire graphs are the same. A counterexample is given by $f(x, y) = x^2 + y^2$ and $g(x, y) = 20 - x^2 - y^2$.

41. False. Any two-variable function that is missing one variable (e.g. $f(x, y) = x^2$) will have parallel lines for contours. Linear functions have the additional property of *evenly-spaced* parallel lines for contours.

45. True. Since the graph of a linear function is a plane, any vertical slice parallel to the yz-plane will yield a line.

49. False. All of the columns have to have the same slope, as do the rows, but the row slopes can differ from the column slopes.

53. True. The graph of $f(x, y)$ is the set of all points (x, y, z) satisfying $z = f(x, y)$. If we define the three-variable function g by $g(x, y, z) = f(x, y) - z$, then the level surface $g = 0$ is exactly the same as the graph of $f(x, y)$.

57. False. The level surface $g = 0$ of the function $g(x, y, z) = x^2 + y^2 + z^2$ consists of only the origin.

CHAPTER THIRTEEN

Solutions for Section 13.1

Exercises

1. $\|\vec{z}\| = \sqrt{(1)^2 + (-3)^2 + (-1)^2} = \sqrt{1 + 9 + 1} = \sqrt{11}$.

5. $\|\vec{y}\| = \sqrt{(4)^2 + (-7)^2} = \sqrt{16 + 49} = \sqrt{65}$.

9. $-4\vec{i} + 8\vec{j} - 0.5\vec{i} + 0.5\vec{k} = -4.5\vec{i} + 8\vec{j} + 0.5\vec{k}$

13. $\|\vec{v}\| = \sqrt{1.2^2 + (-3.6)^2 + 4.1^2} = \sqrt{31.21} \approx 5.6$.

17. **(a)** See Figure 13.1.
    (b) $\|\vec{v}\| = \sqrt{5^2 + 7^2} = \sqrt{74} = 8.602$.
    (c) We see in Figure 13.2 that $\tan\theta = \frac{7}{5}$ and so $\theta = 54.46°$.

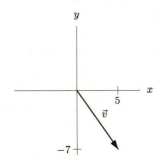

Figure 13.1

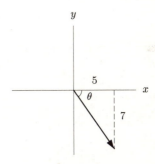

Figure 13.2

Problems

21. **(a)** True, since vectors $\vec{c}$ and $\vec{f}$ point in the same direction and have the same length.
    (b) False, since vectors $\vec{a}$ and $\vec{d}$ point in opposite directions. We have $\vec{a} = -\vec{d}$.
    (c) False, since $-\vec{b}$ points in the opposite direction to $\vec{b}$, the vectors $-\vec{b}$ and $\vec{a}$ are perpendicular.
    (d) True. The vector $\vec{f}$ can be "moved" to point directly up the z-axis.
    (e) True. We move in the positive x-direction following vector $\vec{a}$ and then in the positive y-direction following vector $-\vec{b}$. The resulting sum is the vector $\vec{e}$.
    (f) False, vector $\vec{d}$ is the negative of the vector $\vec{g} - \vec{c}$. It is true that $\vec{d} = \vec{c} - \vec{g}$.

25.

$$\text{Displacement} = \text{Cat's coordinates} - \text{Bottom of the tree's coordinates}$$
$$= (1 - 2)\vec{i} + (4 - 4)\vec{j} + (0 - 0)\vec{k} = -\vec{i}.$$

29. Since the component of $\vec{v}$ in the $\vec{i}$-direction is 3, we have $\vec{v} = 3\vec{i} + b\vec{j}$ for some b. Since $\|\vec{v}\| = 5$, we have $\sqrt{3^2 + b^2} = 5$, so $b = 4$ or $b = -4$. There are two vectors satisfying the properties given: $\vec{v} = 3\vec{i} + 4\vec{j}$ and $\vec{v} = 3\vec{i} - 4\vec{j}$.

33. In Figure 13.3 let O be the origin, points A, B, and C be the vertices of the triangle, point D be the midpoint of $\overline{BC}$, and Q be the point in the line segment $\overline{DA}$ that is $\frac{1}{3}|DA|$ away from D.

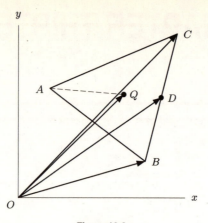

Figure 13.3

From Figure 13.3 we see that

$$\overrightarrow{OQ} = \overrightarrow{OD} + \overrightarrow{DQ} = \overrightarrow{OD} + \frac{1}{3}\overrightarrow{DA}$$

$$= \overrightarrow{OD} + \frac{1}{3}(\overrightarrow{OA} - \overrightarrow{OD})$$

$$= \overrightarrow{OD} + \frac{1}{3}\overrightarrow{OA} - \frac{1}{3}\overrightarrow{OD}$$

$$= \frac{1}{3}\overrightarrow{OA} + \frac{2}{3}\overrightarrow{OD}.$$

Because the diagonals of a parallelogram meet at their midpoint, and $2\overrightarrow{OD}$ is a diagonal of the parallelogram formed by $\overrightarrow{OB}$ and $\overrightarrow{OC}$, we have:

$$\overrightarrow{OD} = \frac{1}{2}(\overrightarrow{OB} + \overrightarrow{OC}),$$

so we can write:

$$\overrightarrow{OQ} = \frac{1}{3}\overrightarrow{OA} + \frac{2}{3}\left(\frac{1}{2}\right)(\overrightarrow{OB} + \overrightarrow{OC}) = \frac{1}{3}(\overrightarrow{OA} + \overrightarrow{OB} + \overrightarrow{OC}).$$

Thus a vector from the origin to a point $\frac{1}{3}$ of the way along median AD from D, the midpoint, is given by $\frac{1}{3}(\overrightarrow{OA} + \overrightarrow{OB} + \overrightarrow{OC})$.

In a similar manner we can show that the vector from the origin to the point $\frac{1}{3}$ of the way along any median from the midpoint of the side it bisects is also $\frac{1}{3}(\overrightarrow{OA} + \overrightarrow{OB} + \overrightarrow{OC})$. See Figure 13.4 and 13.5.

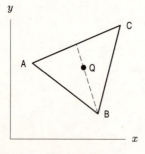

Figure 13.4

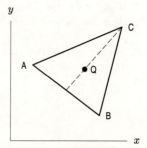

Figure 13.5

Thus the medians of a triangle intersect at a point $\frac{1}{3}$ of the way along each median from the side that each bisects.

Solutions for Section 13.2

Exercises

1. Scalar

5. Writing $\vec{P} = (P_1, P_2, \cdots, P_{50})$ where P_i is the population of the i-th state, shows that $\vec{P}$ can be thought of as a vector with 50 components.

9. We need to calculate the length of each vector.

$$\|21\vec{i} + 35\vec{j}\| = \sqrt{21^2 + 35^2} = \sqrt{1666} \approx 40.8,$$
$$\|40\vec{i}\| = \sqrt{40^2} = 40.$$

So the first car is faster.

Problems

13. (a) The velocity vector for the boat is $\vec{b} = 25\vec{i}$ and the velocity vector for the current is

$$\vec{c} = -10\cos(45°)\vec{i} - 10\sin(45°)\vec{j} = -7.07\vec{i} - 7.07\vec{j}.$$

The actual velocity of the boat is

$$\vec{b} + \vec{c} = 17.93\vec{i} - 7.07\vec{j}.$$

(b) $\|\vec{b} + \vec{c}\| = 19.27$ km/hr.

(c) We see in Figure 13.6 that $\tan\theta = \dfrac{7.07}{17.93}$, so $\theta = 21.52°$ south of east.

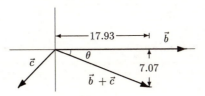

Figure 13.6

17. Let the x-axis point east and the y-axis point north. Since the wind is blowing from the northeast at a speed of 50 km/hr, the velocity of the wind is

$$\vec{w} = -50\cos 45°\vec{i} - 50\sin 45°\vec{j} \approx -35.4\vec{i} - 35.4\vec{j}.$$

Let $\vec{a}$ be the velocity of the airplane, relative to the air, and let ϕ be the angle from the x-axis to $\vec{a}$; since $\|\vec{a}\| = 600$ km/hr, we have $\vec{a} = 600\cos\phi\vec{i} + 600\sin\phi\vec{j}$. (See Figure 13.7.)

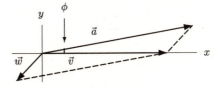

Figure 13.7

Now the resultant velocity, $\vec{v}$, is given by

$$\vec{v} = \vec{a} + \vec{w} = (600\cos\phi\vec{i} + 600\sin\phi\vec{j}) + (-35.4\vec{i} - 35.4\vec{j})$$
$$= (600\cos\phi - 35.4)\vec{i} + (600\sin\phi - 35.4)\vec{j}.$$

Since the airplane is to fly due east, i.e., in the x direction, then the y-component of the velocity must be 0, so we must have

$$600 \sin \phi - 35.4 = 0$$
$$\sin \phi = \frac{35.4}{600}.$$

Thus $\phi = \arcsin(35.4/600) \approx 3.4°$.

21. We want the total force on the object to be zero. We must choose the third force $\vec{F}_3$ so that $\vec{F}_1 + \vec{F}_2 + \vec{F}_3 = 0$. Since $\vec{F}_1 + \vec{F}_2 = 11\vec{i} - 4\vec{j}$, we need $\vec{F}_3 = -11\vec{i} + 4\vec{j}$.

25.

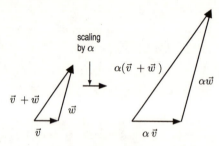

Figure 13.8

The effect of scaling the left-hand picture in Figure 13.8 is to stretch each vector by a factor of α (shown with $\alpha > 1$). Since, after scaling up, the three vectors $\alpha\,\vec{v}$, $\alpha\vec{w}$, and $\alpha(\vec{v} + \vec{w})$ form a similar triangle, we know that $\alpha(\vec{v} + \vec{w})$ is the sum of the other two: that is

$$\alpha(\vec{v} + \vec{w}) = \alpha\,\vec{v} + \alpha\vec{w}.$$

29. According to the definition of scalar multiplication, $1 \cdot \vec{v}$ has the same direction and magnitude as $\vec{v}$, so it is the same as $\vec{v}$.

Solutions for Section 13.3

Exercises

1. $\vec{c} \cdot \vec{y} = (\vec{i} + 6\vec{j}) \cdot (4\vec{i} - 7\vec{j}) = (1)(4) + (6)(-7) = 4 - 42 = -38$.

5. Since $\vec{a} \cdot \vec{y}$ and $\vec{c} \cdot \vec{z}$ are both scalars, the answer to this equation is the product of two numbers and therefore a number. We have

$$\vec{a} \cdot \vec{y} = (2\vec{j} + \vec{k}) \cdot (4\vec{i} - 7\vec{j}) = 0(4) + 2(-7) + 1(0) = -14$$
$$\vec{c} \cdot \vec{z} = (\vec{i} + 6\vec{j}) \cdot (\vec{i} - 3\vec{j} - \vec{k}) = 1(1) + 6(-3) + 0(-1) = -17$$

Thus,

$$(\vec{a} \cdot \vec{y})(\vec{c} \cdot \vec{z}) = 238$$

9. Writing the equation in the form

$$3x + 4y - z = 7$$

shows that a normal vector is

$$\vec{n} = 3\vec{i} + 4\vec{j} - \vec{k}$$

Problems

13. Since $3\vec{i} + \sqrt{3}\vec{j} = \sqrt{3}(\sqrt{3}\vec{i} + \vec{j})$, we know that $3\vec{i} + \sqrt{3}\vec{j}$ and $\sqrt{3}\vec{i} + \vec{j}$ are scalar multiples of one another, and therefore parallel.

Since $(\sqrt{3}\vec{i} + \vec{j}) \cdot (\vec{i} - \sqrt{3}\vec{j}) = \sqrt{3} - \sqrt{3} = 0$, we know that $\sqrt{3}\vec{i} + \vec{j}$ and $\vec{i} - \sqrt{3}\vec{j}$ are perpendicular.

Since $3\vec{i} + \sqrt{3}\vec{j}$ and $\sqrt{3}\vec{i} + \vec{j}$ are parallel, $3\vec{i} + \sqrt{3}\vec{j}$ and $\vec{i} - \sqrt{3}\vec{j}$ are perpendicular, too.

17. Since the plane is normal to the vector $5\vec{i} + \vec{j} - 2\vec{k}$ and passes through the point $(0, 1, -1)$, an equation for the plane is

$$5x + y - 2z = 5 \cdot 0 + 1 \cdot 1 + (-2) \cdot (-1) = 3$$
$$5x + y - 2z = 3.$$

21. (a) The plane can be written as $5x - 2y - z + 7 = 0$, so the vector $5\vec{i} - 2\vec{j} - \vec{k}$ is normal to the plane. The vector $\lambda\vec{i} + \vec{j} + 0.5\vec{k}$ is parallel to $5\vec{i} - 2\vec{j} - \vec{k}$ if one is a scalar multiple of the other. This occurs if the coeffients are in proportion:

$$\frac{\lambda}{5} = \frac{1}{-2} = \frac{0.5}{-1}.$$

Solving gives $\lambda = -2.5$.

(b) Substituting $x = a + 1$, $y = a$, $z = a - 1$ into the equation of the plane gives

$$a - 1 = 5(a + 1) - 2a + 7$$
$$a - 1 = 5a + 5 - 2a + 7$$
$$-13 = 2a$$
$$a = -6.5.$$

25. Let

$$\vec{a} = \vec{a}_{\text{parallel}} + \vec{a}_{\text{perp}}$$

where $\vec{a}_{\text{parallel}}$ is parallel to $\vec{d}$, and $\vec{a}_{\text{perp}}$ is perpendicular to $\vec{d}$. Then $\vec{a}_{\text{parallel}}$ is the projection of $\vec{a}$ in the direction of $\vec{d}$:

$$\vec{a}_{\text{parallel}} = \left(\vec{a} \cdot \frac{\vec{d}}{\|\vec{d}\|} \right) \frac{\vec{d}}{\|\vec{d}\|}$$
$$= \left((3\vec{i} + 2\vec{j} - 6\vec{k}) \cdot \frac{(2\vec{i} - 4\vec{j} + \vec{k})}{\sqrt{2^2 + 4^2 + 1^2}} \right) \frac{(2\vec{i} - 4\vec{j} + \vec{k})}{\sqrt{2^2 + 4^2 + 1^2}}$$
$$= -\frac{8}{21}(2\vec{i} - 4\vec{j} + \vec{k})$$
$$= -\frac{8}{21}\vec{d}$$

Since we now know $\vec{a}$ and $\vec{a}_{\text{parallel}}$, we can solve for $\vec{a}_{\text{perp}}$:

$$\vec{a}_{\text{perp}} = \vec{a} - \vec{a}_{\text{parallel}}$$
$$= (3\vec{i} + 2\vec{j} - 6\vec{k}) - \left(-\frac{8}{21} \right)(2\vec{i} - 4\vec{j} + \vec{k})$$
$$= \frac{79}{21}\vec{i} + \frac{10}{21}\vec{j} - \frac{118}{21}\vec{k}.$$

Thus we can now write $\vec{a}$ as the sum of two vectors, one parallel to $\vec{d}$, the other perpendicular to $\vec{d}$:

$$\vec{a} = -\frac{8}{21}\vec{d} + \left(\frac{79}{21}\vec{i} + \frac{10}{21}\vec{j} - \frac{118}{21}\vec{k} \right)$$

29. We have

$$\vec{p} \cdot \vec{q} = (1.00)(43) + (3.50)(57) + (4.00)(12) + (2.75)(78) + (5.00)(20) + (3.00)(35)$$
$$= 710 \text{ dollars.}$$

The vendor took in \$710 in from sales. The quantity $\vec{p} \cdot \vec{q}$ represents the total revenue earned.

33. (a) The geometric definition of the dot product says that

$$\vec{n} \cdot \overrightarrow{P_0 P} = \|\vec{n}\| \|\overrightarrow{P_0 P}\| \cos\theta,$$

where θ is the angle between $\vec{n}$ and $\overrightarrow{P_0 P}$ with $0 \le \theta \le \pi$. To say that the dot product $\vec{n} \cdot \overrightarrow{P_0 P}$ is positive means that the angle between $\vec{n}$ and $\overrightarrow{P_0 P}$ is between 0 and $\pi/2$, and strictly less than $\pi/2$. Hence $\vec{n}$ and $\overrightarrow{P_0 P}$ are both pointing to the same side of the plane. Thus, all the points satisfying $\vec{n} \cdot \overrightarrow{P_0 P} > 0$ are on the same side of the plane, the side which $\vec{n}$ points to. To say that the dot product is negative is to say that $\pi/2 < \theta \le \pi$, and this means that $\overrightarrow{P_0 P}$ and $\vec{n}$ are pointing to opposite sides of the plane. Thus, all points satisfying $\vec{n} \cdot \overrightarrow{P_0 P} < 0$ are on the side of the plane opposite to $\vec{n}$.

(b) Suppose the normal vector is $\vec{n} = a\vec{i} + b\vec{j} + c\vec{k}$, let $P_0 = (x_0, y_0, z_0)$ be a point in the plane and let $P = (x, y, z)$ be a variable point. Then $\overrightarrow{P_0 P} = (x - x_0)\vec{i} + (y - y_0)\vec{j} + (z - z_0)\vec{k}$. Then $\vec{n} \cdot \overrightarrow{P_0 P} > 0$ means

$$a(x - x_0) + b(y - y_0) + c(z - z_0) > 0$$

and $\vec{n} \cdot \overrightarrow{P_0 P} < 0$ means

$$a(x - x_0) + b(y - y_0) + c(z - z_0) < 0$$

If the equation of the plane is written $ax + by + cz = d$ (with $d = ax_0 + by_0 + cz_0$) then the inequalities become

$$ax + by + cz > d \quad \text{and} \quad ax + by + cz < d.$$

(c) We test each of the points $P = (-1, -1, 1)$, $Q = (-1, -1, -1)$ and $R = (1, 1, 1)$, using the coordinate version of the inequalites in part (b):

$$\begin{aligned}
P : \quad & 2 \cdot (-1) - 3 \cdot (-1) + 4 \cdot 1 = 5 > 4 \\
Q : \quad & 2 \cdot (-1) - 3 \cdot (-1) + 4 \cdot (-1) = -3 < 4 \\
R : \quad & 2 \cdot 1 - 3 \cdot 1 + 4 \cdot 1 = 3 < 4
\end{aligned}$$

Therefore Q and R are on the same side of the plane as each other; P is on the other side.

37. Since $\vec{u} \cdot \vec{w} = \vec{v} \cdot \vec{w}$, $(\vec{u} - \vec{v}) \cdot \vec{w} = 0$. This equality holds for any $\vec{w}$, so we can take $\vec{w} = \vec{u} - \vec{v}$. This gives

$$\|\vec{u} - \vec{v}\|^2 = (\vec{u} - \vec{v}) \cdot (\vec{u} - \vec{v}) = 0,$$

that is,

$$\|\vec{u} - \vec{v}\| = 0.$$

This implies $\vec{u} - \vec{v} = 0$, that is, $\vec{u} = \vec{v}$.

41. We substitute $\vec{u} = u_1\vec{i} + u_2\vec{j} + u_3\vec{k}$ and by the result of Problem 38, we expand as follows:

$$\begin{aligned}
(\vec{u} \cdot \vec{v})_{\text{geom}} &= (u_1\vec{i} + u_2\vec{j} + u_3\vec{k}) \cdot \vec{v} \\
&= (u_1\vec{i}) \cdot \vec{v} + (u_2\vec{j}) \cdot \vec{v} + (u_3\vec{k}) \cdot \vec{v}
\end{aligned}$$

where all the dot products are defined geometrically By the result of Problem 39 we can write

$$(\vec{u} \cdot \vec{v})_{\text{geom}} = u_1(\vec{i} \cdot \vec{v})_{\text{geom}} + u_2(\vec{j} \cdot \vec{v})_{\text{geom}} + u_3(\vec{k} \cdot \vec{v})_{\text{geom}}.$$

Now substitute $\vec{v} = v_1\vec{i} + v_2\vec{j} + v_3\vec{k}$ and expand, again using Problem 38 and the geometric definition of the dot product:

$$\begin{aligned}
(\vec{u} \cdot \vec{v})_{\text{geom}} = \; & u_1 \left(\vec{i} \cdot (v_1\vec{i} + v_2\vec{j} + v_3\vec{k})\right)_{\text{geom}} \\
& + u_2 \left(\vec{j} \cdot (v_1\vec{i} + v_2\vec{j} + v_3\vec{k})\right)_{\text{geom}} \\
& + u_3 \left(\vec{k} \cdot (v_1\vec{i} + v_2\vec{j} + v_3\vec{k})\right)_{\text{geom}} \\
= \; & u_1 v_1 (\vec{i} \cdot \vec{i})_{\text{geom}} + u_1 v_2 (\vec{i} \cdot \vec{j})_{\text{geom}} + u_1 v_3 (\vec{i} \cdot \vec{k})_{\text{geom}} \\
& + u_2 v_1 (\vec{i} \cdot \vec{i})_{\text{geom}} + u_2 v_2 (\vec{i} \cdot \vec{j})_{\text{geom}} + u_2 v_3 (\vec{i} \cdot \vec{k})_{\text{geom}} \\
& + u_3 v_1 (\vec{i} \cdot \vec{i})_{\text{geom}} + u_3 v_2 (\vec{i} \cdot \vec{j})_{\text{geom}} + u_3 v_3 (\vec{i} \cdot \vec{k})_{\text{geom}}
\end{aligned}$$

The geometric definition of the dot product shows that

$$\vec{i} \cdot \vec{i} = \|\vec{i}\| \, \|\vec{i}\| \cos 0 = 1$$
$$\vec{i} \cdot \vec{j} = \|\vec{i}\| \, \|\vec{j}\| \cos \frac{\pi}{2} = 0.$$

Similarly $\vec{j} \cdot \vec{j} = \vec{k} \cdot \vec{k} = 1$ and $\vec{i} \cdot \vec{k} = \vec{j} \cdot \vec{k} = 0$. Thus, the expression for $(\vec{u} \cdot \vec{v})_{\text{geom}}$ becomes

$$(\vec{u} \cdot \vec{v})_{\text{geom}} = u_1 v_1 (1) + u_1 v_2 (0) + u_1 v_3 (0)$$
$$+ u_2 v_1 (0) + u_2 v_2 (1) + u_2 v_3 (0)$$
$$+ u_3 v_1 (0) + u_3 v_2 (0) + u_3 v_3 (1)$$
$$= u_1 v_1 + u_2 v_2 + u_3 v_3.$$

Solutions for Section 13.4

Exercises

1. $\vec{v} \times \vec{w} = \vec{k} \times \vec{j} = -\vec{i}$ (remember $\vec{i}, \vec{j}, \vec{k}$ are unit vectors along the axes, and you must use the right hand rule.)

5. $\vec{v} = 2\vec{i} - 3\vec{j} + \vec{k}$, and $\vec{w} = \vec{i} + 2\vec{j} - \vec{k}$

$$\vec{v} \times \vec{w} = \begin{vmatrix} \vec{i} & \vec{j} & \vec{k} \\ 2 & -3 & 1 \\ 1 & 2 & -1 \end{vmatrix} = \vec{i} + 3\vec{j} + 7\vec{k}$$

9.

$$(\vec{i} + \vec{j}) \times (\vec{i} \times \vec{j}) = (\vec{i} + \vec{j}) \times \vec{k}$$
$$= (\vec{i} \times \vec{k}) + (\vec{j} \times \vec{k})$$
$$= -\vec{j} + \vec{i} = \vec{i} - \vec{j}.$$

Problems

13. Since $\vec{v} \times \vec{w}$ is perpendicular to both $\vec{v}$ and $\vec{w}$, we can conclude that $\vec{v} \times \vec{w}$ is parallel to the z-axis.

17. (a) If we let $\overrightarrow{PQ}$ in Figure 13.9 be the vector from point P to point Q and $\overrightarrow{PR}$ be the vector from P to R, then

$$\overrightarrow{PQ} = -\vec{i} + 2\vec{k}$$

$$\overrightarrow{PR} = 2\vec{i} - \vec{k},$$

then the area of the parallelogram determined by $\overrightarrow{PQ}$ and $\overrightarrow{PR}$ is:

$$\begin{aligned} \text{Area of} \\ \text{parallelogram} \end{aligned} = \|\overrightarrow{PQ} \times \overrightarrow{PR}\| = \left\| \begin{vmatrix} \vec{i} & \vec{j} & \vec{k} \\ -1 & 0 & 2 \\ 2 & 0 & -1 \end{vmatrix} \right\| = \|3\vec{j}\| = 3.$$

Thus, the area of the triangle PQR is

$$\left(\begin{array}{c} \text{Area of} \\ \text{triangle} \end{array} \right) = \frac{1}{2} \left(\begin{array}{c} \text{Area of} \\ \text{parallelogram} \end{array} \right) = \frac{3}{2} = 1.5.$$

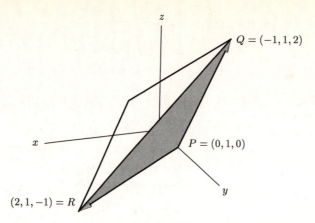

Figure 13.9

(b) Since $\vec{n} = \overrightarrow{PQ} \times \overrightarrow{PR}$ is perpendicular to the plane PQR, and from above, we have $\vec{n} = 3\vec{j}$, the equation of the plane has the form $3y = C$. At the point $(0, 1, 0)$ we get $3 = C$, therefore $3y = 3$, i.e., $y = 1$.

21. We use the same normal vector $\vec{n} = 4\vec{i} + 26\vec{j} + 14\vec{k}$ and the point $(4, 5, 6)$ to get $4(x-4) + 26(y-5) + 14(z-6) = 0$.

25. If $\lambda = 0$, then all three cross products are $\vec{0}$, since the cross product of the zero vector with any other vector is always 0.

If $\lambda > 0$, then $\lambda \vec{v}$ and $\vec{v}$ are in the same direction and $\vec{w}$ and $\lambda \vec{w}$ are in the same direction. Therefore the unit normal vector $\vec{n}$ is the same in all three cases. In addition, the angles between $\lambda \vec{v}$ and $\vec{w}$, and between $\vec{v}$ and $\vec{w}$, and between $\vec{v}$ and $\lambda \vec{w}$ are all θ. Thus,

$$(\lambda \vec{v}) \times \vec{w} = \|\lambda \vec{v}\| \|\vec{w}\| \sin \theta \vec{n}$$
$$= \lambda \|\vec{v}\| \|\vec{w}\| \sin \theta \vec{n}$$
$$= \lambda (\vec{v} \times \vec{w})$$
$$= \|\vec{v}\| \|\lambda \vec{w}\| \sin \theta \vec{n}$$
$$= \vec{v} \times (\lambda \vec{w})$$

If $\lambda < 0$, then $\lambda \vec{v}$ and $\vec{v}$ are in opposite directions, as are $\vec{w}$ and $\lambda \vec{w}$ in opposite directions. Therefore if $\vec{n}$ is the normal vector in the definition of $\vec{v} \times \vec{w}$, then the right-hand rule gives $-\vec{n}$ for $(\lambda \vec{v}) \times \vec{w}$ and $\vec{v} \times (\lambda \vec{w})$. In addition, if the angle between $\vec{v}$ and $\vec{w}$ is θ, then the angle between $\lambda \vec{v}$ and $\vec{w}$ and between $\vec{v}$ and $\lambda \vec{w}$ is $(\pi - \theta)$. Since if $\lambda < 0$, we have $|\lambda| = -\lambda$, so

$$(\lambda \vec{v}) \times \vec{w} = \|\lambda \vec{v}\| \|\vec{w}\| \sin(\pi - \theta)(-\vec{n})$$
$$= |\lambda| \|\vec{v}\| \|\vec{w}\| \sin(\pi - \theta)(-\vec{n})$$
$$= -\lambda \|\vec{v}\| \|\vec{w}\| \sin \theta (-\vec{n})$$
$$= \lambda \|\vec{v}\| \|\vec{w}\| \sin \theta \vec{n}$$
$$= \lambda (\vec{v} \times \vec{w}).$$

Similarly,

$$\vec{v} \times (\lambda \vec{w}) = \|\vec{v}\| \|\lambda \vec{w}\| \sin(\pi - \theta)(-\vec{n})$$
$$= -\lambda \|\vec{v}\| \|\vec{w}\| \sin \theta (-\vec{n})$$
$$= \lambda (\vec{v} \times \vec{w}).$$

29. Problem 26 tells us that $(\vec{u} \times \vec{v}) \cdot \vec{w} = \vec{u} \cdot (\vec{v} \times \vec{w})$. Using this result on the triple product of $(\vec{a} + \vec{b}) \times \vec{c}$ with any vector $\vec{d}$ together with the fact that the dot product distributes over addition gives us:

$$[(\vec{a} + \vec{b}) \times \vec{c}] \cdot \vec{d} = (\vec{a} + \vec{b}) \cdot (\vec{c} \times \vec{d})$$
$$= \vec{a} \cdot (\vec{c} \times \vec{d}) + \vec{b} \cdot (\vec{c} \times \vec{d}) \quad \text{(dot product is distributive)}$$
$$= (\vec{a} \times \vec{c}) \cdot \vec{d} + (\vec{b} \times \vec{c}) \cdot \vec{d} \quad \text{(using Problem 26 again)}$$
$$= [(\vec{a} \times \vec{c}) + (\vec{b} \times \vec{c})] \cdot \vec{d}. \quad \text{(dot product is distributive)}$$

So, since $[(\vec{a} + \vec{b}) \times \vec{c}] \cdot \vec{d} = [(\vec{a} \times \vec{c}) + (\vec{b} \times \vec{c})] \cdot \vec{d}$, then

$$[(\vec{a} + \vec{b}) \times \vec{c}] \cdot \vec{d} - [(\vec{a} \times \vec{c}) + (\vec{b} \times \vec{c})] \cdot \vec{d} = 0,$$

Since the dot product is distributive, we have

$$[((\vec{a} + \vec{b}) \times \vec{c}) - (\vec{a} \times \vec{c}) - (\vec{b} \times \vec{c})] \cdot \vec{d} = 0.$$

Since this equation is true for all vectors $\vec{d}$, by letting

$$\vec{d} = ((\vec{a} + \vec{b}) \times \vec{c}) - (\vec{a} \times \vec{c}) - (\vec{b} \times \vec{c}),$$

we get

$$\|(\vec{a} + \vec{b}) \times \vec{c} - \vec{a} \times \vec{c} - \vec{b} \times \vec{c}\|^2 = 0$$

and hence

$$(\vec{a} + \vec{b}) \times \vec{c} - (\vec{a} \times \vec{c}) - (\vec{b} \times \vec{c}) = \vec{0}.$$

Thus

$$(\vec{a} + \vec{b}) \times \vec{c} = (\vec{a} \times \vec{c}) + (\vec{b} \times \vec{c}).$$

Solutions for Chapter 13 Review

Exercises

1. $\vec{v} + 2\vec{w} = 2\vec{i} + 3\vec{j} - \vec{k} + 2(\vec{i} - \vec{j} + 2\vec{k}) = 4\vec{i} + \vec{j} + 3\vec{k}$.

5. Since $\vec{v} \cdot \vec{w} = 2 \cdot 1 + 3(-1) + (-1)2 = -3$, we have $(\vec{v} \cdot \vec{w})\vec{v} = -6\vec{i} - 9\vec{j} + 3\vec{k}$.

9. (a) We have $\vec{v} \cdot \vec{w} = 3 \cdot 4 + 2 \cdot (-3) + (-2) \cdot 1 = 4$.
   (b) We have $\vec{v} \times \vec{w} = -4\vec{i} - 11\vec{j} - 17\vec{k}$.
   (c) A vector of length 5 parallel to $\vec{v}$ is

   $$\frac{5}{\|\vec{v}\|}\vec{v} = \frac{5}{\sqrt{17}}(3\vec{i} + 2\vec{j} - 2\vec{k}) = 3.64\vec{i} + 2.43\vec{j} - 2.43\vec{k}.$$

   (d) The angle between vectors $\vec{v}$ and $\vec{w}$ is found using

   $$\cos\theta = \frac{\vec{v} \cdot \vec{w}}{\|\vec{v}\|\|\vec{w}\|} = \frac{4}{\sqrt{17}\sqrt{26}} = 0.190,$$

   so $\theta = 79.0°$.
   (e) The component of vector $\vec{v}$ in the direction of vector $\vec{w}$ is

   $$\frac{\vec{v} \cdot \vec{w}}{\|\vec{w}\|} = \frac{4}{\sqrt{26}} = 0.784.$$

   (f) The answer is any vector $\vec{a}$ such that $\vec{a} \cdot \vec{v} = 0$. One possible answer is $2\vec{i} - 2\vec{j} + \vec{k}$.
   (g) A vector perpendicular to both is the cross product:

   $$\vec{v} \times \vec{w} = -4\vec{i} - 11\vec{j} - 17\vec{k}.$$

13. The vector $\vec{w}$ we want is shown in Figure 13.10, where the given vector is $\vec{v} = 4\vec{i} + 3\vec{j}$. The vectors $\vec{v}$ and $\vec{w}$ are the same length and the two angles marked α are equal, so the two right triangles shown are congruent. Thus

$$a = -3 \qquad \text{and} \qquad b = 4.$$

Therefore

$$\vec{w} = -3\vec{i} + 4\vec{j}.$$

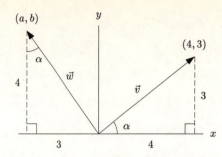

Figure 13.10

Problems

17. (a) On the x-axis, $y = z = 0$, so $5x = 21$, giving $x = \frac{21}{5}$. So the only such point is $\left(\frac{21}{5}, 0, 0\right)$.
   (b) Other points are $(0, -21, 0)$, and $(0, 0, 3)$. There are many other possible answers.
   (c) $\vec{n} = 5\vec{i} - \vec{j} + 7\vec{k}$. It is the normal vector.
   (d) The vector between two points in the plane is parallel to the plane. Using the points from part (b), the vector $3\vec{k} - (-21\vec{j}) = 21\vec{j} + 3\vec{k}$ is parallel to the plane.

21. The speed is a scalar which equals 30 times the circumference of the circle per minute. So it is a constant. The velocity is a vector. Since the direction of the motion changes all the time, the velocity is not constant. This implies that the acceleration is nonzero.

25. Let the x-axis point east and the y-axis point north. Denote the forces exerted by Charlie, Sam and Alice by $\vec{F}_C, \vec{F}_S$ and $\vec{F}_A$ (see Figure 13.11).

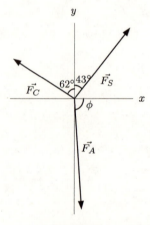

Figure 13.11

Since $\|\vec{F}_C\| = 175$ newtons and the angle θ from the x-axis to $\vec{F}_C$ is $90° + 62° = 152°$, we have

$$\vec{F}_C = 175 \cos 152° \vec{i} + 175 \sin 152° \vec{j} \approx -154.52\vec{i} + 82.16\vec{j}.$$

Similarly,

$$\vec{F}_S = 200 \cos 47° \vec{i} + 200 \sin 47° \vec{j} \approx 136.4\vec{i} + 146.27\vec{j}.$$

Now Alice is to counterbalance Sam and Charlie, so the resultant force of the three forces $\vec{F}_C, \vec{F}_S$ and $\vec{F}_A$ must be 0, that is,

$$\vec{F}_C + \vec{F}_S + \vec{F}_A = 0.$$

Thus, we have

$$\vec{F}_A = -\vec{F}_C - \vec{F}_S$$
$$\approx -(-154.52\vec{i} + 82.16\vec{j}) - (136.4\vec{i} + 146.27\vec{j})$$
$$= 18.12\vec{i} - 228.43\vec{j}$$

and, $\|\vec{F}_A\| = \sqrt{18.12^2 + (-228.43)^2} \approx 229.15$ newtons.

If ϕ is the angle from the x-axis to $\vec{F}_A$, then

$$\phi = \arctan \frac{-228.43}{18.12} \approx -85.5°.$$

29. The displacement from $(1, 1, 1)$ to $(1, 4, 5)$ is

$$\vec{r_1} = (1 - 1)\vec{i} + (4 - 1)\vec{j} + (5 - 1)\vec{k} = 3\vec{j} + 4\vec{k}.$$

The displacement from $(-3, -2, 0)$ to $(1, 4, 5)$ is

$$\vec{r_2} = (1 + 3)\vec{i} + (4 + 2)\vec{j} + (5 - 0)\vec{k} = 4\vec{i} + 6\vec{j} + 5\vec{k}.$$

A normal vector is

$$\vec{n} = \vec{r_1} \times \vec{r_2} = \begin{vmatrix} \vec{i} & \vec{j} & \vec{k} \\ 0 & 3 & 4 \\ 4 & 6 & 5 \end{vmatrix} = (15 - 24)\vec{i} - (-16)\vec{j} + (-12)\vec{k} = -9\vec{i} + 16\vec{j} - 12\vec{k}.$$

The equation of the plane is

$$-9x + 16y - 12z = -9 \cdot 1 + 16 \cdot 1 - 12 \cdot 1 = -5$$
$$9x - 16y + 12z = 5.$$

We pick a point A on the plane, $A = \left(\frac{5}{9}, 0, 0\right)$ and let $P = (0, 0, 0)$. (See Figure 13.12.) Then $\vec{PA} = (5/9)\vec{i}$.

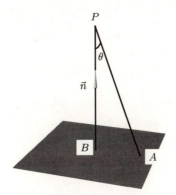

Figure 13.12

So the distance d from the point P to the plane is

$$d = \|\vec{PB}\| = \|\vec{PA}\| \cos \theta$$
$$= \frac{\vec{PA} \cdot \vec{n}}{\|\vec{n}\|} \quad \text{since } \vec{PA} \cdot \vec{n} = \|\vec{PA}\|\|\vec{n}\| \cos \theta)$$
$$= \left| \frac{\left(\frac{5}{9}\vec{i}\right) \cdot (-9\vec{i} + 16\vec{j} - 12\vec{k})}{\sqrt{9^2 + 16^2 + 12^2}} \right|$$
$$= \frac{5}{\sqrt{481}} = 0.23.$$

CAS Challenge Problems

33. $(\vec{a} \times \vec{b}) \cdot \vec{c} = 0$, $(\vec{a} \times \vec{b}) \times (\vec{a} \times \vec{c}) = \vec{0}$

Since $\vec{c}$ is the sum of a scalar multiple of $\vec{a}$ and a scalar multiple of $\vec{b}$, it lies in the plane containing $\vec{a}$ and $\vec{b}$. On the other hand, $\vec{a} \times \vec{b}$ is perpendicular to this plane, so $\vec{a} \times \vec{b}$ is perpendicular to $\vec{c}$. Therefore, $(\vec{a} \times \vec{b}) \cdot \vec{c} = 0$. Also, $\vec{a} \times \vec{c}$ is also perpendicular to the plane, thus parallel to $\vec{a} \times \vec{b}$, and thus $(\vec{a} \times \vec{b}) \times (\vec{a} \times \vec{c}) = \vec{0}$.

37. (a) $\overrightarrow{PQ} \times \overrightarrow{PR}$ is perpendicular to the plane containing P, Q, R, and therefore parallel to the normal vector $a\vec{i} + b\vec{j} + c\vec{k}$.

(b)

$$\overrightarrow{PQ} \times \overrightarrow{PR} = (tv - sw - ty + wy + sz - vz)\vec{i} +$$
$$(-tu + rw + tx - wx - rz + uz)\vec{j} + (su - rv - sx + vx + ry - uy)\vec{k}$$

(c) After substituting $z = (d - ax - by)/c$, $w = (d - au - bv)/c$, $t = (d - ar - bs)/c$ into the result of part (a), and simplifying the expression, we obtain:

$$\overrightarrow{PQ} \times \overrightarrow{PR} = \frac{a(s(u - x) + vx - uy + r(-v + y))}{c}\vec{i} +$$
$$\frac{b(s(u - x) + vx - uy + r(-v + y))}{c}\vec{j} + (s(u - x) + vx - uy + r(-v + y))\vec{k}$$
$$= \frac{(s(u - x) + vx - uy + r(-v + y))}{c}(a\vec{i} + b\vec{j} + c\vec{k}).$$

Thus $\overrightarrow{PQ} \times \overrightarrow{PR}$ is a scalar multiple of $a\vec{i} + b\vec{j} + c\vec{k}$, and hence parallel to it.

CHECK YOUR UNDERSTANDING

1. False. There are exactly two unit vectors: one in the same direction as $\vec{v}$ and the other in the opposite direction. Explicitly, the unit vectors parallel to $\vec{v}$ are $\pm\dfrac{1}{\|\vec{v}\|}\vec{v}$.

5. False. If $\vec{v}$ and $\vec{w}$ are not parallel, the three vectors $\vec{v}$, $\vec{w}$ and $\vec{v} - \vec{w}$ can be thought of as three sides of a triangle. (If the tails of $\vec{v}$ and $\vec{w}$ are placed together, then $\vec{v} - \vec{w}$ is a vector from the head of $\vec{w}$ to the head of $\vec{v}$.) The length of one side of a triangle is less than the sum of the lengths of the other two sides. Alternatively, a counterexample is $\vec{v} = \vec{i}$ and $\vec{w} = \vec{j}$. Then $\|\vec{i} - \vec{j}\| = \sqrt{2}$ but $\|\vec{i}\| - \|\vec{j}\| = 0$.

9. False. To find the displacement vector *from* $(1, 1, 1)$ *to* $(1, 2, 3)$ we subtract $\vec{i} + \vec{j} + \vec{k}$ from $\vec{i} + 2\vec{j} + 3\vec{k}$ to get $(1 - 1)\vec{i} + (2 - 1)\vec{j} + (3 - 1)\vec{k} = \vec{j} + 2\vec{k}$.

13. True. The cosine of the angle between the vectors is negative when the angle is between $\pi/2$ and π.

17. False. If the vectors are nonzero and perpendicular, the dot product will be zero (e.g. $\vec{i} \cdot \vec{j} = 0$).

21. True. The cross product yields a vector.

25. False. If $\vec{u}$ and $\vec{w}$ are two different vectors both of which are parallel to $\vec{v}$, then $\vec{v} \times \vec{u} = \vec{v} \times \vec{w} = \vec{0}$, but $\vec{u} \neq \vec{w}$. A counterexample is $\vec{v} = \vec{i}$, $\vec{u} = 2\vec{i}$ and $\vec{w} = 3\vec{i}$.

29. True. Any vector $\vec{w}$ that is parallel to $\vec{v}$ will give $\vec{v} \times \vec{w} = \vec{0}$.

CHAPTER FOURTEEN

Solutions for Section 14.1

Exercises

1. If h is small, then

$$f_x(3, 2) \approx \frac{f(3 + h, 2) - f(3, 2)}{h}.$$

With $h = 0.01$, we find

$$f_x(3, 2) \approx \frac{f(3.01, 2) - f(3, 2)}{0.01} = \frac{\frac{3.01^2}{(2+1)} - \frac{3^2}{(2+1)}}{0.01} = 2.00333.$$

With $h = 0.0001$, we get

$$f_x(3, 2) \approx \frac{f(3.0001, 2) - f(3, 2)}{0.0001} = \frac{\frac{3.0001^2}{(2+1)} - \frac{3^2}{(2+1)}}{0.0001} = 2.0000333.$$

Since the difference quotient seems to be approaching 2 as h gets smaller, we conclude

$$f_x(3, 2) \approx 2.$$

To estimate $f_y(3, 2)$, we use

$$f_y(3, 2) \approx \frac{f(3, 2 + h) - f(3, 2)}{h}.$$

With $h = 0.01$, we get

$$f_y(3, 2) \approx \frac{f(3, 2.01) - f(3, 2)}{0.01} = \frac{\frac{3^2}{(2.01+1)} - \frac{3^2}{(2+1)}}{0.01} = -0.99668.$$

With $h = 0.0001$, we get

$$f_y(3, 2) \approx \frac{f(3, 2.0001) - f(3, 2)}{0.0001} = \frac{\frac{3^2}{(2.0001+1)} - \frac{3^2}{(2+1)}}{0.0001} = -0.9999667.$$

Thus, it seems that the difference quotient is approaching -1, so we estimate

$$f_y(3, 2) \approx -1.$$

5. (a) We expect f_p to be negative because if the price of the product increases, the sales usually decrease.
(b) If the price of the product is $8 per unit and if $12000 has been spent on advertising, sales increase by approximately 150 units if an additional $1000 is spent on advertising.

9. For $f_w(10, 25)$ we get

$$f_w(10, 25) \approx \frac{f(10 + h, 25) - f(10, 25)}{h}.$$

Choosing $h = 5$ and reading values from Table 12.3 on page 567 of the text, we get

$$f_w(10, 25) \approx \frac{f(15, 25) - f(10, 25)}{5} = \frac{2 - 10}{5} = -1.6$$

This means that when the wind speed is 10 mph and the true temperature is $25°$F, as the wind speed increases from 10 mph by 1 mph we feel a $1.6°$F drop in temperature. This rate is negative because the temperature you feel drops as the wind speed increases.

Problems

13. (a) For points near the point $(0, 5, 3)$, moving in the positive x direction, the surface is sloping down and the function is decreasing. Thus, $f_x(0, 5) < 0$.

(b) Moving in the positive y direction near this point the surface slopes up as the function increases, so $f_y(0, 5) > 0$.

17. (a) Estimate $\partial P/\partial r$ and $\partial P/\partial L$ by using difference quotients and reading values of P from the graph:

$$\frac{\partial P}{\partial r}(8, 4000) \approx \frac{P(16, 4000) - P(8, 4000)}{16 - 8}$$

$$= \frac{100 - 80}{8} = 2.5,$$

and

$$\frac{\partial P}{\partial L} \approx \frac{P(8, 5000) - P(8, 4000)}{5000 - 4000}$$

$$= \frac{100 - 80}{1000} = 0.02.$$

$P_r(8, 4000) \approx 2.5$ means that at an interest rate of 8% and a loan amount of \$4000 the monthly payment increases by approximately \$2.50 for every one percent increase of the interest rate. $P_L(8, 4000) \approx 0.02$ means the monthly payment increases by approximately \$0.02 for every \$1 increase in the loan amount at an 8% rate and a loan amount of \$4000.

(b) Using difference quotients and reading from the graph

$$\frac{\partial P}{\partial r}(8, 6000) \approx \frac{P(14, 6000) - P(8, 6000)}{14 - 8}$$

$$= \frac{140 - 120}{6} = 3.33,$$

and

$$\frac{\partial P}{\partial L}(8, 6000) \approx \frac{P(8, 7000) - P(8, 6000)}{7000 - 6000}$$

$$= \frac{140 - 120}{1000} = 0.02.$$

Again, we see that the monthly payment increases with increases in interest rate and loan amount. The interest rate is $r = 8\%$ as in part (a), but here the loan amount is $L = \$6000$. Since $P_L(8, 4000) \approx P_L(8, 6000)$, the increase in monthly payment per unit increase in loan amount remains the same as in part a). However, in this case, the effect of the interest rate is different: here the monthly payment increases by approximately \$3.33 for every one percent increase of interest rate at $r = 8\%$ and loan amount of \$6000.

(c)

$$\frac{\partial P}{\partial r}(13, 7000) \approx \frac{P(19, 7000) - P(13, 7000)}{19 - 13}$$

$$= \frac{180 - 160}{6} = 3.33,$$

and

$$\frac{\partial P}{\partial L}(13, 7000) \approx \frac{P(13, 8000) - P(13, 7000)}{8000 - 7000}$$

$$= \frac{180 - 160}{1000} = 0.02.$$

The figures show that the rates of change of the monthly payment with respect to the interest rate and loan amount are roughly the same for $(r, L) = (8, 6000)$ and $(r, L) = (13, 7000)$.

21.

Table 14.1 *Estimated values of*
$H(T, w)$ *(in calories/meter^3)*

		w (gm/m^3)			
		0.1	0.2	0.3	0.4
	10	110	240	330	450
T (°C)	20	100	180	260	350
	30	70	150	220	300
	40	65	140	200	270

Values of H from the graph are given in Table 14.1. In order to compute $H_w(T, w)$ for $w = 0.3$, it is useful to have the column corresponding to $w = 0.4$. The row corresponding to $T = 40$ is not used in this problem. The partial derivative $H_w(T, w)$ can be approximated by

$$H_w(10, 0.1) \approx \frac{H(10, 0.1 + h) - H(10, 0.1)}{h} \quad \text{for small } h.$$

We choose $h = 0.1$ because we can read off a value for $H(10, 0.2)$ from the graph. If we take $H(10, 0.2) = 240$, we get the approximation

$$H_w(10, 0.1) \approx \frac{H(10, 0.2) - H(10, 0.1)}{0.1} = \frac{240 - 110}{0.1} = 1300.$$

In practical terms, we have found that for fog at $10°$ C containing 0.1 g/m^3 of water, an increase in the water content of the fog will increase the heat requirement for dissipating the fog at the rate given by $H_w(10, 0.1)$. Specifically, a 1 g/m^3 increase in the water content will increase the heat required to dissipate the fog by about 1300 calories per cubic meter of fog. Wetter fog is harder to dissipate. Other values of $H_w(T, w)$ in Table 14.2 are computed using the formula

$$H_w(T, w) \approx \frac{H(T, w + 0.1) - H(T, w)}{0.1},$$

where we have used Table 14.1 to evaluate H.

Table 14.2 *Table of values of $H_w(T, w)$ (in cal/gm)*

		w (gm/m^3)		
		0.1	0.2	0.3
	10	1300	900	1200
T (°C)	20	800	800	900
	30	800	700	800

Solutions for Section 14.2

Exercises

1. $f_x(x, y) = 10xy^3 + 8y^2 - 6x$ and $f_y(x, y) = 15x^2y^2 + 16xy$.

5. $V_r = \frac{2}{3}\pi r h$

9. $g_x(x, y) = \dfrac{\partial}{\partial x} \ln(ye^{xy}) = (ye^{xy})^{-1} \dfrac{\partial}{\partial x}(ye^{xy}) = (ye^{xy})^{-1} \cdot y\dfrac{\partial}{\partial x}(e^{xy}) = (ye^{xy})^{-1} \cdot y \cdot y \cdot e^{xy} = y$

13. $\dfrac{\partial}{\partial m}\left(\dfrac{1}{2}mv^2\right) = \dfrac{1}{2}v^2$

17. $\dfrac{\partial V}{\partial r} = \dfrac{8}{3}\pi r h$ and $\dfrac{\partial V}{\partial h} = \dfrac{4}{3}\pi r^2$.

21. $\dfrac{\partial}{\partial T}\left(\dfrac{2\pi r}{T}\right) = -\dfrac{2\pi r}{T^2}$

25. $\dfrac{\partial T}{\partial l} = \dfrac{2\pi}{\sqrt{g}} \cdot \dfrac{1}{2}l^{-1/2} = \dfrac{\pi}{\sqrt{lg}}.$

29. $\dfrac{\partial}{\partial M}\left(\dfrac{2\pi r^{3/2}}{\sqrt{GM}}\right) = 2\pi r^{3/2}(-\dfrac{1}{2})(GM)^{-3/2}(G) = -\pi r^{3/2}\cdot\dfrac{G}{GM\sqrt{GM}} = -\dfrac{\pi r^{3/2}}{M\sqrt{GM}}$

33. We regard x as constant and differentiate with respect to y using the product rule:

$$\frac{\partial z}{\partial y} = 2e^{x+2y}\sin y + e^{x+2y}\cos y$$

Substituting $x = 1$, $y = 0.5$ gives

$$\left.\frac{\partial z}{\partial y}\right|_{(1,0.5)} = 2e^2\sin(0.5) + e^2\cos(0.5) = 13.6.$$

Problems

37. Substituting $w = 65$ and $h = 160$, we have

$$f(65, 160) = 0.01(65^{0.25})(160^{0.75}) = 1.277\text{ m}^2.$$

This tells us that a person who weighs 65 kg and is 160 cm tall has a surface area of about 1.277 m². Since

$$f_w(w, h) = 0.01(0.25w^{-0.75})h^{0.75}\text{ m}^2\text{/kg},$$

we have $f_w(65, 160) = 0.005$ m²/kg. Thus, an increase of 1 kg in weight increases surface area by about 0.005 m². Since

$$f_h(w, h) = 0.01w^{0.25}(0.75h^{-0.25})\text{ m}^2\text{/cm},$$

we have $f_h(65, 160) = 0.006$ m²/cm. Thus, an increase of 1 cm in height increases surface area by about 0.006 m².

41. We compute the partial derivatives:

$$\frac{\partial Q}{\partial K} = b\alpha K^{\alpha-1}L^{1-\alpha} \quad\text{so}\quad K\frac{\partial Q}{\partial K} = b\alpha K^{\alpha}L^{1-\alpha}$$

$$\frac{\partial Q}{\partial L} = b(1-\alpha)K^{\alpha}L^{-\alpha} \quad\text{so}\quad L\frac{\partial Q}{\partial L} = b(1-\alpha)K^{\alpha}L^{1-\alpha}$$

Adding these two results, we have:

$$K\frac{\partial Q}{\partial K} + L\frac{\partial Q}{\partial L} = b(\alpha + 1 - \alpha)K^{\alpha}L^{1-\alpha} = Q.$$

Solutions for Section 14.3

Exercises

1. We have

$$z = e^y + x + x^2 + 6.$$

The partial derivatives are

$$\left.\frac{\partial z}{\partial x}\right|_{(x,y)=(1,0)} = (2x+1)\bigg|_{(x,y)=(1,0)} = 3$$

$$\left.\frac{\partial z}{\partial y}\right|_{(x,y)=(1,0)} = e^y\bigg|_{(x,y)=(1,0)} = 1.$$

So the equation of the tangent plane is

$$z = 9 + 3(x - 1) + y = 6 + 3x + y.$$

5. Since $z_x = -e^{-x}\cos(y)$ and $z_y = -e^{-x}\sin(y)$, we have

$$dz = -e^{-x}\cos(y)dx - e^{-x}\sin(y)dy.$$

9. We have $dg = g_x\,dx + g_t\,dt$. Finding the partial derivatives, we have $g_x = 2x\sin(2t)$ so $g_x(2, \frac{\pi}{4}) = 4\sin(\pi/2) = 4$, and $g_t = 2x^2\cos(2t)$ so $g_t(2, \frac{\pi}{4}) = 8\cos(\frac{\pi}{2}) = 0$. Thus $dg = 4\,dx$.

Problems

13. (a) Since the equation of a tangent plane should be linear, this answer is wrong.

(b) The student didn't substitute the values $x = 2$, $y = 3$ into the formulas for the partial derivatives used in the formula of a tangent plane.

(c) Let $f(x, y) = z = x^3 - y^2$. Since $f_x(x, y) = 3x^2$ and $f_y(x, y) = -2y$, substituting $x = 2$, $y = 3$ gives $f_x(2, 3) = 12$ and $f_y(2, 3) = -6$. Then the equation of the tangent plane is

$$z = 12(x - 2) - 6(y - 3) - 1, \quad \text{or} \quad z = 12x - 6y - 7.$$

17. Making use of the values of P_r and P_L from the solution to Problem 17 on page 218, we have the local linearizations:

For $(r, L) = (8, 4000)$,

$$P(r, L) \approx 80 + 2.5(r - 8) + 0.02(L - 4000),$$

For $(r, L) = (8, 6000)$,

$$P(r, L) \approx 120 + 3.33(r - 8) + 0.02(L - 6000),$$

For $(r, L) = (13, 7000)$,

$$P(r, L) \approx 160 + 3.33(r - 13) + 0.02(L - 7000).$$

21. The error in η is approximated by $d\eta$, where

$$d\eta = \frac{\partial \eta}{\partial r} dr + \frac{\partial \eta}{\partial p} dp.$$

We need to find

$$\frac{\partial \eta}{\partial r} = \frac{\pi}{8} \frac{p 4 r^3}{v}$$

and

$$\frac{\partial \eta}{\partial p} = \frac{\pi}{8} \frac{r^4}{v}.$$

For $r = 0.005$ and $p = 10^5$ we get

$$\frac{\partial \eta}{\partial r}(0.005, 10^5) = 3.14159 \cdot 10^7, \quad \frac{\partial \eta}{\partial p}(0.005, 10^5) = 0.39270,$$

so that

$$d\eta = \frac{\partial \eta}{\partial r} dr + \frac{\partial \eta}{\partial p} dp.$$

is largest when we take all positive values to give

$$d\eta = 3.14159 \cdot 10^7 \cdot 0.00025 + 0.39270 \cdot 1000 = 8246.68.$$

This seems quite large but $\eta(0.005, 10^5) = 39269.9$ so the maximum error represents about 20% of any value computed by the given formula. Notice also the relative error in r is $\pm 5\%$, which means the relative error in r^4 is $\pm 20\%$.

25. (a) The area of a circle of radius r is given by

$$A = \pi r^2$$

and the perimeter is

$$L = 2\pi r.$$

Thus we get

$$r = \frac{L}{2\pi}$$

and

$$A = \pi \left(\frac{L}{2\pi}\right)^2 = \frac{L^2 \pi}{4\pi^2} = \frac{L^2}{4\pi}.$$

Thus we get

$$\pi = \frac{L^2}{4A}.$$

(b) We will treat π as a function of L and A.

$$d\pi = \frac{\partial \pi}{\partial L}dL + \frac{\partial \pi}{\partial A}dA = \frac{2L}{4A}dL - \frac{L^2}{4A^2}dA.$$

If L is in error by a factor λ, then $\Delta L = \lambda L$, and if A is in error by a factor μ, then $\Delta A = \mu A$. Therefore,

$$\Delta \pi \approx \frac{2L}{4A}\Delta L - \frac{L^2}{4A^2}\Delta A$$

$$= \frac{2L}{4A}\lambda L - \frac{L^2}{4A^2}\mu A$$

$$= \frac{2\lambda L^2}{4A} - \frac{\mu L^2}{4A} = (2\lambda - \mu)\frac{L^2}{4A} = (2\lambda - \mu)\pi,$$

so π is in error by a factor of $2\lambda - \mu$.

Solutions for Section 14.4

Exercises

1. Since the partial derivatives are

$$\frac{\partial f}{\partial x} = \frac{15}{2}x^4 - 0 = \frac{15}{2}x^4$$

$$\frac{\partial f}{\partial y} = 0 - \frac{24}{7}y^5 = -\frac{24}{7}y^5$$

we have

$$\text{grad } f = \frac{\partial f}{\partial x}\vec{i} + \frac{\partial f}{\partial y}\vec{j} = \left(\frac{15}{2}x^4\right)\vec{i} - \left(\frac{24}{7}y^5\right)\vec{j}.$$

5. Since the partial derivatives are

$$f_x = \frac{x}{\sqrt{x^2 + y^2}} \quad \text{and} \quad f_y = \frac{y}{\sqrt{x^2 + y^2}},$$

we have

$$\nabla f = \frac{1}{\sqrt{x^2 + y^2}}(x\vec{i} + y\vec{j}).$$

9. Since the partial derivatives are

$$z_x = (2x)\cos(x^2 + y^2), \quad \text{and} \quad z_y = (2y)\cos(x^2 + y^2),$$

$$\nabla z = 2x\cos(x^2 + y^2)\vec{i} + 2y\cos(x^2 + y^2)\vec{j}.$$

13. Since the partial derivatives are

$$z_x = \frac{e^y(x + y) - xe^y}{(x + y)^2} = \frac{ye^y}{(x + y)^2}$$

$$z_y = \frac{xe^y(x + y) - xe^y}{(x + y)^2} = \frac{e^y(x^2 + xy - x)}{(x + y)^2}$$

we have

$$\nabla z = \frac{ye^y}{(x + y)^2}\vec{i} + \frac{e^y(x^2 + xy - x)}{(x + y)^2}\vec{j}$$

17. Since the partial derivatives are

$$f_r = 2\pi(h + r) \quad \text{and} \quad f_h = 2\pi r,$$

we have

$$\nabla f(2, 3) = 10\pi\vec{i} + 4\pi\vec{j}.$$

21. Since the values of z decrease as we move in direction $\vec{i}$ from the point $(-2, 2)$, the directional derivative is negative.

25. Since the values of z decrease as we move in direction $\vec{i} + 2\vec{j}$ from the point $(0, -2)$, the directional derivative is negative.

29. The approximate direction of the gradient vector at point $(-2, 0)$ is $-\vec{i}$, since the gradient vector is perpendicular to the contour and points in the direction of increasing z-values. Answers may vary since answers are approximate and any positive multiple of the vector given is also correct.

33. The approximate direction of the gradient vector at point $(-2, 2)$ is $-\vec{i} + \vec{j}$, since the gradient vector is perpendicular to the contour and points in the direction of increasing z-values. Answers may vary since answers are approximate and any positive multiple of the vector given is also correct.

37. Since $f_x = 2\cos(2x - y)$ and $f_y = -\cos(2x - y)$, at $(1, 2)$ we have $\text{grad } f = 2\cos(0)\vec{i} - \cos(0)\vec{j} = 2\vec{i} - \vec{j}$. Thus

$$f_{\vec{u}}(1, 2) = \text{grad } f \cdot \left(\frac{3}{5}\vec{i} - \frac{4}{5}\vec{j}\right) = \frac{2 \cdot 3 - 1(-4)}{5} = \frac{10}{5} = 2.$$

Problems

41. Since $\vec{u} = (\vec{i} - \vec{j})/\sqrt{2}$, we head away from the point $(3, 1)$ toward the point $(4, 0)$.
From the graph, we see that $f(3, 1) = 1$ and $f(4, 0) = 4$. Since the points $(3, 1)$ and $(4, 0)$ are a distance $\sqrt{2}$ apart, we have

$$f_{\vec{u}}(3, 1) \approx \frac{f(4, 0) - f(3, 1)}{\sqrt{2}} = \frac{4 - 1}{\sqrt{2}} = 2.12.$$

45. The directional derivative is approximately the change in z (as we move in direction $\vec{v}$) divided by the horizontal change in position. In the direction of $\vec{v}$, the directional derivative $\approx \dfrac{2 - 1}{\|\vec{i} + \vec{j}\|} = \dfrac{1}{\sqrt{2}} \approx 0.7$.

49. **(a)** In the $\vec{i} - \vec{j}$ direction the function is decreasing, so the value of $g_{\vec{u}}(2, 5)$ is negative.
(b) In the $\vec{i} + \vec{j}$ direction the function is decreasing, so the value of $g_{\vec{u}}(2, 5)$ negative as well.

53. **(a)** First we will find a unit vector in the same direction as the vector $\vec{v} = 3\vec{i} - 2\vec{j}$. Since this vector has magnitude $\sqrt{13}$, a unit vector is

$$\vec{u}_1 = \frac{1}{\|\vec{v}\|}\vec{v} = \frac{3}{\sqrt{13}}\vec{i} - \frac{2}{\sqrt{13}}\vec{j}.$$

The partial derivatives are

$$f_x(x, y) = \frac{(1 + x^2) - (x + y) \cdot 2x}{(1 + x^2)^2} = \frac{1 - x^2 - 2xy}{(1 + x^2)^2},$$

$$\text{and} \quad f_y(x, y) = \frac{1}{1 + x^2},$$

then, at the point P, we have

$$f_x(P) = f_x(1, -2) = \frac{1 - 1^2 - 2 \cdot 1 \cdot (-2)}{(1 + 1^2)^2} = 1,$$

$$f_y(P) = f_y(1, -2) = f_y(1, -2) = \frac{1}{1 + 1^2} = \frac{1}{2}.$$

Thus

$$f_{\vec{u}_1}(P) = \text{grad } f(P) \cdot \vec{u}_1$$
$$= \left(\vec{i} + \frac{1}{2}\vec{j}\right) \cdot \left(\frac{3}{\sqrt{13}}\vec{i} - \frac{2}{\sqrt{13}}\vec{j}\right)$$
$$= \frac{3}{\sqrt{13}} - \frac{1}{\sqrt{13}} = \frac{2}{\sqrt{13}}.$$

(b) The unit vector in the same direction as the vector $\vec{v} = -\vec{i} + 4\vec{j}$ is

$$\vec{u}_2 = \frac{1}{\|\vec{v}\|}\vec{v} = \frac{1}{\sqrt{(-1)^2 + 4^2}}(-\vec{i} + 4\vec{j})$$

$$= -\frac{1}{\sqrt{17}}\vec{i} + \frac{4}{\sqrt{17}}\vec{j}.$$

Since we have calculated from part (a) that $f_x(P) = 1$ and $f_y(P) = 1/2$,

$$f_{\vec{u}_2}(P) = \text{grad } f(P) \cdot \vec{u}_2$$

$$= (\vec{i} + \frac{1}{2}\vec{j}) \cdot (-\frac{1}{\sqrt{17}}\vec{i} + \frac{4}{\sqrt{17}}\vec{j})$$

$$= -\frac{1}{\sqrt{17}} + \frac{2}{\sqrt{17}} = \frac{1}{\sqrt{17}}.$$

(c) The direction of greatest increase is grad f at P. By part (a) we have found that

$$f_x(P) = 1 \quad \text{and} \quad f_y(P) = \frac{1}{2}.$$

Therefore the direction of greatest increase is

$$\text{grad } f(P) = \vec{i} + \frac{1}{2}\vec{j}.$$

57. Let's put a coordinate plane on the area you are hiking, with your trail along the x-axis and the second trail branching off at the origin as in Figure 14.1. You are moving in the positive x direction. Let $h(x, y)$ be the elevation at the point (x, y) on the mountain.

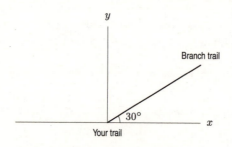

Figure 14.1: Two trails

Since the trail along the x-axis ascends at a 20° angle, we have $h_x(0,0) = \tan 20°$. Since the trail is the steepest path, grad h must point along your trail in the positive x direction. Thus

$$\text{grad } h = h_x\vec{i} + 0\vec{j} = \tan 20°\vec{i}.$$

We must compute the rate of change of elevation in the direction of the branch trail. The unit vector in this direction is $\vec{u} = \cos 30°\vec{i} + \sin 30°\vec{j}$, and thus the directional derivative is

$$h_{\vec{u}} = (\text{grad } h) \cdot \vec{u} = (\tan 20°\vec{i}) \cdot (\cos 30°\vec{i} + \sin 30°\vec{j}) = (\tan 20°)(\cos 30°) = 0.3152.$$

The angle of ascent of the branch trail is thus $\tan^{-1}(0.3152) = 17.5°$.

61. First, we check that $(2)(3) = 6$. Then let $f(x, y) = xy$ so that the given curve is the contour $f(x, y) = 6$. Since $f_x = y$ and $f_y = x$, we have grad $f(2, 3) = 3\vec{i} + 2\vec{j}$. Since gradients are perpendicular to contours, a vector normal to the curve at $(2, 3)$ is $\vec{n} = 3\vec{i} + 2\vec{j}$. Using the normal vector to a line the same way we use the normal vector to a plane, we get that the equation of the tangent line is $3(x - 2) + 2(y - 3) = 0$.

65. At the point $(1.2, 0)$, the value of the function is 4.2. Nearby, the largest value is 8.9 at the point $(1.4, -1)$. Since the gradient vector points in the direction of maximum increase, it points into the fourth quadrant.

69. **(a)** P corresponds to greatest rate of increase of f and Q corresponds to greatest rate of decrease of f. See Figure 14.2.
    (b) The points are marked in Figure 14.3.
    (c) Amplitude is $\| \operatorname{grad} f \|$. The equation is

$$f_{\vec{u}} = \| \operatorname{grad} f \| \cos \theta.$$

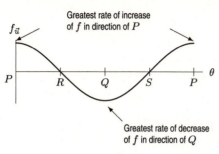

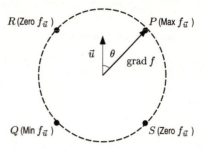

Figure 14.2	**Figure 14.3**

Solutions for Section 14.5

Exercises

1. We have $f_x = 2x$, $f_y = 3y^2$, and $f_z = -4z^3$. Thus

$$\operatorname{grad} f = 2x\vec{i} + 3y^2\vec{j} - 4z^3\vec{k}.$$

5. We have $f_x = -2x/(x^2 + y^2 + z^2)^2$, $f_y = -2y/(x^2 + y^2 + z^2)^2$, and $f_z = -2z/(x^2 + y^2 + z^2)^2$. Thus

$$\operatorname{grad} f = \frac{-2}{(x^2 + y^2 + z^2)^2}(x\vec{i} + y\vec{j} + z\vec{k}).$$

9. We have $f_x = yz$, $f_y = xz$, and $f_z = yz$. Thus $f_x(1, 2, 3) = 6$, $f_y(1, 2, 3) = 3$, and $f_z(1, 2, 3) = 2$, so

$$\operatorname{grad} f = 6\vec{i} + 3\vec{j} + 2\vec{k}.$$

13. We have $\operatorname{grad} f = y\vec{i} + x\vec{j} + 2z\vec{k}$, so $\operatorname{grad} f(1, 1, 0) = \vec{i} + \vec{j}$. A unit vector in the direction we want is $u = (1/\sqrt{2})(-\vec{i} + \vec{k})$. Therefore, the directional derivative is

$$\operatorname{grad} f(1, 1, 0) \cdot \vec{u} = \frac{1(-1) + 1 \cdot 0 + 0 \cdot 1}{\sqrt{2}} = \frac{-1}{\sqrt{2}}.$$

17. First, we check that $(-1)^2 - (1)^2 + 2^2 = 4$. Then let $f(x, y, z) = x^2 - y^2 + z^2$ so that the given surface is the level surface $f(x, y, z) = 4$. Since $f_x = 2x$, $f_y = -2y$, and $f_z = 2z$, we have $\operatorname{grad} f(-1, 1, 2) = -2\vec{i} - 2\vec{j} + 4\vec{k}$. Since gradients are perpendicular to level surfaces, a vector normal to the surface at $(-1, 1, 2)$ is $\vec{n} = -2\vec{i} - 2\vec{j} + 4\vec{k}$. Thus an equation for the tangent plane is

$$-2(x + 1) - 2(y - 1) + 4(z - 2) = 0.$$

21. First, we check that $1 = 2^2 - 3$. Then we let $f(x, y, z) = y^2 - z^2 + 3$, so that the given surface is the level surface $f(x, y, z) = 0$. Since $f_x = 0$, $f_y = 2y$, and $f_z = -2z$, we have $\operatorname{grad} f(-1, 1, 2) = 2\vec{j} - 4\vec{k}$. Since gradients are perpendicular to level surfaces, a vector normal to the surface at $(-1, 1, 2)$ is $\vec{n} = 2\vec{j} - 4\vec{k}$. Thus an equation for the tangent plane is

$$2(y - 1) - 4(z - 2) = 0.$$

Problems

25. **(a)** The function $T(x, y, z) =$ constant where $x^2 + y^2 + z^2 =$ constant. These surfaces are spheres centered at the origin.
  (b) Calculating the partial derivative with respect to x gives

$$\frac{\partial T}{\partial x} = -2xe^{-(x^2+y^2+z^2)}.$$

Similar calculations for the other variables shows that

$$\text{grad } T = (-2x\vec{i} - 2y\vec{j} - 2z\vec{k})e^{-(x^2+y^2+z^2)}.$$

  (c) At the point $(1, 0, 0)$

$$\text{grad } T(1, 0, 0) = -2e^{-1}\vec{i}.$$

Moving from the point $(1, 0, 0)$ to $(2, 1, 0)$, you move in the direction

$$(2 - 1)\vec{i} + (1 - 0)\vec{j} = \vec{i} + \vec{j}.$$

A unit vector in this direction is

$$\vec{u} = \frac{\vec{i} + \vec{j}}{\sqrt{2}}.$$

The directional derivative of $T(x, y, z)$ in this direction at the point $(1, 0, 0)$ is

$$T_{\vec{u}}(1, 0, 0) = -2e^{-1}\vec{i} \cdot \frac{\vec{i} + \vec{j}}{\sqrt{2}} = -\sqrt{2}e^{-1}.$$

Since you are moving at a speed of 3 units per second,

$$\text{Rate of change of temperature } = -\sqrt{2}e^{-1} \cdot 3 = -3\sqrt{2}e^{-1} \text{ degrees/second.}$$

29. **(a)** The surface is the level surface $F(x, y, z) = 7$, where $F(x, y, z) = x^2 + y^2 - xyz$. Thus the normal vector to the tangent plane is grad $F = (2x - yz)\vec{i} + (2y - xz)\vec{j} + (-xy)\vec{k}$. Evaluated at $(2, 3, 1)$, we get the normal to the plane

$$\vec{n} = \vec{i} + 4\vec{j} - 6\vec{k}.$$

Thus the equation of the plane is

$$(x - 2) + 4(y - 3) - 6(z - 1) = 0.$$

  (b) Solving $x^2 + y^2 - xyz = 7$ for z, we get

$$z = \frac{x^2 + y^2 - 7}{xy}.$$

Thus, we have

$$f(x, y) = \frac{x^2 + y^2 - 7}{xy} = \frac{x}{y} + \frac{y}{x} - \frac{7}{xy}.$$

We have

$$f_x(x, y) = \frac{1}{y} - \frac{y}{x^2} + \frac{7}{x^2 y}$$

$$f_y(x, y) = -\frac{x}{y^2} + \frac{1}{x} + \frac{7}{xy^2}.$$

Thus $f_x(2, 3) = 1/3 - 3/4 + 7/12 = 1/6$ and $f_y(2, 3) = -2/9 + 1/2 + 7/18 = 2/3$. Thus the equation of the tangent plane is

$$z = 1 + (1/6)(x - 2) + (2/3)(y - 3).$$

This is the same as the answer to part (a) when that equation is solved for z.

33. If write $\vec{r} = x\vec{i} + y\vec{j} + z\vec{k}$, then we know

$$\text{grad } f(x, y, z) = g(x, y, z)(x\vec{i} + y\vec{j} + z\vec{k}) = g(x, y, z)\vec{r}$$

so grad f is everywhere radially outward, and therefore perpendicular to a sphere centered at the origin. If f were not constant on such a sphere, then grad f would have a component tangent to the sphere. Thus, f must be constant on any sphere centered at the origin.

Solutions for Section 14.6

Exercises

1. Using the chain rule we see:

$$\frac{dz}{dt} = \frac{\partial z}{\partial x}\frac{dx}{dt} + \frac{\partial z}{\partial y}\frac{dy}{dt}$$

$$= -y^2 e^{-t} + 2xy\cos t$$

$$= -(\sin t)^2 e^{-t} + 2e^{-t}\sin t\cos t$$

$$= \sin(t)e^{-t}(2\cos t - \sin t)$$

We can also solve the problem using one variable methods:

$$z = e^{-t}(\sin t)^2$$

$$\frac{dz}{dt} = \frac{d}{dt}(e^{-t}(\sin t)^2)$$

$$= \frac{de^{-t}}{dt}(\sin t)^2 + e^{-t}\frac{d(\sin t)^2}{dt}$$

$$= -e^{-t}(\sin t)^2 + 2e^{-t}\sin t\cos t$$

$$= e^{-t}\sin t(2\cos t - \sin t)$$

5. Substituting into the chain rule gives

$$\frac{dz}{dt} = \frac{\partial z}{\partial x}\frac{dx}{dt} + \frac{\partial z}{\partial y}\frac{dy}{dt} = e^y(2) + xe^y(-2t)$$

$$= 2e^y(1 - xt) = 2e^{1-t^2}(1 - 2t^2).$$

9. Since z is a function of two variables x and y which are functions of two variables u and v, the two chain rule identities which apply are:

$$\frac{\partial z}{\partial u} = \frac{\partial z}{\partial x}\frac{\partial x}{\partial u} + \frac{\partial z}{\partial y}\frac{\partial y}{\partial u} = e^y\left(\frac{1}{u}\right) + xe^y \cdot 0 = \frac{e^v}{u}.$$

$$\frac{\partial z}{\partial v} = \frac{\partial z}{\partial x}\frac{\partial x}{\partial v} + \frac{\partial z}{\partial y}\frac{\partial y}{\partial v} = e^y(0) + xe^y \cdot 1 = e^v\ln u.$$

13. Since z is a function of two variables x and y which are functions of two variables u and v, the two chain rule identities which apply are:

$$\frac{\partial z}{\partial u} = \frac{\partial z}{\partial x}\frac{\partial x}{\partial u} + \frac{\partial z}{\partial y}\frac{\partial y}{\partial u} = \left(\cos\left(\frac{x}{y}\right)\right)\left(\frac{1}{y}\right)\frac{1}{u} + \left(\cos\left(\frac{x}{y}\right)\right)\left(\frac{-x}{y^2}\right) \cdot 0$$

$$= \frac{1}{vu}\cos\left(\frac{\ln u}{v}\right).$$

$$\frac{\partial z}{\partial v} = \frac{\partial z}{\partial x}\frac{\partial x}{\partial v} + \frac{\partial z}{\partial y}\frac{\partial y}{\partial v} = \left(\cos\left(\frac{x}{y}\right)\right)\left(\frac{1}{y}\right) \cdot 0 + \left(\cos\left(\frac{x}{y}\right)\right)\left(\frac{-x}{y^2}\right) \cdot 1 = -\frac{\ln u}{v^2}\cos\left(\frac{\ln u}{v}\right).$$

Problems

17.

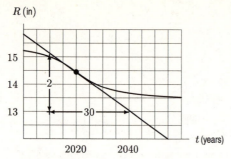

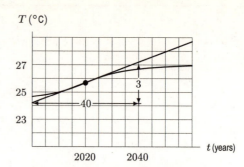

Figure 14.4: Global warming predictions: Rainfall as a function of time

Figure 14.5: Global warming predictions: Temperature as a function of time

We know that, as long as the temperature and rainfall stay close to their current values of $R = 15$ inches and $T = 30°$C, a change, ΔR, in rainfall and a change, ΔT, in temperature produces a change, ΔC, in corn production given by

$$\Delta C \approx 3.3\Delta R - 5\Delta T.$$

Now both R and T are functions of time t (in years), and we want to find the effect of a small change in time, Δt, on R and T. Figure 14.4 shows that the slope of the graph for R versus t is about $-2/30 \approx -0.07$ in/year when $t = 2020$. Similarly, Figure 14.5 shows the slope of the graph of T versus t is about $3/40 \approx 0.08°$C/year when $t = 2020$. Thus, around the year 2020,

$$\Delta R \approx -0.07\Delta t \quad \text{and} \quad \Delta T \approx 0.08\Delta t.$$

Substituting these into the equation for ΔC, we get

$$\Delta C \approx (3.3)(-0.07)\Delta t - (5)(0.08)\Delta t \approx -0.6\Delta t.$$

Since at present $C = 100$, corn production will decline by about 0.6 % between the years 2020 and 2021. Now $\Delta C \approx -0.6\Delta t$ tells us that when $t = 2020$,

$$\frac{\Delta C}{\Delta t} \approx -0.6, \quad \text{and therefore, that} \quad \frac{dC}{dt} \approx -0.6.$$

21. All are done using the chain rule.

   (a) We have $u = x$, $v = 3$. Thus $du/dx = 1$ and $dv/dx = 0$ so

$$f'(x) = F_u(x, 3)(1) + F_v(x, 3)(0) = F_u(x, 3).$$

   (b) We have $u = 3$, $v = x$. Thus $du/dx = 0$ and $dv/dx = 1$ so

$$f'(x) = F_u(3, x)(0) + F_v(3, x)(1) = F_v(3, x).$$

   (c) We have $u = x$, $v = x$. Thus $du/dx = dv/dx = 1$ so

$$f'(x) = F_u(x, x)(1) + F_v(x, x)(1) = F_u(x, x) + F_v(x, x).$$

   (d) We have $u = 5x$, $v = x^2$. Thus $du/dx = 5$ and $dv/dx = 2x$ so

$$f'(x) = F_u(5x, x^2)(5) + F_v(5x, x^2)(2x).$$

25. Since $\left(\dfrac{\partial U}{\partial P}\right)_V$ involves the variables P and V, we are viewing U as a function of these two variables, so $U = U_3(P, V)$. Then

$$\left(\frac{\partial U}{\partial P}\right)_V = \frac{\partial U_3(P, V)}{\partial P}.$$

29. We will use analysis similar to that in Example 7. Since V is a function of P and T, we have

$$dV = \left(\frac{\partial V}{\partial T}\right)_P dT + \left(\frac{\partial V}{\partial P}\right)_T dP$$

We are interested in $\left(\frac{\partial U}{\partial V}\right)_T$ so we use the formula for dU corresponding to U_2. Substituting g for dV into this formula for dU gives

$$dU = \left(\frac{\partial U}{\partial T}\right)_V dT + \left(\frac{\partial U}{\partial V}\right)_T \left(\left(\frac{\partial V}{\partial T}\right)_P dT + \left(\frac{\partial V}{\partial P}\right)_T dP\right)$$

$$= \left(\left(\frac{\partial U}{\partial T}\right)_V + \left(\frac{\partial U}{\partial V}\right)_T \left(\frac{\partial V}{\partial T}\right)_P\right) dT + \left(\frac{\partial U}{\partial V}\right)_T \left(\frac{\partial V}{\partial P}\right)_T dP$$

But we are also interested in $\left(\frac{\partial U}{\partial P}\right)_T$ so we compare with the formula for dU corresponding to U_1.

$$dU = \left(\frac{\partial U}{\partial T}\right)_P dT + \left(\frac{\partial U}{\partial P}\right)_T dP.$$

Since the coefficients of dP must be identical, we get

$$\left(\frac{\partial U}{\partial P}\right)_T = \left(\frac{\partial U}{\partial V}\right)_T \left(\frac{\partial V}{\partial P}\right)_T.$$

33. Use chain rule for the equation $0 = F(x, y, f(x, y))$. Differentiating both sides with respect to x, remembering $z = f(x, y)$ and regarding y as a constant gives:

$$0 = \frac{\partial F}{\partial x}\frac{dx}{dx} + \frac{\partial F}{\partial z}\frac{dz}{dx}.$$

Since $dx/dx = 1$, we get

$$-\frac{\partial F}{\partial x} = \frac{\partial F}{\partial z}\frac{\partial z}{\partial x},$$

so

$$\frac{\partial z}{\partial x} = -\frac{\partial F/\partial x}{\partial F/\partial z}.$$

Similarly, differentiating both sides of the equation $0 = F(x, y, f(x, y))$ with respect to y gives:

$$0 = \frac{\partial F}{\partial y}\frac{dy}{dy} + \frac{\partial F}{\partial z}\frac{dz}{dy}.$$

Since $dy/dy = 1$, we get

$$-\frac{\partial F}{\partial y} = \frac{\partial F}{\partial z}\frac{\partial z}{\partial y},$$

so

$$\frac{\partial z}{\partial y} = -\frac{\partial F/\partial y}{\partial F/\partial z}.$$

Solutions for Section 14.7

Exercises

1. We have $f_x = 6xy + 5y^3$ and $f_y = 3x^2 + 15xy^2$, so $f_{xx} = 6y$, $f_{xy} = 6x + 15y^2$, $f_{yx} = 6x + 15y^2$, and $f_{yy} = 30xy$.

5. Since $f = (x + y)e^y$, the partial derivatives are

$$f_x = e^y, \quad f_y = e^y(x + 1 + y)$$
$$f_{xx} = 0, \quad f_{yx} = e^y = f_{xy}$$
$$f_{yy} = xe^y + e^y + e^y + ye^y = e^y(x + 2 + y).$$

9. We have $f_x = 6\cos 2x \cos 5y$ and $f_y = -15\sin 2x \sin 5y$, so $f_{xx} = -12\sin 2x \cos 5y$, $f_{xy} = -30\cos 2x \sin 5y$, $f_{yx} = -30\cos 2x \sin 5y$, and $f_{yy} = -75\sin 2x \cos 5y$.

13. The quadratic Taylor expansion about $(0, 0)$ is given by

$$f(x, y) \approx Q(x, y) = f(0,0) + f_x(0,0)x + f_y(0,0)y + \frac{1}{2}f_{xx}(0,0)x^2 + f_{xy}(0,0)xy + \frac{1}{2}f_{yy}(0,0)y^2.$$

First we find all the relevant derivatives

$$f(x, y) = \cos(x + 3y)$$
$$f_x(x, y) = -\sin(x + 3y)$$
$$f_y(x, y) = -3\sin(x + 3y)$$
$$f_{xx}(x, y) = -\cos(x + 3y)$$
$$f_{yy}(x, y) = -9\cos(x + 3y)$$
$$f_{xy}(x, y) = -3\cos(x + 3y)$$

Now we evaluate each of these derivatives at $(0, 0)$ and substitute into the formula to get as our final answer:

$$Q(x, y) = 1 - \frac{1}{2}x^2 - 3xy - \frac{9}{2}y^2$$

Notice this is the same as what you get if you substitute $x + 3y$ for u in the single variable quadratic approximation $Q(u) = 1 - u^2/2$ for $\cos u$.

17. The quadratic Taylor expansion about $(0, 0)$ is given by

$$f(x, y) \approx Q(x, y) = f(0,0) + f_x(0,0)x + f_y(0,0)y + \frac{1}{2}f_{xx}(0,0)x^2 + f_{xy}(0,0)xy + \frac{1}{2}f_{yy}(0,0)y^2$$

So first we find all the relevant derivatives

$$f(x, y) = \sin 2x + \cos y$$
$$f_x(x, y) = 2\cos 2x$$
$$f_y(x, y) = -\sin y$$
$$f_{xx}(x, y) = -4\sin 2x$$
$$f_{yy}(x, y) = -\cos y$$
$$f_{xy}(x, y) = 0$$

We substitute into the formula to get for our answer:

$$Q(x, y) = 1 + 2x - \frac{1}{2}y^2$$

21. (a) $f_x(P) < 0$ because f decreases as you go to the right.
   (b) $f_y(P) = 0$ because f does not change as you go up.
   (c) $f_{xx}(P) < 0$ because f_x decreases as you go to the right (f_x changes from a small negative number to a large negative number).
   (d) $f_{yy}(P) = 0$ because f_y does not change as you go up.
   (e) $f_{xy}(P) = 0$ because f_x does not change as you go up.

25. (a) $f_x(P) < 0$ because f decreases as you go to the right.
   (b) $f_y(P) < 0$ because f decreases as you go up.
   (c) $f_{xx}(P) = 0$ because f_x does not change as you go to the right. (Notice that the level curves are equidistant and parallel, so the partial derivatives of f do not change if you move horizontally or vertically.)
   (d) $f_{yy}(P) = 0$ because f_y does not change as you go up.
   (e) $f_{xy}(P) = 0$ because f_x does not change as you go up.

29. We have $f(1, 0) = 1$ and the relevant derivatives are:

$$f_x = \frac{1}{2}(x + 2y)^{-1/2} \quad \text{so} \quad f_x(1, 0) = \frac{1}{2}$$
$$f_y = (x + 2y)^{-1/2} \quad \text{so} \quad f_y(1, 0) = 1$$
$$f_{xx} = -\frac{1}{4}(x + 2y)^{-3/2} \quad \text{so} \quad f_{xx}(1, 0) = -\frac{1}{4}$$
$$f_{xy} = -\frac{1}{2}(x + 2y)^{-3/2} \quad \text{so} \quad f_{xy}(1, 0) = -\frac{1}{2}$$
$$f_{yy} = -(x + 2y)^{-3/2} \quad \text{so} \quad f_{yy}(1, 0) = -1 \, .$$

Thus the linear approximation, $L(x, y)$ to $f(x, y)$ at $(1, 0)$, is given by:

$$f(x, y) \approx L(x, y) = f(1, 0) + f_x(1, 0)(x - 1) + f_y(1, 0)(y - 0)$$
$$= 1 + \frac{1}{2}(x - 1) + y.$$

The quadratic approximation, $Q(x, y)$ to $f(x, y)$ near $(1, 0)$, is given by:

$$f(x, y) \approx Q(x, y) = f(1, 0) + f_x(1, 0)(x - 1) + f_y(1, 0)(y - 0) + \frac{1}{2}f_{xx}(1, 0)(x - 1)^2$$
$$+ f_{xy}(1, 0)(x - 1)(y - 0) + \frac{1}{2}f_{yy}(1, 0)(y - 0)^2$$
$$= 1 + \frac{1}{2}(x - 1) + y - \frac{1}{8}(x - 1)^2 - \frac{1}{2}(x - 1)y - \frac{1}{2}y^2.$$

The values of the approximations are

$$L(0.9, 0.2) = 1 - 0.05 + 0.2 = 1.15$$
$$Q(0.9, 0.2) = 1 - 0.05 + 0.2 - 0.00125 + 0.01 - 0.02 = 1.13875$$

and the exact value is

$$f(0.9, 0.2) = \sqrt{1.3} \approx 1.14018.$$

Observe that the quadratic approximation is closer to the exact value.

Problems

33. (a) $z_{yx} = z_{xy} = 4y$

(b) $z_{xyx} = \frac{\partial}{\partial x}(z_{xy}) = \frac{\partial}{\partial x}(4y) = 0$

(c) $z_{xyy} = z_{yxy} = \frac{\partial}{\partial y}(4y) = 4$

37. (a) Calculate the partial derivatives:

$$
\begin{array}{lll}
f(x, y) = \sin x \sin y & f(0, 0) = 0 & f(\frac{\pi}{2}, \frac{\pi}{2}) = 1 \\
f_x(x, y) = \cos x \sin y & f_x(0, 0) = 0 & f_x(\frac{\pi}{2}, \frac{\pi}{2}) = 0 \\
f_y(x, y) = \sin x \cos y & f_y(0, 0) = 0 & f_y(\frac{\pi}{2}, \frac{\pi}{2}) = 0 \\
f_{xx}(x, y) = -\sin x \sin y & f_{xx}(0, 0) = 0 & f_{xx}(\frac{\pi}{2}, \frac{\pi}{2}) = -1 \\
f_{xy}(x, y) = \cos x \cos y & f_{xy}(0, 0) = 1 & f_{xy}(\frac{\pi}{2}, \frac{\pi}{2}) = 0 \\
f_{yy}(x, y) = -\sin x \sin y & f_{yy}(0, 0) = 0 & f_{yy}(\frac{\pi}{2}, \frac{\pi}{2}) = -1
\end{array}
$$

Thus, the Taylor polynomial about $(0, 0)$ is

$$f(x, y) \approx Q_1(x, y) = xy.$$

The Taylor polynomial about $(\frac{\pi}{2}, \frac{\pi}{2})$ is

$$f(x, y) \approx Q_2(x, y) = 1 - \frac{1}{2}\left(x - \frac{\pi}{2}\right)^2 - \frac{1}{2}\left(y - \frac{\pi}{2}\right)^2.$$

(b)

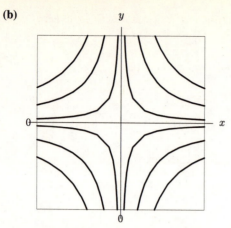

Figure 14.6: $f(x, y) \approx xy$: Quadratic approximation about $(0, 0)$

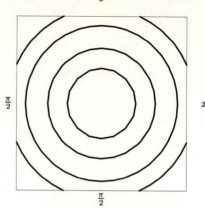

Figure 14.7: $f(x, y) \approx$ $1 - \frac{1}{2}(x - \frac{\pi}{2})^2 - \frac{1}{2}(y - \frac{\pi}{2})^2$: Quadratic approximation about $\left(\frac{\pi}{2}, \frac{\pi}{2}\right)$

41. Letting $G(t) = f(t, 0)$ so that $G'(t) = f_x(t, 0)$, the Fundamental Theorem tells us that $\int_{t=0}^{a} G'(t)dt = G(a) - G(0)$. Thus

$$\int_{t=0}^{a} f_x(t, 0)dt = f(a, 0) - f(0, 0)$$

Letting $H(t) = f(a, t)$ so that $H'(t) = f_y(a, t)$, the Fundamental Theorem tells us that $\int_{t=0}^{b} H'(t)dt = H(b) - H(0)$. Thus

$$\int_{t=0}^{b} f_y(a, t)dt = f(a, b) - f(a, 0)$$

Thus

$$f(0, 0) + \int_{t=0}^{a} f_x(t, 0)dt + \int_{t=0}^{b} f_y(a, t)dt$$
$$= f(0, 0) + (f(a, 0) - f(0, 0)) + (f(a, b) - f(a, 0))$$
$$= f(a, b).$$

Solutions for Section 14.8

Exercises

1. Not differentiable at $(0, 0)$.

5. Not differentiable at all points on the x or y axes.

Problems

9. (a) The contour diagram for $f(x, y) = xy/\sqrt{x^2 + y^2}$ is shown in Figure 14.8.

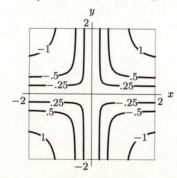

Figure 14.8

(b) By the chain rule, f is differentiable at all points (x, y) where $x^2 + y^2 \neq 0$, and so at all points $(x, y) \neq (0, 0)$.

(c) The partial derivatives of f are given by

$$f_x(x, y) = \frac{y^3}{(x^2 + y^2)^{3/2}}, \qquad \text{for} \quad (x, y) \neq (0, 0),$$

and

$$f_y(x, y) = \frac{x^3}{(x^2 + y^2)^{3/2}}, \qquad \text{for} \quad (x, y) \neq (0, 0).$$

Both f_x and f_y are continuous at $(x, y) \neq (0, 0)$.

(d) If f were differentiable at $(0, 0)$, the chain rule would imply that the function

$$g(t) = \begin{cases} f(t, t), & t \neq 0 \\ 0, & t = 0 \end{cases}$$

would be differentiable at $t = 0$. But

$$g(t) = \frac{t^2}{\sqrt{2t^2}} = \frac{1}{\sqrt{2}} \cdot \frac{t^2}{|t|} = \frac{1}{\sqrt{2}} \cdot |t|,$$

which is not differentiable at $t = 0$. Hence, f is not differentiable at $(0, 0)$.

(e) The partial derivatives of f at $(0, 0)$ are given by

$$f_x(0, 0) = \lim_{x \to 0} \frac{f(x, 0) - f(0, 0)}{x} = \lim_{x \to 0} \frac{\frac{x \cdot 0}{\sqrt{x^2 + 0^2}} - 0}{x} = \lim_{x \to 0} \frac{0 - 0}{x} = 0,$$

$$f_y(0, 0) = \lim_{y \to 0} \frac{f(0, y) - f(0, 0)}{y} = \lim_{y \to 0} \frac{\frac{0 \cdot y}{\sqrt{0^2 + y^2}} - 0}{y} = \lim_{y \to 0} \frac{0 - 0}{y} = 0.$$

The limit $\lim\limits_{(x,y) \to (0,0)} f_x(x, y)$ doesn't exist since if we choose $x = y = t, t \neq 0$, then

$$f_x(x, y) = f_x(t, t) = \frac{t^3}{(2t^2)^{3/2}} = \frac{t^3}{2\sqrt{2} \cdot |t|^3} = \begin{cases} \frac{1}{2\sqrt{2}}, & t > 0, \\ -\frac{1}{2\sqrt{2}}, & t < 0. \end{cases}$$

Thus, f_x is not continuous at $(0, 0)$. Similarly, f_y is not continuous at $(0, 0)$.

13. (a) The contour diagram of $f(x, y) = \sqrt{|xy|}$ is shown in Figure 14.9.

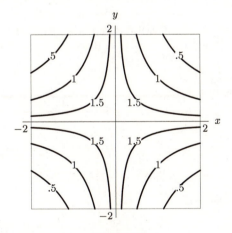

Figure 14.9

(b) The graph of $f(x, y) = \sqrt{|xy|}$ is shown in Figure 14.10.

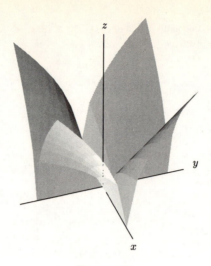

Figure 14.10

(c) f is clearly differentiable at (x, y) where $x \neq 0$ and $y \neq 0$. So we need to look at points $(x_0, 0)$, $x_0 \neq 0$ and $(0, y_0)$, $y_0 \neq 0$. At $(x_0, 0)$:

$$f_x(x_0, 0) = \lim_{x \to x_0} \frac{f(x, 0) - f(x_0, 0)}{x - x_0} = 0$$

$$f_y(x_0, 0) = \lim_{y \to 0} \frac{f(x_0, y) - f(x_0, 0)}{y} = \lim_{y \to 0} \frac{\sqrt{|x_0 y|}}{y}$$

which doesn't exist. So f is not differentiable at the points $(x_0, 0)$, $x_0 \neq 0$. Similarly, f is not differentiable at the points $(0, y_0)$, $y_0 \neq 0$.

(d)

$$f_x(0, 0) = \lim_{x \to 0} \frac{f(x, 0) - f(0, 0)}{x} = 0$$

$$f_y(0, 0) = \lim_{y \to 0} \frac{f(0, y) - f(0, 0)}{y} = 0$$

(e) Let $\vec{u} = (\vec{i} + \vec{j})/\sqrt{2}$:

$$f_{\vec{u}}(0, 0) = \lim_{t \to 0^+} \frac{f(\frac{t}{\sqrt{2}}, \frac{t}{\sqrt{2}}) - f(0, 0)}{t} = \lim_{t \to 0^+} \frac{\sqrt{\frac{t^2}{2}}}{t} = \frac{1}{\sqrt{2}}.$$

We know that $\nabla f(0, 0) = \vec{0}$ because both partial derivatives are 0. But if f were differentiable, $f_{\vec{u}}(0, 0) = \nabla f(0, 0) \cdot \vec{u} = f_x(0, 0) \cdot \frac{1}{\sqrt{2}} + f_y(0, 0) \cdot \frac{1}{\sqrt{2}} = 0$. But since, in fact, $f_{\vec{u}}(0, 0) = 1/\sqrt{2}$, we conclude that f is not differentiable.

Solutions for Chapter 14 Review

Exercises

1. $\frac{\partial z}{\partial x} = \frac{\partial}{\partial x}\left[(x^2 + x - y)^7\right] = 7(x^2 + x - y)^6(2x + 1) = (14x + 7)(x^2 + x - y)^6.$

$\frac{\partial z}{\partial y} = \frac{\partial}{\partial y}\left[(x^2 + x - y)^7\right] = -7(x^2 + x - y)^6.$

5. The first order partial derivative f_x is

$$f_x = -x(x^2 + y^2)^{-3/2}$$

Thus the second order partials are

$$f_{xx} = -(x^2 + y^2)^{-3/2} + 3x^2(x^2 + y^2)^{-5/2} = (2x^2 - y^2)(x^2 + y^2)^{-5/2}$$

and

$$f_{xy} = 3xy(x^2 + y^2)^{-5/2}.$$

9. Since the partial derivatives are

$$f_x = \frac{2x}{x^2 + y^2} \quad \text{and} \quad f_y = \frac{2y}{x^2 + y^2},$$

we have

$$\nabla f = \frac{2}{x^2 + y^2}(x\vec{i} + y\vec{j}).$$

13. We have grad $f = 3x^2\vec{i} - 3y^2\vec{j}$, so grad $f(2,-1) = 12\vec{i} - 3\vec{j}$. A unit vector in the direction we want is $\vec{u} = (1/\sqrt{2})(\vec{i} - \vec{j})$. Therefore, the directional derivative is

$$\text{grad } f(-2, 1) \cdot \vec{u} = \frac{12 \cdot 1 - 3(-1)}{\sqrt{2}} = \frac{15}{\sqrt{2}}.$$

17. The unit vector $\vec{u}$ in the direction of $\vec{v} = \vec{i} - \vec{k}$ is $\vec{u} = \frac{1}{\sqrt{2}}\vec{i} - \frac{1}{\sqrt{2}}\vec{k}$. We have

$$f_x(x, y, z) = 6xy^2, \qquad \text{and } f_x(-1, 0, 4) = 0$$
$$f_y(x, y, z) = 6x^2y + 2z, \quad \text{and } f_y(-1, 0, 4) = 8$$
$$f_z(x, y, z) = 2y, \qquad \text{and } f_z(-1, 0, 4) = 0.$$

So,

$$f_{\vec{u}}(-1, 0, 4) = f_x(-1, 0, 4)\left(\frac{1}{\sqrt{2}}\right) + f_y(-1, 0, 4)(0) + f_z(-1, 0, 4)\left(-\frac{1}{\sqrt{2}}\right)$$

$$= 0\left(\frac{1}{\sqrt{2}}\right) + 8(0) + 0\left(-\frac{1}{\sqrt{2}}\right)$$

$$= 0.$$

21. The given surface is the level surface for $f(x, y, z) = z^2 - 2xyz - x^2 - y^2$ passing through the point $(1, 2, -1)$. Thus a normal vector is grad $f(1, 2, -1)$. We have

$$\text{grad } f = (-2x - 2yz)\vec{i} + (-2y - 2xz)\vec{j} + (2z - 2xy)\vec{k},$$

so a normal vector is grad $f(1, 2, -1) = 2\vec{i} - 2\vec{j} - 6\vec{k}$.

Problems

25. (a) The difference quotient for evaluating $f_w(2, 2)$ is

$$f_w(2, 2) \approx \frac{f(2 + 0.01, 2) - f(2, 2)}{h} = \frac{e^{(2.01)\ln 2} - e^{2\ln 2}}{0.01} = \frac{e^{\ln(2^{2.01})} - e^{\ln(2^2)}}{0.01}$$

$$= \frac{2^{(2.01)} - 2^2}{0.01} \approx 2.78$$

The difference quotient for evaluating $f_z(2, 2)$ is

$$f_z(2, 2) \approx \frac{f(2, 2 + 0.01) - f(2, 2)}{h}$$

$$= \frac{e^{2\ln(2.01)} - e^{2\ln 2}}{0.01} = \frac{(2.01)^2 - 2^2}{0.01} = 4.01$$

(b) Using the derivative formulas we get

$$f_w = \frac{\partial f}{\partial w} = \ln z \cdot e^{w\ln z} = z^w \cdot \ln z$$

$$f_z = \frac{\partial f}{\partial z} = e^{w\ln z} \cdot \frac{w}{z} = w \cdot z^{w-1}$$

so

$$f_w(2, 2) = 2^2 \cdot \ln 2 \approx 2.773$$
$$f_z(2, 2) = 2 \cdot 2^{2-1} = 4.$$

29. Using local linearity with the slope in the x direction of -3 and slope in the y direction of 4, we get the values in Table 14.3.

Table 14.3

		0.9	1.0	1.1
	1.8	7.2	7.6	8.0
x	2.0	6.6	7.0	7.4
	2.2	6.0	6.4	6.8

33. To increase f as much as possible, we should head in the direction of the gradient from the point $(2, 1)$. The rate of increase of f in the direction of the gradient is the magnitude of the gradient. Since at the point $(2, 1)$, we know grad $f = -3\vec{i} + 4\vec{j}$, the magnitude is 5. The furthest we can go from $(2, 1)$, inside the circle, is 0.1 units, so the most we can increase f is $(0.1)(5) = 0.5$. Thus

$$\text{Largest value of function} \approx f(2, 1) + 0.5 = 7.5.$$

This value is achieved at the point obtained from $(2, 1)$ by a displacement of 0.1 units in the direction of grad f, that is, a displacement by the vector

$$(0.1)\frac{\text{grad } f}{\|\text{grad } f\|} = (0.1)((-3/5)\vec{i} + (4/5)\vec{j}) = -0.06\vec{i} + 0.08\vec{j}.$$

Thus, the largest value of f is achieved at the point $(2 - 0.06, 1 + 0.08) = (1.94, 1.08)$.

37. The sign of $\partial f / \partial P_1$ tells you whether f (the number of people who ride the bus) increases or decreases when P_1 is increased. Since P_1 is the price of taking the bus, as it increases, f should decrease. This is because fewer people will be willing to pay the higher price, and more people will choose to ride the train. On the other hand, the sign of $\dfrac{\partial f}{\partial P_2}$ tells you the change in f as P_2 increases. Since P_2 is the cost of riding the train, as it increases, f should increase. This is because fewer people will be willing to pay the higher fares for the train, and more people will choose to ride the bus.

Therefore, $\dfrac{\partial f}{\partial P_1} < 0$ and $\dfrac{\partial f}{\partial P_2} > 0$.

41. The differential is

$$dP = \frac{\partial P}{\partial L}\, dL + \frac{\partial P}{\partial K}\, dK = 10L^{-0.75}K^{0.75}\, dL + 30L^{0.25}K^{-0.25}\, dK.$$

When $L = 2$ and $K = 16$, this is

$$dP \approx 47.6\, dL + 17.8\, dK.$$

45. $f_x = 2x$, $f_y = -2y$, so grad $f(3, -1) = 6\vec{i} + 2\vec{j}$. For the direction $\theta = \pi/4$, the direction is $\vec{u} = \frac{1}{\sqrt{2}}\vec{i} + \frac{1}{\sqrt{2}}\vec{j}$, so $f_{\vec{u}}(3, -1) = (6\vec{i} + 2\vec{j}) \cdot (\frac{1}{\sqrt{2}}\vec{i} + \frac{1}{\sqrt{2}}\vec{j}) = \frac{8}{\sqrt{2}} = 4\sqrt{2}$.

The directional derivative is largest in the direction of the gradient vector grad $f(3, -1) = 6\vec{i} + 2\vec{j}$.

49. (a) Fix $y = 3$. When x changes from 2.00 to 2.01, $f(x, 3)$ decreases from 7.56 to 7.42. So

$$\left.\frac{\partial f}{\partial x}\right|_{(2,3)} \approx \left.\frac{\Delta f}{\Delta x}\right|_{(2,3)} = \frac{7.42 - 7.56}{2.01 - 2.00} = \frac{-0.14}{0.01} = -14.$$

Fix $x = 2$, when y changes from 3.00 to 3.02, $f(2, y)$ increases from 7.56 to 7.61. So

$$\left.\frac{\partial f}{\partial y}\right|_{(2,3)} \approx \left.\frac{\Delta f}{\Delta y}\right|_{(2,3)} = \frac{7.61 - 7.56}{3.02 - 3.00} = \frac{0.05}{0.02} = 2.5.$$

(b) Since the unit vector $\vec{u}$ of the direction $\vec{i} + 3\vec{j}$ is

$$\vec{u} = \frac{\vec{i} + 3\vec{j}}{\|\vec{i} + 3\vec{j}\|} = \frac{1}{\sqrt{10}}\vec{i} + \frac{3}{\sqrt{10}}\vec{j},$$

$$f_{\vec{u}}(2,3) = \operatorname{grad} f(2,3) \cdot \vec{u} \approx \left(\frac{\Delta f}{\Delta x}\bigg|_{(2,3)} \vec{i} + \frac{\Delta f}{\Delta y}\bigg|_{(2,3)} \vec{j} \right) \cdot \vec{u}$$

$$= (-14\vec{i} + 2.5\vec{j}) \cdot \left(\frac{1}{\sqrt{10}}\vec{i} + \frac{3}{\sqrt{10}}\vec{j} \right) = -\frac{6.5}{\sqrt{10}} \approx -2.055.$$

(c) Maximum rate equals $\| \operatorname{grad} f\| \approx \sqrt{(-14)^2 + (2.5)^2} \approx 14.221$ in the direction of the gradient which is approximately equal to $-14\vec{i} + 2.5\vec{j}$.

(d) The equation of the level curve is

$$f(x, y) = f(2, 3) = 7.56.$$

(e) The vector must be perpendicular to grad f, so $\vec{v} = 2.5\vec{i} + 14\vec{j}$ is a possible answer. (There are many others.).

(f) The differential at the point $(2, 3)$ is

$$df = -14\, dx + 2.5\, dy.$$

If $dx = 0.03$, $dy = 0.04$, we get

$$df = -14(0.03) + 2.5(0.04) = -0.32.$$

The df approximates the change in f when (x, y) changes from $(2, 3)$ to $(2.03, 3.04)$.

53. We know from Problem 52 that $V(x, y) = xF(2x + y)$ will solve the equation for any function F, so it is enough to find an F such that $xF(2x + y) = y^2$ when $x = 1$.
In other words, $1 \cdot F(2 + y) = y^2$, i.e. $F(y) = (y - 2)^2$ for all y. So one solution is $V(x, y) = xF(2x + y) = x(2x + y - 2)^2 = 4x^3 + xy^2 + 4x + 4x^2y - 8x^2 - 4xy$.

CAS Challenge Problems

57. (a) We have

$$f(1, 2) = A_0 + A_1 + 2A_2 + A_3 + 2A_4 + 4A_5,$$
$$f_x(1, 2) = A_1 + 2A_3 + 2A_4,$$
$$f_y(1, 2) = A_2 + A_4 + 4A_5,$$
$$f_{xx}(1, 2) = 2A_3,$$
$$f_{xy}(1, 2) = A_4,$$
$$f_{yy}(1, 2) = 2A_5.$$

Thus $Q(x, y) = (A_0 + A_1 + 2A_2 + A_3 + 2A_4 + 4_A5) + (A_1 + 2A_3 + 2A_4)(x - 1) + (A_2 + A_4 + 4A_5)(y - 2) + A_3(x - 1)^2 + A_4(x - 1)(y - 2) + A_5(y - 2)^2$. Expanding this expression in powers of x and y, we find

$$Q(x, y) = A_0 + A_1x + A_2y + A_3x^2 + A_4xy + A_5y^2 = f(x, y)$$

(b) The quadratic expansion for f about $(1, 2)$ is equal to f itself. This is because f is already a quadratic function, and is true about any point (a, b).

(c) The linear approximation of $f(x, y)$ at $(1, 2)$ is:

$$A_0 + A_1 + 2A_2 + A_3 + 2A_4 + 4A_5 + (A_1 + 2A_3 + 2A_4)(x - 1) + (A_2 + A_4 + 4A_5)(y - 2).$$

When we expand this we get

$$A_0 - A_3 + A_4 - 4A_5 + (A_1 + 2A_3 + 2A_4)x + (A_2 + A_4 + 4A_5)y.$$

This is not the same as $f(x, y)$, and it is not even the same as the linear part of $f(x, y)$, namely $A_0 + A_1x + A_2y$.

CHECK YOUR UNDERSTANDING

1. True. This is the instantaneous rate of change of f in the x-direction at the point $(10, 20)$.

5. True. The property $f_x(a, b) > 0$ means that f increases in the positive x-direction near (a, b), so f must decrease in the negative x-direction near (a, b).

9. True. A function with constant $f_x(x, y)$ and $f_y(x, y)$ has constant x-slope and constant y-slope, and therefore has a graph which is a plane.

13. True. Since g is a function of x only, it can be treated like a constant when taking the y partial derivative.

17. True. The definition of $f_x(x, y)$ is the limit of the difference quotient

$$f_x(x, y) = \lim_{h \to 0} \frac{f(x + h, y) - f(x, y)}{h}.$$

The symmetry of f gives

$$\frac{f(x + h, y) - f(x, y)}{h} = \frac{f(y, x + h) - f(y, x)}{h}.$$

The definition of $f_y(y, x)$ is the limit of the difference quotient

$$f_y(y, x) = \lim_{h \to 0} \frac{f(y, x + h) - f(y, x)}{h}.$$

Thus $f_x(x, y) = f_y(y, x)$.

21. False. The equation $z = 2 + 2x(x - 1) + 3y^2(y - 1)$ is not linear. The correct equation is $z = 2 + 2(x - 1) + 3(y - 1)$, which is obtained by evaluating the partial derivatives at the point $(1, 1)$.

25. False. The graph of f is a paraboloid, opening upward. The tangent plane to this surface at any point lies completely under the surface (except at the point of tangency). So the local linearization *underestimates* the value of f at nearby points.

29. True. If f is linear, then $f(x, y) = mx + ny + c$ for some m, n and c. So $f_x = m$ and $f_y = n$ giving $df = m\, dx + n\, dy$, which is linear in the variables dx and dy.

33. True. If $\vec{u}$ is a unit vector, then the directional derivative is given by the formula $f_{\vec{u}}(a, b) = \operatorname{grad} f(a, b) \cdot \vec{u}$.

37. True. The gradient points in the direction of maximal increase of f, and the opposite direction gives the direction of maximum decrease for f.

41. True. It is the rate of change of f in the direction of $\vec{u}$ at the point (x_0, y_0).

45. Is never true. If $\| \operatorname{grad} f \| = 0$, then $\operatorname{grad} f = 0$, so $\operatorname{grad} f \cdot \vec{u} = 0$ for any unit vector $\vec{u}$. Thus the directional derivative must be zero.

CHAPTER FIFTEEN

Solutions for Section 15.1

Exercises

1. The point A is not a critical point and the contour lines look like parallel lines. The point B is a critical point and is a local maximum; the point C is a saddle point.

5. The partial derivatives are $f_x(x, y) = 3x^2 - 3$ which vanishes for $x = \pm 1$ and $f_y(x, y) = 3y^2 - 3$ which vanishes for $y = \pm 1$. The points $(1, 1), (1, -1), (-1, 1), (-1, -1)$ where both partials vanish are the critical points. To determine the nature of these critical points we calculate their discriminant and use the second derivative test. The discriminant is

$$D = f_{xx}(x, y)f_{yy}(x, y) - f_{xy}^2(x, y) = (6x)(6y) - 0 = 36xy.$$

At $(1, 1)$ and $(-1, -1)$ the discriminant is positive. Since $f_{xx}(1, 1) = 6$ is positive, $(1, 1)$ is a local minimum. And since $f_{xx}(-1, -1) = -6$ is negative, $(-1, -1)$ is a local maximum. The remaining two points, $(1, -1)$ and $(-1, 1)$ are saddle points since the discriminant is negative.

9. To find the critical points, we solve $f_x = 0$ and $f_y = 0$ for x and y. Solving

$$f_x = 3x^2 - 3 = 0$$
$$f_y = 3y^2 - 12y = 0$$

shows that $x = -1$ or $x = 1$ and $y = 0$ or $y = 4$. There are four critical points: $(-1, 0)$, $(1, 0)$, $(-1, 4)$, and $(1, 4)$.
We have
$$D = (f_{xx})(f_{yy}) - (f_{xy})^2 = (6x)(6y - 12) - (0)^2 = (6x)(6y - 12).$$
At critical point $(-1, 0)$, we have $D > 0$ and $f_{xx} < 0$, so f has a local maximum at $(-1, 0)$.
At critical point $(1, 0)$, we have $D < 0$, so f has a saddle point at $(1, 0)$.
At critical point $(-1, 4)$, we have $D < 0$, so f has a saddle point at $(-1, 4)$.
At critical point $(1, 4)$, we have $D > 0$ and $f_{xx} > 0$, so f has a local minimum at $(1, 4)$.

13. At a critical point, $f_x = 0$, $f_y = 0$.

$$f_x = 8y - (x + y)^3 = 0, \text{ we know } 8y = (x + y)^3.$$

$$f_y = 8x - (x + y)^3 = 0, \text{ we know } 8x = (x + y)^3.$$

Therefore we must have $x = y$. Since $(x+y)^3 = (2y)^3 = 8y^3$, this tells us that $8y - 8y^3 = 0$. Solving gives $y = 0, \pm 1$. Thus the critical points are $(0, 0), (1, 1), (-1, -1)$.
$f_{yy} = f_{xx} = -3(x + y)^2$, and $f_{xy} = 8 - 3(x + y)^2$.
The discriminant is

$$D(x, y) = f_{xx}f_{yy} - f_{xy}^2$$
$$= 9(x + y)^4 - (64 - 48(x + y)^2 + 9(x + y)^4)$$
$$= -64 + 48(x + y)^2.$$

$D(0, 0) = -64 < 0$, so $(0, 0)$ is a saddle point.
$D(1, 1) = -64 + 192 > 0$ and $f_{xx}(1, 1) = -12 < 0$, so $(1, 1)$ is a local maximum.
$D(-1, -1) = -64 + 192 > 0$ and $f_{xx}(-1, -1) = -12 < 0$, so $(-1, -1)$ is a local maximum.

Problems

17. To find critical points, set partial derivatives equal to zero:

$$E_x = \sin x = 0 \quad \text{when} \quad x = 0, \pm\pi, \pm 2\pi, \cdots$$

$$E_y = y = 0 \quad \text{when} \quad y = 0.$$

The critical points are

$$\cdots (-2\pi, 0), (-\pi, 0), (0, 0), (\pi, 0), (2\pi, 0), (3\pi, 0) \cdots$$

To classify, calculate $D = E_{xx}E_{yy} - (E_{xy})^2 = \cos x$.
At the points $(0,0), (\pm 2\pi, 0), (\pm 4\pi, 0), (\pm 6\pi, 0), \cdots$

$$D = (1) > 0 \quad \text{and} \quad E_{xx} > 0 \quad (\text{Since} E_{xx}(0, 2k\pi) = \cos(2k\pi) = 1).$$

Therefore $(0,0), (\pm 2\pi, 0), (\pm 4\pi, 0), (\pm 6\pi, 0), \cdots$ are local minima.
At the points $(\pm \pi, 0), (\pm 3\pi, 0), (\pm 5\pi, 0), (\pm 7\pi, 0), \cdots$, we have $\cos(2k+1)\pi = -1$, so

$$D = (-1) < 0.$$

Therefore $(\pm \pi, 0), (\pm 3\pi, 0), (\pm 5\pi, 0), (\pm 7\pi, 0), \cdots$ are saddle points.

21. (a) $(1, 3)$ is a critical point. Since $f_{xx} > 0$ and the discriminant

$$D = f_{xx}f_{yy} - f_{xy}^2 = f_{xx}f_{yy} - 0^2 = f_{xx}f_{yy} > 0,$$

the point $(1, 3)$ is a minimum.

(b)

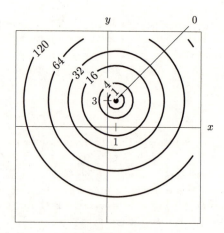

Figure 15.1

25. The first order partial derivatives are

$$f_x(x, y) = 2kx - 2y \quad \text{and} \quad f_y(x, y) = 2ky - 2x.$$

And the second order partial derivatives are

$$f_{xx}(x, y) = 2k \qquad f_{xy}(x, y) = -2 \qquad f_{yy}(x, y) = 2k$$

Since $f_x(0,0) = f_y(0,0) = 0$, the point $(0,0)$ is a critical point. The discriminant is

$$D = (2k)(2k) - 4 = 4(k^2 - 1).$$

For $k = \pm 2$, the discriminant is positive, $D = 12$. When $k = 2$, $f_{xx}(0,0) = 4$ which is positive so we have a local minimum at the origin. When $k = -2$, $f_{xx}(0,0) = -4$ so we have a local maximum at the origin. In the case $k = 0$, $D = -4$ so the origin is a saddle point.

Lastly, when $k = \pm 1$ the discriminant is zero, so the second derivative test can tell us nothing. Luckily, we can factor $f(x, y)$ when $k = \pm 1$. When $k = 1$,

$$f(x, y) = x^2 - 2xy + y^2 = (x - y)^2.$$

This is always greater than or equal to zero. So $f(0,0) = 0$ is a minimum and the surface is a trough-shaped parabolic cylinder with its base along the line $x = y$.

When $k = -1$,

$$f(x, y) = -x^2 - 2xy - y^2 = -(x + y)^2.$$

This is always less than or equal to zero. So $f(0,0) = 0$ is a maximum. The surface is a parabolic cylinder, with its top ridge along the line $x = -y$.

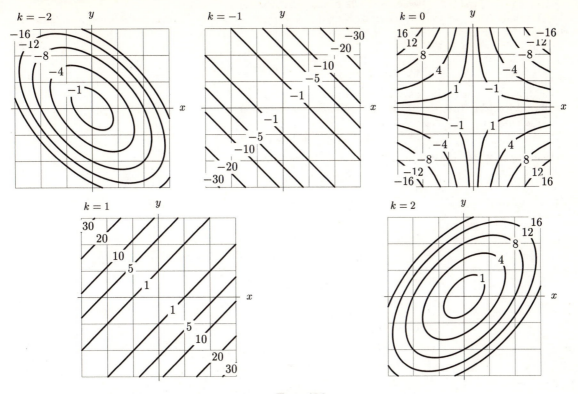

Figure 15.2

Solutions for Section 15.2

Exercises

1. Mississippi lies entirely within a region designated as 80s so we expect both the maximum and minimum daily high temperatures within the state to be in the 80s. The southwestern-most corner of the state is close to a region designated as 90s, so we would expect the temperature here to be in the high 80s, say 87-88. The northern-most portion of the state is located near the center of the 80s region. We might expect the high temperature there to be between 83-87.

Alabama also lies completely within a region designated as 80s so both the high and low daily high temperatures within the state are in the 80s. The southeastern tip of the state is close to a 90s region so we would expect the temperature here to be about 88-89 degrees. The northern-most part of the state is near the center of the 80s region so the temperature there is 83-87 degrees.

Pennsylvania is also in the 80s region, but it is touched by the boundary line between the 80s and a 70s region. Thus we expect the low daily high temperature to occur there and be about 80 degrees. The state is also touched by a boundary line of a 90s region so the high will occur there and be 89-90 degrees.

New York is split by a boundary between an 80s and a 70s region, so the northern portion of the state is likely to be about 74-76 while the southern portion is likely to be in the low 80s, maybe 81-84 or so.

California contains many different zones. The northern coastal areas will probably have the daily high as low as 65-68, although without another contour on that side, it is difficult to judge how quickly the temperature is dropping off to the west. The tip of Southern California is in a 100s region, so there we expect the daily high to be 100-101.

Arizona will have a low daily high around 85-87 in the northwest corner and a high in the 100s, perhaps 102-107 in its southern regions.

Massachusetts will probably have a high daily high around 81-84 and a low daily high of 70.

5. To maximize $z = x^2 + y^2$, it suffices to maximize x^2 and y^2. We can maximize both of these at the same time by taking the point $(1, 1)$, where $z = 2$. It occurs on the boundary of the square. (Note: We also have maxima at the points $(-1, -1), (-1, 1)$ and $(1, -1)$ which are on the boundary of the square.)

To minimize $z = x^2 + y^2$, we choose the point $(0, 0)$, where $z = 0$. It does not occur on the boundary of the square.

9. The maximum value, which is slightly above 30, say 30.5, occurs approximately at the origin. The minimum value, which is about 20.5, occurs at $(2.5, 5)$.

Problems

13. Let the sides be x, y, z cm. Then the volume is given by $V = xyz = 32$.

The surface area S is given by

$$S = 2xy + 2xz + 2yz.$$

Substituting $z = 32/(xy)$ gives

$$S = 2xy + \frac{64}{y} + \frac{64}{x}.$$

At a critical point,

$$\frac{\partial S}{\partial x} = 2y - \frac{64}{x^2} = 0$$

$$\frac{\partial S}{\partial y} = 2x - \frac{64}{y^2} = 0,$$

The symmetry of the equations (or by dividing the equations) tells us that $x = y$ and

$$2x - \frac{64}{x^2} = 0$$

$$x^3 = 32$$

$$x = 32^{1/3} = 3.17 \text{ cm}.$$

Thus the only critical point is $x = y = (32)^{1/3}$ cm and $z = 32/\left((32)^{1/3} \cdot (32)^{1/3}\right) = (32)^{1/3}$ cm. At the critical point

$$S_{xx}S_{yy} - (S_{xy})^2 = \frac{128}{x^3} \cdot \frac{128}{y^3} - 2^2 = \frac{(128)^2}{x^3 y^3} - 4.$$

Since $D > 0$ and $S_{xx} > 0$ at this critical point, the critical point $x = y = z = (32)^{1/3}$ is a local minimum. Since $S \to \infty$ as $x, y \to \infty$, the local minimum is a global minimum.

17. We minimize the square of the distance from the point (x, y, z) to the origin:

$$S = x^2 + y^2 + z^2.$$

Since $z^2 = 9 - xy - 3x$, we have

$$S = x^2 + y^2 + 9 - xy - 3x.$$

At a critical point

$$\frac{\partial S}{\partial x} = 2x - y - 3 = 0$$

$$\frac{\partial S}{\partial y} = 2y - x = 0,$$

so $x = 2y$, and

$$2(2y) - y - 3 = 0$$

giving $y = 1$, so $x = 2$ and $z^2 = 9 - 2 \cdot 1 - 3 \cdot 2 = 1$, so $z = \pm 1$. We have

$$D = S_{xx}S_{yy} - (S_{xy})^2 = 2 \cdot 2 - (-1)^2 = 4 - 1 > 0,$$

so, since $D > 0$ and $S_{xx} > 0$, the critical points are local minima. Since $S \to \infty$ as $x, y \to \pm\infty$, the local minima are global minima.

If $x = 2, y = 1, z = \pm 1$, we have $S = 2^2 + 1^2 + 1^2 = 6$, so the shortest distance to the origin is $\sqrt{6}$.

21. We calculate the partial derivatives and set them to zero.

$$\frac{\partial \text{(range)}}{\partial t} = -10t - 6h + 400 = 0$$

$$\frac{\partial \text{(range)}}{\partial h} = -6t - 6h + 300 = 0.$$

$$10t + 6h = 400$$
$$6t + 6h = 300$$

solving we obtain

$$4t = 100$$

so

$$t = 25$$

Solving for h, we obtain $6h = 150$, yielding $h = 25$. Since the range is quadratic in h and t, the second derivative test tells us this is a local and global maximum. So the optimal conditions are $h = 25\%$ humidity and $t = 25°$C.

25. Let $P(K, L)$ be the profit obtained using K units of capital and L units of labor. The cost of production is given by

$$C(K, L) = kK + \ell L,$$

and the revenue function is given by

$$R(K, L) = pQ = pAK^a L^b.$$

Hence, the profit is

$$P = R - C = pAK^a L^b - (kK + \ell L).$$

In order to find local maxima of P, we calculate the partial derivatives and see where they are zero. We have:

$$\frac{\partial P}{\partial K} = apAK^{a-1} L^b - k,$$

$$\frac{\partial P}{\partial L} = bpAK^a L^{b-1} - \ell.$$

The critical points of the function $P(K, L)$ are solutions (K, L) of the simultaneous equations:

$$\frac{k}{a} = pAK^{a-1} L^b,$$

$$\frac{\ell}{b} = pAK^a L^{b-1}.$$

Multiplying the first equation by K and the second by L, we get

$$\frac{kK}{a} = \frac{\ell L}{b},$$

and so

$$K = \frac{\ell a}{kb} L.$$

Substituting for K in the equation $k/a = pAK^{a-1} L^b$, we get:

$$\frac{k}{a} = pA \left(\frac{\ell a}{kb} \right)^{a-1} L^{a-1} L^b.$$

We must therefore have

$$L^{1-a-b} = pA \left(\frac{a}{k} \right)^a \left(\frac{\ell}{b} \right)^{a-1}.$$

Hence, if $a + b \neq 1$,

$$L = \left[pA \left(\frac{a}{k} \right)^a \left(\frac{\ell}{b} \right)^{(a-1)} \right]^{1/(1-a-b)},$$

and

$$K = \frac{\ell a}{kb} L = \frac{\ell a}{kb} \left[pA \left(\frac{a}{k} \right)^a \left(\frac{\ell}{b} \right)^{(a-1)} \right]^{1/(1-a-b)}.$$

To see if this is really a local maximum, we apply the second derivative test. We have:

$$\frac{\partial^2 P}{\partial K^2} = a(a-1)pAK^{a-2}L^b,$$

$$\frac{\partial^2 P}{\partial L^2} = b(b-1)pAK^aL^{b-2},$$

$$\frac{\partial^2 P}{\partial K\partial L} = abpAK^{a-1}L^{b-1}.$$

Hence,

$$D = \frac{\partial^2 P}{\partial K^2}\frac{\partial^2 P}{\partial L^2} - \left(\frac{\partial^2 P}{\partial K\partial L}\right)^2$$

$$= ab(a-1)(b-1)p^2A^2K^{2a-2}L^{2b-2} - a^2b^2p^2A^2K^{2a-2}L^{2b-2}$$

$$= ab((a-1)(b-1) - ab)p^2A^2K^{2a-2}L^{2b-2}$$

$$= ab(1-a-b)p^2A^2K^{2a-2}L^{2b-2}.$$

Now a, b, p, A, K, and L are positive numbers. So, the sign of this last expression is determined by the sign of $1 - a - b$.

(a) We assumed that $a + b < 1$, so $D > 0$, and as $0 < a < 1$, then $\partial^2 P/\partial K^2 < 0$ and so we have a unique local maximum. To verify that the local maximum is a global maximum, we focus on the cost. Let $C = kK + \ell L$. Since $K \geq 0$ and $L \geq 0$, $K \leq C/k$ and $L \leq C/\ell$. Therefore the profit satisfies:

$$P = pAK^aL^b - (kK + \ell L)$$

$$\leq pA\left(\frac{C}{k}\right)^a\left(\frac{C}{\ell}\right)^b - C$$

$$= mC^{a+b} - C$$

where $m = pA(1/k)^a(1/\ell)^b$. Since $a + b < 1$, the profit is negative for large costs C, say $C \geq C_0$ ($C_0 = m^{1-a-b}$ will do). Therefore, in the KL-plane for $K \geq 0$ and $L \geq 0$, the profit is less than or equal to zero everywhere on or above the line $kK + \ell L = C_o$. Thus the global maximum must occur inside the triangle bounded by this line and the K and L axes. Since $P \leq 0$ on the K and L axes as well, the global maximum must be in the interior of the triangle at the unique local maximum we found.

In the case $a + b < 1$, we have decreasing returns to scale. That is, if the amount of capital and labor used is multiplied by a constant $\lambda > 0$, we get less than λ times the production.

(b) Now suppose $a + b \geq 1$. If we multiply K and L by λ for some $\lambda > 0$, then

$$Q(\lambda K, \lambda L) = A(\lambda K)^a(\lambda L)^b = \lambda^{a+b}Q(K, L).$$

We also see that

$$C(\lambda K, \lambda L) = \lambda C(K, L).$$

So if $a + b = 1$, we have

$$P(\lambda K, \lambda L) = \lambda P(K, L).$$

Thus, if $\lambda = 2$, so we are doubling the inputs K and L, then the profit P is doubled and hence there can be no maximum profit.

If $a + b > 1$, we have increasing returns to scale and there can again be no maximum profit: doubling the inputs will more than double the profit. In this case, the profit increases without bound as K, L go toward infinity.

29. (a) The function f is continuous in the region R, but R is not closed and bounded so a special analysis is required.

Notice that $f(x, y)$ tends to ∞ as (x, y) tends farther and farther from the origin or tends toward any point on the x or y axis. This suggests that a minimum for f, if it exists, can not be too far from the origin or too close to the axes. For example, if $x > 10$ then $f(x, y) > 4x > 40$, and if $y > 10$ then $f(x, y) > 5y > 50$. If $0 < x < 0.1$ then $f(x, y) > 2/x > 20$, and if $0 < y < 0.1$ then $f(x, y) > 3/y > 30$.

Since $f(1, 1) = 14$, a global minimum for f if it exists must be in the smaller region $R' : 0.1 \leq x \leq 10$, $0.1 \leq y \leq 10$. The region R' is closed and bounded and so f does have a minimum value at some point in R', and since that value is at most 14, it is also a global minimum for all of R.

(b) Since the region R has no boundary, the minimum value must occur at a critical point of f. At a critical point we have

$$f_x = -\frac{2}{x^2} + 4 = 0 \qquad f_y = -\frac{3}{y^2} + 5 = 0.$$

The only critical point is $(\sqrt{1/2}, \sqrt{3/5}) \approx (0.7071, 0.7746)$, at which f achieves the minimum value $f(\sqrt{1/2}, \sqrt{3/5}) = 4\sqrt{2} + 2\sqrt{15} \approx 13.403$.

Solutions for Section 15.3

Exercises

1. Our objective function is $f(x, y) = x + y$ and our equation of constraint is $g(x, y) = x^2 + y^2 = 1$. To optimize $f(x, y)$ with Lagrange multipliers, we solve $\nabla f(x, y) = \lambda \nabla g(x, y)$ subject to $g(x, y) = 1$. The gradients of f and g are

$$\nabla f(x, y) = \vec{i} + \vec{j},$$
$$\nabla g(x, y) = 2x\vec{i} + 2y\vec{j}.$$

So the equation $\nabla f = \lambda \nabla g$ becomes

$$\vec{i} + \vec{j} = \lambda(2x\vec{i} + 2y\vec{j})$$

Solving for λ gives

$$\lambda = \frac{1}{2x} = \frac{1}{2y},$$

which tells us that $x = y$. Going back to our equation of constraint, we use the substitution $x = y$ to solve for y:

$$g(y, y) = y^2 + y^2 = 1$$
$$2y^2 = 1$$
$$y^2 = \frac{1}{2}$$
$$y = \pm\sqrt{\frac{1}{2}} = \pm\frac{\sqrt{2}}{2}.$$

Since $x = y$, our critical points are $(\frac{\sqrt{2}}{2}, \frac{\sqrt{2}}{2})$ and $(-\frac{\sqrt{2}}{2}, -\frac{\sqrt{2}}{2})$. Since the constraint is closed and bounded, maximum and minimum values of f subject to the constraint exist. Evaluating f at the critical points we find that the maximum value is $f(\frac{\sqrt{2}}{2}, \frac{\sqrt{2}}{2}) = \sqrt{2}$ and the minimum value is $f(-\frac{\sqrt{2}}{2}, -\frac{\sqrt{2}}{2}) = -\sqrt{2}$.

5. Our objective function is $f(x, y) = xy$ and our equation of constraint is $g(x, y) = 4x^2 + y^2 = 8$. Their gradients are

$$\nabla f(x, y) = y\vec{i} + x\vec{j},$$
$$\nabla g(x, y) = 8x\vec{i} + 2y\vec{j}.$$

So the equation $\nabla f = \lambda \nabla g$ becomes $y\vec{i} + x\vec{j} = \lambda(8x\vec{i} + 2y\vec{j})$. This gives

$$8x\lambda = y \quad \text{and} \quad 2y\lambda = x.$$

Multiplying, we get

$$8x^2\lambda = 2y^2\lambda.$$

If $\lambda = 0$, then $x = y = 0$, which doesn't satisfy the constraint equation. So $\lambda \neq 0$ and we get

$$2y^2 = 8x^2$$
$$y^2 = 4x^2$$
$$y = \pm 2x.$$

To find x, we substitute for y in our equation of constraint.

$$4x^2 + y^2 = 8$$
$$4x^2 + 4x^2 = 8$$
$$x^2 = 1$$
$$x = \pm 1$$

So our critical points are $(1, 2)$, $(1, -2)$, $(-1, 2)$ and $(-1, -2)$. Since the constraint is closed and bounded, maximum and minimum values of f subject to the constraint exist. Evaluating $f(x, y)$ at the critical points, we have

$$f(1, 2) = f(-1, -2) = 2$$
$$f(1, -2) = f(1, -2) = -2.$$

So the maximum value of f on $g(x, y) = 8$ is 2, and the minimum value is -2.

9. Our objective function is $f(x, y, z) = 2x + y + 4z$ and our equation of constraint is $g(x, y, z) = x^2 + y + z^2 = 16$. Their gradients are

$$\nabla f(x, y, z) = 2\vec{i} + 1\vec{j} + 4\vec{k},$$
$$\nabla g(x, y, z) = 2x\vec{i} + 1\vec{j} + 2z\vec{k}.$$

So the equation $\nabla f = \lambda \nabla g$ becomes $2\vec{i} + 1\vec{j} + 4\vec{k} = \lambda(2x\vec{i} + 1\vec{j} + 2z\vec{k})$. Solving for λ we find

$$\lambda = \frac{2}{2x} = \frac{1}{1} = \frac{4}{2z}$$
$$\lambda = \frac{1}{x} = 1 = \frac{2}{z}.$$

Which tells us that $x = 1$ and $z = 2$. Going back to our equation of constraint, we can solve for y.

$$g(1, y, 2) = 16$$
$$1^2 + y + 2^2 = 16$$
$$y = 11.$$

So our one critical point is at $(1, 11, 2)$. The value of f at this point is $f(1, 11, 2) = 2 + 11 + 8 = 21$. This is the maximum value of $f(x, y, z)$ on $g(x, y, z) = 16$. To see this, note that for $y = 16 - x^2 - z^2$,

$$f(x, y, z) = 2x + 16 - x^2 - z^2 + 4z = 21 - (x - 1)^2 - (z - 2)^2 \le 21.$$

As $y \to -\infty$, the point $(-\sqrt{16 - y}, y, 0)$ is on the constraint and $f(-\sqrt{16 - y}, y, 0) \to -\infty$, so there is no minimum value for $f(x, y, z)$ on $g(x, y, z) = 16$.

13. The region $x^2 + y^2 \le 2$ is the shaded disk of radius $\sqrt{2}$ centered at the origin (including the circle $x^2 + y^2 = 2$) as shown in Figure 15.3.

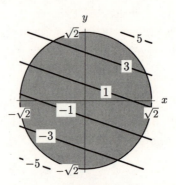

Figure 15.3

We first find the local maxima and minima of f in the interior of our disk. So we need to find the extrema of

$$f(x, y) = x + 3y, \quad \text{in the region} \quad x^2 + y^2 < 2.$$

As

$$f_x = 1$$
$$f_y = 3$$

f doesn't have critical points. Now let's find the local extrema of f on the boundary of the disk. We want to find the extrema of $f(x, y) = x + 3y$ subject to the constraint $g(x, y) = x^2 + y^2 - 2 = 0$. We use Lagrange multipliers

$$\text{grad } f = \lambda \text{ grad } g \quad \text{and} \quad x^2 + y^2 = 2,$$

which give

$$1 = 2\lambda x$$
$$3 = 2\lambda y$$
$$x^2 + y^2 = 2.$$

As λ cannot be zero, we solve for x and y in the first two equations and get $x = \frac{1}{2\lambda}$ and $y = \frac{3}{2\lambda}$. Plugging into the third equation gives

$$8\lambda^2 = 10$$

so $\lambda = \pm\frac{\sqrt{5}}{2}$ and we get the solutions $(\frac{1}{\sqrt{5}}, \frac{3}{\sqrt{5}})$ and $(-\frac{1}{\sqrt{5}}, -\frac{3}{\sqrt{5}})$. Evaluating f at these points gives

$$f(\frac{1}{\sqrt{5}}, \frac{3}{\sqrt{5}}) = 2\sqrt{5} \quad \text{and}$$

$$f(-\frac{1}{\sqrt{5}}, -\frac{3}{\sqrt{5}}) = -2\sqrt{5}$$

The region $x^2 + y^2 \leq 2$ is closed and bounded, so maximum and minimum values of f in the region exist. Therefore $(\frac{1}{\sqrt{5}}, \frac{3}{\sqrt{5}})$ is a global maximum of f and $(-\frac{1}{\sqrt{5}}, -\frac{3}{\sqrt{5}})$ is a global minimum of f on the whole region $x^2 + y^2 \leq 2$.

Problems

17. (a) The contour for $z = 1$ is the line $1 = 2x + y$, or $y = -2x + 1$. The contour for $z = 3$ is the line $3 = 2x + y$, or $y = -2x + 3$. The contours are all lines with slope -2. See Figure 15.4.

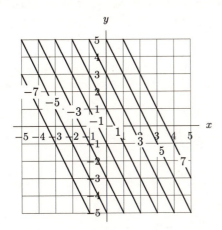

Figure 15.4

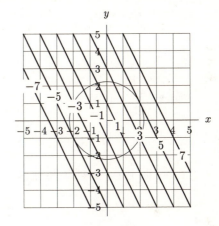

Figure 15.5

(b) The graph of $x^2 + y^2 = 5$ is a circle of radius $\sqrt{5} = 2.236$ centered at the origin. See Figure 15.5.

(c) The circle representing the constraint equation in Figure 15.5 appears to be tangent to the contour close to $z = 5$ at the point $(2, 1)$, and this is the contour with the highest z-value that the circle intersects. The circle is tangent to the contour $z = -5$ approximately at the point $(-2, -1)$, and this is the contour with the lowest z-value that the circle intersects. Therefore, subject to the constraint $x^2 + y^2 = 5$, the function f has a maximum value of about 5 at the point $(2, 1)$ and a minimum value of about -5 at the point $(-2, -1)$.

Since the radius vector, $2\vec{i} + \vec{j}$, at the point $(2, 1)$ is perpendicular to the line $2x + y = 5$, the maximum is exactly 5 and occurs at $(2, 1)$. Similarly, the minimum is exactly -5 and occurs at $(-2, -1)$.

(d) The objective function is $f(x, y) = 2x + y$ and the constraint equation is $g(x, y) = x^2 + y^2 = 5$, and so $\operatorname{grad} f = 2\vec{i} + \vec{j}$ and $\operatorname{grad} g = (2x)\vec{i} + (2y)\vec{j}$. Setting $\operatorname{grad} f = \lambda \operatorname{grad} g$ gives

$$2 = \lambda(2x),$$
$$1 = \lambda(2y).$$

On the constraint, $x \neq 0$ and $y \neq 0$. Thus, from the first equation, we have $\lambda = 1/x$, and from the second equation we have $\lambda = 1/(2y)$. Setting these equal gives

$$x = 2y.$$

Substituting this into the constraint equation $x^2 + y^2 = 5$ gives $(2y)^2 + y^2 = 5$ so $y = -1$ and $y = 1$. Since $x = 2y$, the maximum or minimum values occur at $(2, 1)$ or $(-2, -1)$. Since $f(2, 1) = 5$ and $f(-2, -1) = -5$, the function $f(x, y) = 2x + y$ subject to the constraint $x^2 + y^2 = 5$ has a maximum value of 5 at the point $(2, 1)$ and a minimum value of -5 at the point $(-2, -1)$. This confirms algebraically what we observed graphically in part (c).

21. (a) The problem is to maximize

$$V = 1000 D^{0.6} N^{0.3}$$

subject to the budget constraint in dollars

$$40000 D + 10000 N \leq 600000$$

or (in thousand dollars)

$$40 D + 10 N \leq 600$$

(b) Let $B = 40 D + 10 N = 600$ (thousand dollars) be the budget constraint. At the optimum

$$\nabla V = \lambda \nabla B,$$

$$\text{so} \quad \frac{\partial V}{\partial D} = \lambda \frac{\partial B}{\partial D} = 40\lambda$$

$$\frac{\partial V}{\partial N} = \lambda \frac{\partial B}{\partial N} = 10\lambda.$$

$$\text{Thus} \quad \frac{\frac{\partial V}{\partial D}}{\frac{\partial V}{\partial N}} = 4.$$

Therefore, at the optimum point, the rate of increase in the number of visits to the number of doctors is four times the corresponding rate for nurses. This factor of four is the same as the ratio of the salaries.

(c) Differentiating and setting $\nabla V = \lambda \nabla B$ yields

$$600 D^{-0.4} N^{0.3} = 40\lambda$$

$$300 D^{0.6} N^{-0.7} = 10\lambda$$

Thus, we get

$$\frac{600 D^{-0.4} N^{0.3}}{40} = \lambda = \frac{300 D^{0.6} N^{-0.7}}{10}$$

So

$$N = 2D.$$

To solve for D and N, substitute in the budget constraint:

$$600 - 40 D - 10 N = 0$$

$$600 - 40 D - 10 \cdot (2D) = 0$$

So $D = 10$ and $N = 20$.

$$\lambda = \frac{600 (10^{-0.4})(20^{0.3})}{40} \approx 14.67$$

Thus the clinic should hire 10 doctors and 20 nurses. With that staff, the clinic can provide

$$V = 1000 (10^{0.6})(20^{0.3}) \approx 9{,}779 \text{ visits per year.}$$

(d) From part c), the Lagrange multiplier is $\lambda = 14.67$. At the optimum, the Lagrange multiplier tells us that about 14.67 extra visits can be generated through an increase of \$1,000 in the budget. (If we had written out the constraint in dollars instead of thousands of dollars, the Lagrange multiplier would tell us the number of extra visits per dollar.)

(e) The marginal cost, MC, is the cost of an additional visit. Thus, at the optimum point, we need the reciprocal of the Lagrange multiplier:

$$\text{MC} = \frac{1}{\lambda} \approx \frac{1}{14.67} \approx 0.068 \text{ (thousand dollars)}$$

i.e. at the optimum point, an extra visit costs the clinic 0.068 thousand dollars, or \$68.

This production function exhibits declining returns to scale (e.g. doubling both inputs less than doubles output, because the two exponents add up to less than one). This means that for large V, increasing V will require increasing D and N by more than when V is small. Thus the cost of an additional visit is greater for large V than for small. In other words, the marginal cost will rise with the number of visits.

25. Constraint is $G = P_1 x + P_2 y - K = 0$.
Since $\nabla Q = \lambda \nabla G$, we have
$$cax^{a-1}y^b = \lambda P_1 \quad \text{and} \quad cbx^a y^{b-1} = \lambda P_2.$$

Dividing the two equations yields $\dfrac{cax^{a-1}y^b}{cbx^a y^{b-1}} = \dfrac{\lambda P_1}{\lambda P_2}$, or simplifying, $\dfrac{ay}{bx} = \dfrac{P_1}{P_2}$. Hence, $y = \dfrac{bP_1}{aP_2}x$.

Substitute into the constraint to obtain $P_1 x + P_2 \dfrac{bP_1}{aP_2}x = P_1 \left(\dfrac{a+b}{a}\right)x = K$, giving
$$x = \frac{aK}{(a+b)P_1} \quad \text{and} \quad y = \frac{bK}{(a+b)P_2}.$$

We now check that this is indeed the maximization point. Since $x, y \geq 0$, possible maximization points are $(0, \dfrac{K}{P_2})$,
$(\dfrac{K}{P_1}, 0)$, and $(\dfrac{aK}{(a+b)P_1}, \dfrac{bK}{(a+b)P_2})$. Since $Q = 0$ for the first two points and Q is positive for the last point, it follows
that $(\dfrac{aK}{(a+b)P_1}, \dfrac{bK}{(a+b)P_2})$ gives the maximal value.

29. We want to minimize the function $h(x, y)$ subject to the constraint that
$$g(x, y) = x^2 + y^2 = 1{,}000^2 = 1{,}000{,}000.$$

Using the method of Lagrange multipliers, we obtain the following system of equations:
$$h_x = -\frac{10x + 4y}{10{,}000} = 2\lambda x,$$
$$h_y = -\frac{4x + 4y}{10{,}000} = 2\lambda y,$$
$$x^2 + y^2 = 1{,}000{,}000.$$

Multiplying the first equation by y and the second by x we get
$$\frac{-y(10x + 4y)}{10{,}000} = \frac{-x(4x + 4y)}{10{,}000}.$$

Hence:
$$2y^2 + 3xy - 2x^2 = (2y - x)(y + 2x) = 0,$$
and so the climber either moves along the line $x = 2y$ or $y = -2x$.

We must now choose one of these lines and the direction along that line which will lead to the point of minimum
height on the circle. To do this we find the points of intersection of these lines with the circle $x^2 + y^2 = 1{,}000{,}000$,
compute the corresponding heights, and then select the minimum point.

If $x = 2y$, the third equation gives
$$5y^2 = 1{,}000^2,$$
so that $y = \pm 1{,}000/\sqrt{5} \approx \pm 447.21$ and $x = \pm 894.43$. The corresponding height is $h(\pm 894.43, \pm 447.21) = 2400$ m.
If $y = -2x$, we find that $x = \pm 447.21$ and $y = \mp 894.43$. The corresponding height is $h(\pm 447.21, \mp 894.43) = 2900$ m. Therefore, she should travel along the line $x = 2y$, in either of the two possible directions.

33. **(a)** The objective function $f(x, y) = px + qy$ gives the cost to buy x units of input 1 at unit price p and y units of input
2 at unit price q.

The constraint $g(x, y) = u$ tells us that we are only considering the cost of inputs x and y that can be used to
produce quantity u of the product.

Thus the number $C(p, q, u)$ gives the minimum cost to the company of producing quantity u if the inputs it
needs have unit prices p and q.

(b) The Lagrangian function is
$$\mathcal{L}(x, y, \lambda) = px + qy - \lambda(xy - u).$$
We look for solutions to the system of equations we get from grad $\mathcal{L} = \vec{0}$:
$$\frac{\partial \mathcal{L}}{\partial x} = p - \lambda y = 0$$
$$\frac{\partial \mathcal{L}}{\partial y} = q - \lambda x = 0$$
$$\frac{\partial \mathcal{L}}{\partial \lambda} = -(xy - u) = 0.$$

We see that $\lambda = p/y = q/x$ so $y = px/q$. Substituting for y in the constraint $xy = u$ leads to $x = \sqrt{qu/p}$, $y = \sqrt{pu/q}$ and $\lambda = \sqrt{pq/u}$. The minimum cost is thus

$$C(p, q, u) = p\sqrt{\frac{qu}{p}} + q\sqrt{\frac{pu}{q}} = 2\sqrt{pqu}.$$

Solutions for Chapter 15 Review

Exercises

1. At a critical point

$$f_x(x, y) = 2xy - 2y = 0$$
$$f_y(x, y) = x^2 + 4y - 2x = 0.$$

From the first equation, $2y(x - 1) = 0$, so either $y = 0$ or $x = 1$. If $y = 0$, then $x^2 - 2x = 0$, so $x = 0$ or $x = 2$. Thus $(0, 0)$ and $(2, 0)$ are critical points. If $x = 1$, then $1^2 + 4y - 2 = 0$, so $y = 1/4$. Thus $(1, 1/4)$ is a critical point. Now

$$D = f_{xx}f_{yy} - (f_{xy})^2 = 2y \cdot 4 - (2x - 2)^2 = 8y - 4(x - 1)^2,$$

so

$$D(0, 0) = -4, \quad D(2, 0) = -4, \quad D(1, \tfrac{1}{4}) = 2$$

so $(0, 0)$ and $(2, 0)$ are saddle points. Since $f_{yy} = 4 > 0$, we see that $(1, 1/4)$ is a local minimum.

5. We find critical points:

$$f_x(x, y) = 12 - 6x = 0$$
$$f_y(x, y) = 6 - 2y = 0$$

so $(2, 3)$ is the only critical point. At this point

$$D = f_{xx}f_{yy} - (f_{xy})^2 = (-6)(-2) = 12 > 0,$$

and $f_{xx} < 0$, so $(2, 3)$ is a local maximum. Since this is a quadratic, the local maximum is a global maximum.
Alternatively, we complete the square, giving

$$f(x, y) = 10 - 3(x^2 - 4x) - (y^2 - 6y) = 31 - 3(x - 2)^2 - (y - 3)^2.$$

This expression for f shows that its maximum value (which is 31) occurs where $x = 2, y = 3$.

9. The objective function is $f(x, y) = x^2 + 2y^2$ and the constraint equation is $g(x, y) = 3x + 5y = 200$, so grad $f = (2x)\vec{i} + (4y)\vec{j}$ and grad $g = 3\vec{i} + 5\vec{j}$. Setting grad $f = \lambda$ grad g gives

$$2x = 3\lambda,$$
$$4y = 5\lambda.$$

From the first equation, we have $\lambda = 2x/3$, and from the second equation we have $\lambda = 4y/5$. Setting these equal gives

$$x = 1.2y.$$

Substituting this into the constraint equation $3x + 5y = 200$ gives $y = 23.256$. Since $x = 1.2y$, we have $x = 27.907$. A maximum or minimum value of f can occur only at $(27.907, 23.256)$.
We have $f(27.907, 23.256) = 1860.484$. From Figure 15.6, we see that the point $(27.907, 23.256)$ is a minimum value of f subject to the given constraint.

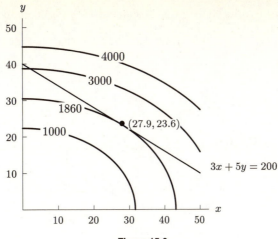

Figure 15.6

13. The region $x^2 \geq y$ is the shaded region in Figure 15.7 which includes the parabola $y = x^2$.

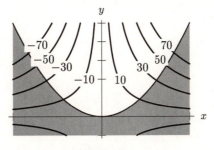

Figure 15.7

We first want to find the local maxima and minima of f in the interior of our region. So we need to find the extrema of
$$f(x, y) = x^2 - y^2, \quad \text{in the region} \quad x^2 > y.$$
For this we compute the critical points:
$$f_x = 2x = 0$$
$$f_y = -2y = 0.$$
As $(0, 0)$ does not belong to the region $x^2 > y$, we have no critical points. Now let's find the local extrema of f on the boundary of our region, hence this time we have to solve a constraint problem. We want to find the extrema of $f(x, y) = x^2 - y^2$ subject to $g(x, y) = x^2 - y = 0$. We use Lagrange multipliers:
$$\operatorname{grad} f = \lambda \operatorname{grad} g \quad \text{and} \quad x^2 = y.$$
This gives
$$2x = 2\lambda x$$
$$2y = \lambda$$
$$x^2 = y.$$
From the first equation we get $x = 0$ or $\lambda = 1$.

If $x = 0$, from the third equation we get $y = 0$, so one solution is $(0, 0)$. If $x \neq 0$, then $\lambda = 1$ and from the second equation we get $y = \frac{1}{2}$. This gives $x^2 = \frac{1}{2}$ so the solutions $\left(\frac{1}{\sqrt{2}}, \frac{1}{2}\right)$ and $\left(-\frac{1}{\sqrt{2}}, \frac{1}{2}\right)$.

So $f(0, 0) = 0$ and $f\left(\frac{1}{\sqrt{2}}, \frac{1}{2}\right) = f\left(-\frac{1}{\sqrt{2}}, \frac{1}{2}\right) = \frac{1}{4}$. From Figure 15.7 showing the level curves of f and the region $x^2 \geq y$, we see that $(0, 0)$ is a local minimum of f on $x^2 = y$, but not a global minimum and that $\left(\frac{1}{\sqrt{2}}, \frac{1}{2}\right)$ and $\left(-\frac{1}{\sqrt{2}}, \frac{1}{2}\right)$ are global maxima of f on $x^2 = y$ but *not* global maxima of f on the whole region $x^2 \geq y$.

So there are no global extrema of f in the region $x^2 \geq y$.

Problems

17. Since $f_{xx} < 0$ and $D = f_{xx}f_{yy} - f_{xy}^2 > 0$, the point $(1, 3)$ is a maximum. See Figure 15.8.

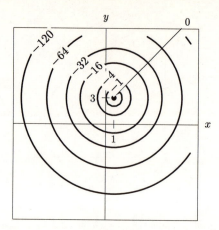

Figure 15.8

21. We want to maximize $f(x, y) = 80x^{0.75}y^{0.25}$ subject to the budget constraint $g(x, y) = 6x + 4y = 8000$. Setting grad $f = \lambda$ grad g gives us

$$80(0.75x^{-0.25})y^{0.25} = 6\lambda,$$
$$80x^{0.75}(0.25y^{-0.75}) = 4\lambda.$$

At the maximum, $x, y \neq 0$. From the first equation we have $\lambda = 10y^{0.25}/x^{0.25}$, and from the second equation we have $\lambda = 5x^{0.75}/y^{0.75}$. Setting these equal gives

$$y = 0.5x.$$

Substituting this into the constraint equation $6x + 4y = 8000$, we see that $x = 1000$. Since $y = 0.5x$, we obtain $y = 500$. That the point $(1000, 500)$ gives the maximum production is suggested by Figure 15.9, since the values of f decrease as we move along the constraint away from $(1000, 500)$. To produce the maximum quantity, the company should use 1000 units of labor and 500 units of capital.

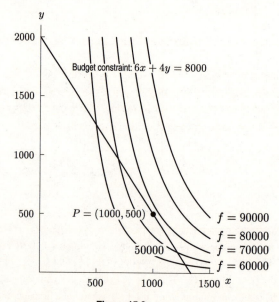

Figure 15.9

25. (a) To be producing the maximum quantity Q under the cost constraint given, the firm should be using K and L values given by

$$\frac{\partial Q}{\partial K} = 0.6aK^{-0.4}L^{0.4} = 20\lambda$$

$$\frac{\partial Q}{\partial L} = 0.4aK^{0.6}L^{-0.6} = 10\lambda$$

$$20K + 10L = 150.$$

Hence $\dfrac{0.6aK^{-0.4}L^{0.4}}{0.4aK^{0.6}L^{-0.6}} = 1.5\dfrac{L}{K} = \dfrac{20\lambda}{10\lambda} = 2$, so $L = \dfrac{4}{3}K$. Substituting in $20K + 10L = 150$, we obtain $20K + 10\left(\dfrac{4}{3}\right)K = 150$. Then $K = \dfrac{9}{2}$ and $L = 6$, so capital should be reduced by $\dfrac{1}{2}$ unit, and labor should be increased by 1 unit.

(b) $\dfrac{\text{New production}}{\text{Old production}} = \dfrac{a4.5^{0.6}6^{0.4}}{a5^{0.6}5^{0.4}} \approx 1.01$, so tell the board of directors, "Reducing the quantity of capital by 1/2 unit and increasing the quantity of labor by 1 unit will increase production by 1% while holding costs to \$150."

29. (a) The objective function is the complementary energy, $\dfrac{f_1^2}{2k_1} + \dfrac{f_2^2}{2k_2}$, and the constraint is $f_1 + f_2 = mg$. The Lagrangian function is

$$\mathcal{L}(f_1, f_2, \lambda) = \frac{f_1^2}{2k_1} + \frac{f_2^2}{2k_2} - \lambda(f_1 + f_2 - mg).$$

We look for solutions to the system of equations we get from grad $\mathcal{L} = \vec{0}$:

$$\frac{\partial \mathcal{L}}{\partial f_1} = \frac{f_1}{k_1} - \lambda = 0$$

$$\frac{\partial \mathcal{L}}{\partial f_2} = \frac{f_2}{k_2} - \lambda = 0$$

$$\frac{\partial \mathcal{L}}{\partial \lambda} = -(f_1 + f_2 - mg) = 0.$$

Combining $\dfrac{\partial \mathcal{L}}{\partial f_1} - \dfrac{\partial \mathcal{L}}{\partial f_2} = \dfrac{f_1}{k_1} - \dfrac{f_2}{k_2} = 0$ with $\dfrac{\partial \mathcal{L}}{\partial \lambda} = 0$ gives the two equation system

$$\frac{f_1}{k_1} - \frac{f_2}{k_2} = 0$$

$$f_1 + f_2 = mg.$$

Substituting $f_2 = mg - f_1$ into the first equation leads to

$$f_1 = \frac{k_1}{k_1 + k_2}mg$$

$$f_2 = \frac{k_2}{k_1 + k_2}mg.$$

(b) Hooke's Law states that for a spring

Force of spring = Spring constant $\cdot$ Distance stretched or compressed from equilibrium.

Since $f_1 = k_1 \cdot \lambda$ and $f_2 = k_2 \cdot \lambda$, the Lagrange multiplier λ equals the distance the mass stretches the top spring and compresses the lower spring.

33. The wetted perimeter of the trapezoid is given by the sum of the lengths of the three walls, so

$$p = w + \frac{2d}{\sin\theta}$$

We want to minimize p subject to the constraint that the area is fixed at 50 m^2. A trapezoid of height h and with parallel sides of lengths b_1 and b_2 has

$$A = \text{Area} = h\frac{(b_1 + b_2)}{2}.$$

In this case, d corresponds to h and b_1 corresponds to w. The b_2 term corresponds to the width of the exposed surface of the canal. We find that $b_2 = w + (2d)/(\tan\theta)$. Substituting into our original equation for the area along with the fact that the area is fixed at 50 m^2, we arrive at the formula:

$$\text{Area} = \frac{d}{2}\left(w + w + \frac{2d}{\tan\theta}\right) = d\left(w + \frac{d}{\tan\theta}\right) = 50$$

We now solve the constraint equation for one of the variables; we will choose w to give

$$w = \frac{50}{d} - \frac{d}{\tan\theta}.$$

Substituting into the expression for p gives

$$p = w + \frac{2d}{\sin\theta} = \frac{50}{d} - \frac{d}{\tan\theta} + \frac{2d}{\sin\theta}.$$

We now take partial derivatives:

$$\frac{\partial p}{\partial d} = -\frac{50}{d^2} - \frac{1}{\tan\theta} + \frac{2}{\sin\theta}$$

$$\frac{\partial p}{\partial\theta} = \frac{d}{\tan^2\theta} \cdot \frac{1}{\cos^2\theta} - \frac{2d}{\sin^2\theta} \cdot \cos\theta$$

From $\partial p/\partial\theta = 0$, we get

$$\frac{d\cdot\cos^2\theta}{\sin^2\theta} \cdot \frac{1}{\cos^2\theta} = \frac{2d}{\sin^2\theta} \cdot \cos\theta.$$

Since $\sin\theta \neq 0$ and $\cos\theta \neq 0$, canceling gives

$$1 = 2\cos\theta$$

so

$$\cos\theta = \frac{1}{2}.$$

$$\text{Since} \quad 0 < \theta < \frac{\pi}{2}, \quad \text{we get} \quad \theta = \frac{\pi}{3}.$$

Substituting into the equation $\partial p/\partial d = 0$ and solving for d gives:

$$\frac{-50}{d^2} - \frac{1}{\sqrt{3}} + \frac{2}{\sqrt{3}/2} = 0$$

which leads to

$$d = \sqrt{\frac{50}{\sqrt{3}}} \approx 5.37\text{m}.$$

Then

$$w = \frac{50}{d} - \frac{d}{\tan\theta} \approx \frac{50}{5.37} - \frac{5.37}{\sqrt{3}} \approx 6.21 \text{ m}.$$

When $\theta = \pi/3$, $w \approx 6.21$ m and $d \approx 5.37$ m, we have $p \approx 18.61$ m.

Since there is only one critical point, and since p increases without limit as d or θ shrink to zero, the critical point must give the global minimum for p.

CAS Challenge Problems

37. **(a)** We have grad $f = 3\vec{i} + 2\vec{j}$ and grad $g = (4x - 4y)\vec{i} + (-4x + 10y)\vec{j}$, so the Lagrange multiplier equations are

$$3 = \lambda(4x - 4y)$$
$$2 = \lambda(-4x + 10y)$$
$$2x^2 - 4xy + 5y^2 = 20$$

Solving these with a CAS we get $\lambda = -0.4005$, $x = -3.9532$, $y = -2.0806$ and $\lambda = 0.4005$, $x = 3.9532$, $y = 2.0806$. We have $f(-3.9532, -2.0806) = -11.0208$, and $f(3, 9532, 2.0806) = 21.0208$. The constraint equation is $2x^2 - 4xy + 5y^2 = 20$, or, completing the square, $2(x - y)^2 + 3y^2 = 20$. This has the shape of a skewed ellipse, so the constraint curve is bounded, and therefore the local maximum is a global maximum. Thus the maximum value is 21.0208.

(b) The maximum value on $g = 20.5$ is $\approx 21.0208 + 0.5(0.4005) = 21.2211$. The maximum value on $g = 20.2$ is $\approx 21.0208 + 0.2(0.4005) = 21.1008$.

(c) We use the same commands in the CAS from part (a), with 20 replaced by 20.5 and 20.2, and get the maximum values 21.2198 for $g = 20.5$ and 21.1007 for $g = 20.2$. These agree with the approximations we found in part (b) to 2 decimal places.

CHECK YOUR UNDERSTANDING

1. True. By definition, a critical point is either where the gradient of f is zero or does not exist.

5. True. The graph of this function is a cone that opens upward with its vertex at the origin.

9. False. For example, the linear function $f(x, y) = x + y$ has no local extrema at all.

13. False. For example, the linear function $f(x, y) = x + y$ has neither a global minimum or global maximum on all of 2-space.

17. False. On the given region the function f is always less than one. By picking points closer and closer to the circle $x^2 + y^2 = 1$ we can make f larger and larger (although never larger than one). There is no point in the open disk that gives f its largest value.

21. True. The point (a, b) must lie on the constraint $g(x, y) = c$, so $g(a, b) = c$.

25. False. Since grad f and grad g point in opposite directions, they are parallel. Therefore (a, b) could be a local maximum or local minimum of f constrained to $g = c$. However the information given is not enough to determine that it is a minimum. If the contours of g near (a, b) increase in the opposite direction as the contours of f, then at a point with grad $f(a, b) = \lambda$grad $g(a, b)$ we have $\lambda \leq 0$, but this can be a local maximum or minimum.

For example, $f(x, y) = 4 - x^2 - y^2$ has a local maximum at $(1, 1)$ on the constraint $g(x, y) = x + y = 2$. Yet at this point, grad $f = -2\vec{i} - 2\vec{j}$ and grad $g = \vec{i} + \vec{j}$, so grad f and grad g point in opposite directions.

29. True. Since $f(a, b) = M$, we must satisfy the Lagrange conditions that $f_x(a, b) = \lambda g(a, b)$ and $f_y(a, b) = \lambda g_y(a, b)$, for some λ. Thus $f_x(a, b)/f_y(a, b) = g_x(a, b)/g_y(a, b)$.

33. False. The value of λ at a minimum point gives the proportional change in m for a change in c. If $\lambda > 0$ and the change in c is positive, the change in m will also be positive.

CHAPTER SIXTEEN

Solutions for Section 16.1

Exercises

1. In the subrectangle in the top left in Figure 1, it appears that $f(x, y)$ has a maximum value of about 9. In the subrectangle in the top middle, $f(x, y)$ has a maximum value of 10. Continuing in this way, and multiplying by Δx and Δy, we have

$$\text{Overestimate} = (9 + 10 + 12 + 7 + 8 + 10 + 5 + 7 + 8)(10)(5) = 3800.$$

Similarly, we find

$$\text{Underestimate} = (7 + 7 + 8 + 4 + 5 + 7 + 1 + 3 + 6)(10)(5) = 2400.$$

Thus, we expect that

$$2400 \leq \int_R f(x, y) dA \leq 3800.$$

5. Partition R into subrectangles with the lines $x = 0$, $x = 0.5$, $x = 1$, $x = 1.5$, and $x = 2$ and the lines $y = 0$, $y = 1$, $y = 2$, $y = 3$, and $y = 4$. Then we have 16 subrectangles, each of which we denote $R_{(a,b)}$, where (a, b) is the location of the lower-left corner of the subrectangle.

   We want to find a lower bound and an upper bound for the volume above each subrectangle. The lower bound for the volume of $R_{(a,b)}$ is

$$0.5(\text{Min of } f \text{ on } R_{(a,b)})$$

because the area of $R_{(a,b)}$ is $0.5 \cdot 1 = 0.5$. The function $f(x, y) = 2 + xy$ increases with both x and y over the whole region R, as shown in Figure 16.1. Thus,

$$\text{Min of } f \text{ on } R_{(a,b)} = f(a, b) = 2 + ab,$$

because the minimum on each subrectangle is at the corner closest to the origin.

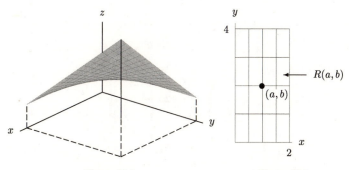

Figure 16.1   **Figure 16.2**

Similarly,

$$\text{Max of } f \text{ on } R_{(a,b)} = f(a + 0.5, b + 1) = 2 + (a + 0.5)(b + 1).$$

So we have

$$\text{Lower sum} = \sum_{(a,b)} 0.5(2 + ab) = 0.5 \sum_{(a,b)} (2 + ab)$$

$$= 16 + 0.5 \sum_{(a,b)} ab$$

Since $a = 0, 0.5, 1, 1.5$ and $b = 0, 1, 2, 3$, expanding this sum gives

$$\begin{aligned}
\text{Lower sum} = 16 + 0.5 \ (\ & 0 \cdot 0 + 0 \cdot 1 + 0 \cdot 2 + 0 \cdot 3 \\
& + 0.5 \cdot 0 + 0.5 \cdot 1 + 0.5 \cdot 2 + 0.5 \cdot 3 \\
& + 1 \cdot 0 + 1 \cdot 1 + 1 \cdot 2 + 1 \cdot 3 \\
& + 1.5 \cdot 0 + 1.5 \cdot 1 + 1.5 \cdot 2 + 1.5 \cdot 3) \\
= 25.
\end{aligned}$$

Similarly, we can compute the upper sum:

$$\begin{aligned}
\text{Upper sum} = \sum_{(a,b)} 0.5(2 + (a + 0.5)(b + 1)) &= 0.5 \sum_{(a,b)} (2 + (a + 0.5)(b + 1)) \\
&= 16 + 0.5 \sum_{(a,b)} (a + 0.5)(b + 1) \\
&= 41.
\end{aligned}$$

Problems

9. Let's break up the room into 25 sections, each of which is 1 meter by 1 meter and has area $\Delta A = 1$.

We shall begin our sum as an upper estimate starting with the lower left corner of the room and continue across the bottom and moving upwards using the highest temperature, T_i, in each case. So the upper Riemann sum becomes

$$\begin{aligned}
\sum_{i=1}^{25} T_i \Delta A = T_1 \Delta A + T_2 \Delta A + T_3 \Delta A + \cdots + T_{25} \Delta A \\
= \Delta A (T_1 + T_2 + T_3 + \cdots + T_{25}) \\
= (1) \ (31 + 29 + 28 + 27 + 27 + \\
29 + 28 + 27 + 27 + 26 + \\
27 + 27 + 26 + 26 + 26 + \\
26 + 26 + 25 + 25 + 25 + \\
25 + 24 + 24 + 24 + 24) \\
= (1)(659) = 659.
\end{aligned}$$

In the same way, the lower Riemann sum is formed by taking the lowest temperature, t_i, in each case:

$$\begin{aligned}
\sum_{i=1}^{25} t_i \Delta A = t_1 \Delta A + t_2 \Delta A + t_3 \Delta A + \cdots + t_{25} \Delta A \\
= \Delta A (t_1 + t_2 + t_3 + \cdots + t_{25}) \\
= (1) \ (27 + 27 + 26 + 26 + 25 + \\
26 + 26 + 25 + 25 + 25 + \\
25 + 24 + 24 + 24 + 24 + \\
24 + 23 + 23 + 23 + 23 + \\
23 + 21 + 20 + 21 + 22) \\
= (1)(602) = 602.
\end{aligned}$$

So, averaging the upper and lower sums we get: 630.5.

To compute the average temperature, we divide by the area of the room, giving

$$\text{Average temperature} = \frac{630.5}{(5)(5)} \approx 25.2°\text{C}.$$

Alternatively we can use the temperature at the central point of each section ΔA. Then the sum becomes

$$\sum_{i=1}^{25} T_i' \Delta A = \Delta A \sum_{i=1}^{25} T_i'$$
$$= (1)(29 + 28 + 27 + 26.5 + 26+$$
$$27 + 27 + 26 + 26 + 25.5+$$
$$26 + 25.5 + 25 + 25 + 25+$$
$$25 + 24 + 24 + 24 + 24+$$
$$24 + 23 + 22 + 22.5 + 23)$$
$$= (1)(630) = 630.$$

Then we get

$$\text{Average temperature} = \frac{\sum_{i=1}^{25} T_i' \Delta A}{\text{Area}} = \frac{630}{(5)(5)} \approx 25.2^\circ\text{C}.$$

13. We use four subrectangles to find an overestimate and underestimate of the integral:

$$\text{Overestimate} = (15 + 9 + 9 + 5)(4)(3) = 456,$$

$$\text{Underestimate} = (5 + 2 + 3 + 1)(4)(3) = 132.$$

A better estimate of the integral is the average of the two:

$$\int_R f(x,y)dA \approx \frac{456 + 132}{2} = 294.$$

The units of the integral are milligrams, and the integral represents the total number of mg of mosquito larvae in this 8 meter by 6 meter section of swamp.

17. The function being integrated is $f(x,y) = 5x$, which is an odd function in x. Since B is symmetric with respect to x, the contributions to the integral cancel out, as $f(x,y) = -f(-x,y)$. Thus, the integral is zero.

21. The region R is symmetric with respect to y and the integrand is an odd function in y, so the integral over R is zero.

25. Since D is a disk of radius 1, in the region D, we have $|y| < 1$. Thus, $-\pi/2 < y < \pi/2$. Thus, $\cos y$ is always positive in the region D and thus its integral is positive.

29. The function $f(x,y)$ is odd with respect to x, and thus the integral is zero in region B, which is symmetric with respect to x.

33. Take a Riemannian sum approximation to

$$\int_R f \, dA \approx \sum_{i,j} f(x_i, y_i)\Delta A.$$

Then, using the fact that $|a + b| \le |a| + |b|$ repeatedly, we have:

$$\left| \int_R f \, dA \right| \approx | \sum_{i,j} f(x_i, y_i)\Delta A| \le \sum |f(x_i, y_i)\Delta A|.$$

Now $|f(x_i, y_j)\Delta A| = |f(x_i, y_j|\Delta A$ since ΔA is non-negative, so

$$\left| \int_R f \, dA \right| \le \sum_{i,j} |f(x_i, y_i)\Delta A| = \sum_{i,j} |f(x_i, y_j)|\Delta A.$$

But the last expression on the right is a Riemann sum approximation to the integral $\int_R |f|dA$, so we have

$$\left| \int_R f \, dA \right| \approx \left| \sum_{i,j} f(x_i, y_j)\Delta A \right| \le \sum_{i,j} |f(x_{i,j})|\Delta A \approx \int_R |f|dA.$$

Thus,

$$\left| \int_R f \, dA \right| \le \int_R |f|dA.$$

Solutions for Section 16.2

Exercises

1. We evaluate the inside integral first:

$$\int_0^3 (x^2 + y^2)\, dy = \left(x^2 y + \frac{y^3}{3}\right)\Bigg|_{y=0}^{y=3} = 3x^2 + 9.$$

Therefore, we have

$$\int_0^2 \int_0^3 (x^2 + y^2)\, dy dx = \int_0^2 (3x^2 + 9)\, dx = (x^3 + 9x)\Big|_0^2 = 26.$$

5. $\displaystyle\int_1^4 \int_1^2 f\, dy\, dx \quad \text{or} \quad \int_1^2 \int_1^4 f\, dx\, dy$

9. The line connecting $(1, 0)$ and $(4, 1)$ is

$$y = \frac{1}{3}(x - 1)$$

So the integral is

$$\int_1^4 \int_{(x-1)/3}^2 f\, dy\, dx$$

13.

$$\int_1^5 \int_x^{2x} \sin x\, dy\, dx = \int_1^5 \sin x \cdot y\Big|_x^{2x}\, dx$$

$$= \int_1^5 \sin x \cdot x\, dx$$

$$= (\sin x - x\cos x)\Big|_1^5$$

$$= (\sin 5 - 5\cos 5) - (\sin 1 - \cos 1) \approx -2.68.$$

See Figure 16.3.

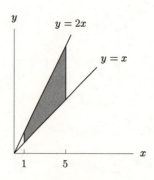

Figure 16.3

17. The region of integration, R, is shown in Figure 16.4.

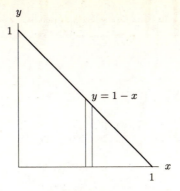

Figure 16.4

Integrating first over y, as shown in the diagram, we obtain

$$\int_R xy\,dA = \int_0^1 \left(\int_0^{1-x} xy\,dy \right) dx = \int_0^1 \frac{xy^2}{2} \bigg|_0^{1-x} dx = \int_0^1 \frac{1}{2}x(1-x)^2\,dx$$

Now integrating with respect to x gives

$$\int_R xy\,dA = \left(\frac{1}{4}x^2 - \frac{1}{3}x^3 + \frac{1}{8}x^4 \right) \bigg|_0^1 = \frac{1}{24}.$$

21. It would be easier to integrate first in the x direction from $x = y - 1$ to $x = -y + 1$, because integrating first in the y direction would involve two separate integrals.

$$\int_R (2x + 3y)^2\,dA = \int_0^1 \int_{y-1}^{-y+1} (2x + 3y)^2\,dx\,dy$$

$$= \int_0^1 \int_{y-1}^{-y+1} (4x^2 + 12xy + 9y^2)\,dx\,dy$$

$$= \int_0^1 \left[\frac{4}{3}x^3 + 6x^2y + 9xy^2 \right]_{y-1}^{-y+1} dy$$

$$= \int_0^1 [\frac{8}{3}(-y+1)^3 + 9y^2(-2y+2)]\,dy$$

$$= \left[-\frac{2}{3}(-y+1)^4 - \frac{9}{2}y^4 + 6y^3 \right]_0^1$$

$$= -\frac{2}{3}(-1) - \frac{9}{2} + 6 = \frac{13}{6}$$

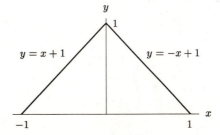

Figure 16.5

25. (a)

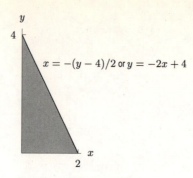

$x = -(y-4)/2$ or $y = -2x + 4$

Figure 16.6

(b) $\int_0^2 \int_0^{-2x+4} g(x,y)\, dy\, dx$.

29. As given, the region of integration is as shown in Figure 16.7.

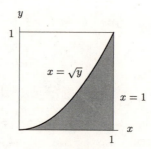

$x = \sqrt{y}$

$x = 1$

Figure 16.7

Reversing the limits gives

$$\int_0^1 \int_0^{x^2} \sqrt{2+x^3}\, dy dx = \int_0^1 \left(y\sqrt{2+x^3}\, \Big|_0^{x^2} \right) dx$$

$$= \int_0^1 x^2 \sqrt{2+x^3}\, dx$$

$$= \frac{2}{9}(2+x^3)^{\frac{3}{2}}\, \Big|_0^1 = \frac{2}{9}(3\sqrt{3} - 2\sqrt{2}).$$

Problems

33. The intersection of the graph of $f(x,y) = 25 - x^2 - y^2$ and xy-plane is a circle $x^2 + y^2 = 25$. The given solid is shown in Figure 16.8.

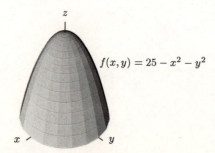

$f(x,y) = 25 - x^2 - y^2$

Figure 16.8

Thus the volume of the solid is

$$V = \int_R f(x, y) \, dA$$

$$= \int_{-5}^{5} \int_{-\sqrt{25-y^2}}^{\sqrt{25-y^2}} (25 - x^2 - y^2) \, dx \, dy.$$

37. The region of integration is shown in Figure 16.9. Thus

$$\text{Volume} = \int_0^1 \int_0^x (x^2 + y^2) \, dy \, dx = \int_0^1 \left(x^2 y + \frac{y^3}{3} \right) \Big|_{y=0}^{y=x} dx = \int_0^1 \frac{4}{3} x^3 \, dx = \frac{x^4}{3} \Big|_0^1 = \frac{1}{3}.$$

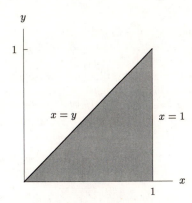

Figure 16.9

41. We want to calculate the volume of the tetrahedron shown in Figure 16.10.

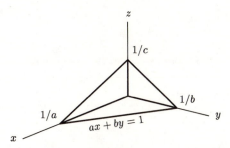

Figure 16.10

We first find the region in the xy-plane where the graph of $ax + by + cz = 1$ is above the xy-plane. When $z = 0$ we have $ax + by = 1$. So the region over which we want to integrate is bounded by $x = 0, y = 0$ and $ax + by = 1$. Integrating with respect to y first, we have

$$\text{Volume} = \int_0^{1/a} \int_0^{(1-ax)/b} z \, dy \, dx = \int_0^{1/a} \int_0^{(1-ax)/b} \frac{1 - by - ax}{c} \, dy \, dx$$

$$= \int_0^{1/a} \left(\frac{y}{c} - \frac{by^2}{2c} - \frac{axy}{c} \right) \Big|_{y=0}^{y=(1-ax)/b} dx$$

$$= \int_0^{1/a} \frac{1}{2bc} (1 - 2ax + a^2 x^2) \, dx$$

$$= \frac{1}{6abc}.$$

45. (a) We have

$$\text{Average value of } f = \frac{1}{\text{Area of Rectangle}} \int_{\text{Rectangle}} f \, dA$$

$$= \frac{1}{6} \int_{x=0}^{2} \int_{y=0}^{3} (ax + by) \, dy dx = \frac{1}{6} \int_{0}^{2} \left(axy + b\frac{y^2}{2} \right) \Big|_{y=0}^{y=3} dx$$

$$= \frac{1}{6} \int_{0}^{2} \left(3ax + \frac{9}{2}b \right) dx = \frac{1}{6} \left(\frac{3}{2}ax^2 + \frac{9}{2}bx \right) \Big|_{0}^{2}$$

$$= \frac{1}{6}(6a + 9b)$$

$$= a + \frac{3}{2}b.$$

The average value will be 20 if and only if $a + (3/2)b = 20$.

This equation can also be expressed as $2a + 3b = 40$, which shows that $f(x, y) = ax + by$ has average value of 20 on the rectangle $0 \leq x \leq 2, 0 \leq y \leq 3$ if and only if $f(2, 3) = 40$.

(b) Since $2a + 3b = 40$, we must have $b = (40/3) - (2/3)a$. Any function $f(x, y) = ax + ((40/3) - (2/3)a)y$ where a is any real number is a correct solution. For example, $a = 1$ leads to the function $f(x, y) = x + (38/3)y$, and $a = -3$ leads to the function $f(x, y) = -3x + (46/3)y$, both of which have average value 20 on the given rectangle. See Figure 16.11 and 16.12.

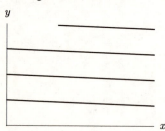

Figure 16.11: $f(x, y) = x + \frac{38}{3}y$

Figure 16.12: $f(x, y) = -3x + \frac{46}{3}y$

Solutions for Section 16.3

Exercises

1.

$$\int_{W} f \, dV = \int_{0}^{2} \int_{-1}^{1} \int_{2}^{3} (x^2 + 5y^2 - z) \, dz \, dy \, dx$$

$$= \int_{0}^{2} \int_{-1}^{1} (x^2 z + 5y^2 z - \frac{1}{2}z^2) \Big|_{2}^{3} dy \, dx$$

$$= \int_{0}^{2} \int_{-1}^{1} (x^2 + 5y^2 - \frac{5}{2}) \, dy \, dx$$

$$= \int_{0}^{2} (x^2 y + \frac{5}{3}y^3 - \frac{5}{2}y) \Big|_{-1}^{1} dx$$

$$= \int_{0}^{2} (2x^2 + \frac{10}{3} - 5) \, dx$$

$$= (\frac{2}{3}x^3 - \frac{5}{3}x) \Big|_{0}^{2}$$

$$= \frac{16}{3} - \frac{10}{3} = 2$$

5. The region is the half cylinder in Figure 16.13.

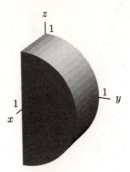

Figure 16.13

9. The region is the half cylinder in Figure 16.14.

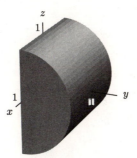

Figure 16.14

13. The region is the quarter sphere in Figure 16.15.

Figure 16.15

Problems

17. The region of integration is shown in Figure 16.16, and the mass of the given solid is given by

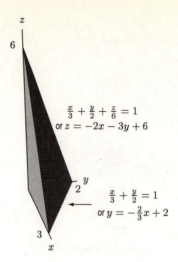

Figure 16.16

$$\text{mass} = \int_R \delta \, dV$$

$$= \int_0^3 \int_0^{-\frac{2}{3}x+2} \int_0^{-2x-3y+6} (x+y) \, dz \, dy \, dx$$

$$= \int_0^3 \int_0^{-\frac{2}{3}x+2} (x+y)z \Big|_0^{-2x-3y+6} \, dy \, dx$$

$$= \int_0^3 \int_0^{-\frac{2}{3}x+2} (x+y)(-2x-3y+6) \, dy \, dx$$

$$= \int_0^3 \int_0^{-\frac{2}{3}x+2} (-2x^2 - 3y^2 - 5xy + 6x + 6y) \, dy \, dx$$

$$= \int_0^3 \left(-2x^2 y - y^3 - \frac{5}{2}xy^2 + 6xy + 3y^2 \right) \Big|_0^{-\frac{2}{3}x+2} \, dx$$

$$= \int_0^3 \left(\frac{14}{27}x^3 - \frac{8}{3}x^2 + 2x + 4 \right) \, dx$$

$$= \left(\frac{7}{54}x^4 - \frac{8}{9}x^3 + x^2 + 4x \right) \Big|_0^3$$

$$= \frac{7}{54} \cdot 3^4 - \frac{8}{9} \cdot 3^3 + 3^2 + 12 = \frac{21}{2} - 3 = \frac{15}{2}.$$

21. Zero. The value of x is positive above the first and fourth quadrants in the xy-plane, and negative (and of equal absolute value) above the second and third quadrants. The integral of x over the entire solid cone is zero because the integrals over the two halves of the cone cancel.

25. Zero. Write the triple integral as an iterated integral, say integrating first with respect to x. For fixed y and z, the x-integral is over an interval symmetric about 0. The integral of x over such an interval is zero. If any of the inner integrals in an iterated integral is zero, then the triple integral is zero.

29. Positive. If (x, y, z) is any point inside the solid W then $\sqrt{x^2 + y^2} < z$. Thus $z - \sqrt{x^2 + y^2} > 0$, and so its integral over the solid W is positive.

33. Zero. You can see this in several ways. One way is to observe that xy is positive on part of the cone above the first quadrant (where x and y are of the same sign) and negative (of equal absolute value) on the part of the cone above the fourth quadrant (where x and y have opposite signs). These add up to zero in the integral of xy over all of W.

Another way to see that the integral is zero is to write the triple integral as an iterated integral, say integrating first with respect to y. For fixed x and z, the y-integral is over an interval symmetric about 0. The integral of y over such an interval is zero. If any of the inner integrals in an iterated integral is zero, then the triple integral is zero.

37. Orient the region as shown in Figure 16.17 and use Cartesian coordinates with origin at the center of the sphere. The equation of the sphere is $x^2 + y^2 + z^2 = 25$, and we want the volume between the planes $z = 3$ and $z = 5$. The plane $z = 3$ cuts the sphere in the circle $x^2 + y^2 + 3^2 = 25$, or $x^2 + y^2 = 16$.

$$\text{Volume} = \int_{-4}^{4} \int_{-\sqrt{16-x^2}}^{\sqrt{16-x^2}} \int_{3}^{\sqrt{25-x^2-y^2}} dz\, dy\, dx.$$

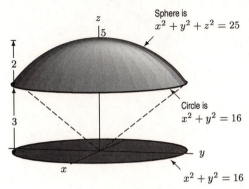

Figure 16.17

41. The volume V of the solid is $1 \cdot 2 \cdot 3 = 6$. We need to compute

$$\frac{m}{6} \int_{W} x^2 + y^2 \, dV = \frac{m}{6} \int_{0}^{1} \int_{0}^{2} \int_{0}^{3} x^2 + y^2 \, dz\, dy\, dx$$

$$= \frac{m}{6} \int_{0}^{1} \int_{0}^{2} 3(x^2 + y^2) \, dy\, dx$$

$$= \frac{m}{2} \int_{0}^{1} (x^2 y + y^3/3) \Big|_{0}^{2} \, dx$$

$$= \frac{m}{2} \int_{0}^{1} (2x^2 + 8/3) \, dx = 5m/3$$

Solutions for Section 16.4

Exercises

1. $\displaystyle\int_{\pi/4}^{3\pi/4} \int_{0}^{2} f \, r\, dr\, d\theta$

5.

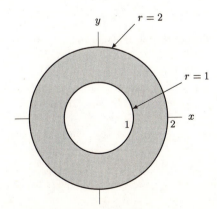

Figure 16.18

9.

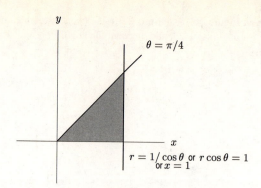

Figure 16.19

13. The region is pictured in Figure 16.20.

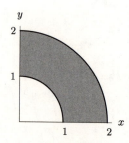

Figure 16.20

Using polar coordinates, we get

$$\int_R (x^2 - y^2)dA = \int_0^{\pi/2} \int_1^2 r^2(\cos^2\theta - \sin^2\theta)r\,dr\,d\theta = \int_0^{\pi/2} (\cos^2\theta - \sin^2\theta) \cdot \frac{1}{4}r^4 \Big|_1^2 d\theta$$

$$= \frac{15}{4} \int_0^{\pi/2} (\cos^2\theta - \sin^2\theta)\,d\theta$$

$$= \frac{15}{4} \int_0^{\pi/2} \cos 2\theta\,d\theta$$

$$= \frac{15}{4} \cdot \frac{1}{2} \sin 2\theta \Big|_0^{\pi/2} = 0.$$

17. From the given limits, the region of integration is in Figure 16.21.

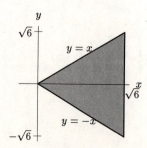

Figure 16.21

In polar coordinates, $-\pi/4 \le \theta \le \pi/4$. Also, $\sqrt{6} = x = r\cos\theta$. Hence, $0 \le r \le \sqrt{6}/\cos\theta$. The integral becomes

$$
\int_0^{\sqrt{6}} \int_{-x}^{x} dy\, dx = \int_{-\pi/4}^{\pi/4} \int_0^{\sqrt{6}/\cos\theta} r\, dr\, d\theta
$$

$$
= \int_{-\pi/4}^{\pi/4} \left(\frac{r^2}{2} \Big|_0^{\sqrt{6}/\cos\theta} \right) d\theta = \int_{-\pi/4}^{\pi/4} \frac{6}{2\cos^2\theta}\, d\theta
$$

$$
= 3\tan\theta \Big|_{-\pi/4}^{\pi/4} = 3 \cdot (1 - (-1)) = 6.
$$

Notice that we can check this answer because the integral gives the area of the shaded triangular region which is $\frac{1}{2} \cdot \sqrt{6} \cdot (2\sqrt{6}) = 6$.

Problems

21. First, let's find where the two surfaces intersect.

$$
\sqrt{8 - x^2 - y^2} = \sqrt{x^2 + y^2}
$$

$$
8 - x^2 - y^2 = x^2 + y^2
$$

$$
x^2 + y^2 = 4
$$

So $z = 2$ at the intersection. See Figure 16.22.

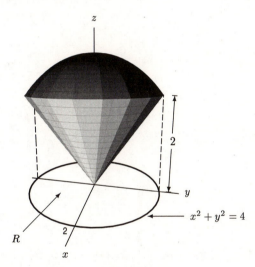

Figure 16.22

The volume of the ice cream cone has two parts. The first part (which is the volume of the cone) is the volume of the solid bounded by the plane $z = 2$ and the cone $z = \sqrt{x^2 + y^2}$. Hence, this volume is given by $\int_R (2 - \sqrt{x^2 + y^2})\, dA$, where R is the disk of radius 2 centered at the origin, in the xy-plane. Using polar coordinates, we have:

$$
\int_R \left(2 - \sqrt{x^2 + y^2} \right) dA = \int_0^{2\pi} \int_0^2 (2 - r) \cdot r\, dr\, d\theta
$$

$$
= \int_0^{2\pi} \left[\left(r^2 - \frac{r^3}{3} \right) \Big|_0^2 \right] d\theta
$$

$$
= \frac{4}{3} \int_0^{2\pi} d\theta
$$

$$
= 8\pi/3
$$

The second part is the volume of the region above the plane $z = 2$ but inside the sphere $x^2 + y^2 + z^2 = 8$, which is given by $\int_R (\sqrt{8 - x^2 - y^2} - 2) \, dA$ where R is the same disk as before. Now

$$\int_R (\sqrt{8 - x^2 - y^2} - 2) \, dA = \int_0^{2\pi} \int_0^2 (\sqrt{8 - r^2} - 2) r \, dr \, d\theta$$

$$= \int_0^{2\pi} \int_0^2 r\sqrt{8 - r^2} \, dr \, d\theta - \int_0^{2\pi} \int_0^2 2r \, dr \, d\theta$$

$$= \int_0^{2\pi} \left(-\frac{1}{3}(8 - r^2)^{3/2} \Big|_0^2 \right) d\theta - \int_0^{2\pi} r^2 \Big|_0^2 \, d\theta$$

$$= -\frac{1}{3} \int_0^{2\pi} (4^{3/2} - 8^{3/2}) \, d\theta - \int_0^{2\pi} 4 \, d\theta$$

$$= -\frac{1}{3} \cdot 2\pi(8 - 16\sqrt{2}) - 8\pi$$

$$= \frac{2\pi}{3}(16\sqrt{2} - 8) - 8\pi$$

$$= \frac{8\pi(4\sqrt{2} - 5)}{3}$$

Thus, the total volume is the sum of the two volumes, which is $32\pi(\sqrt{2} - 1)/3$.

25. (a) We must first decide where to put the origin. We locate the origin at the center of one disk and locate the center of the second disk at the point $(1, 0)$. See Figure 16.23. (Other choices of origin are possible.)

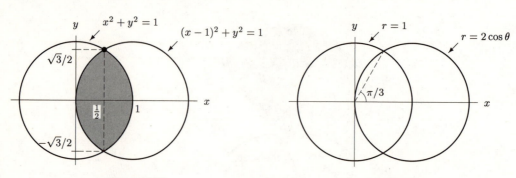

Figure 16.23   Figure 16.24

By symmetry, the points of intersection of the circles are half-way between the centers, at $x = 1/2$. The y-values at these points are given by

$$y = \pm\sqrt{1 - x^2} = \pm\sqrt{1 - \left(\frac{1}{2}\right)^2} = \pm\frac{\sqrt{3}}{2}.$$

We integrate in the x-direction first, so that it is not necessary to set up two integrals. The right-side of the circle $x^2 + y^2 = 1$ is given by

$$x = \sqrt{1 - y^2}.$$

The left side of the circle $(x - 1)^2 + y^2 = 1$ is given by

$$x = 1 - \sqrt{1 - y^2}.$$

Thus the area of overlap is given by

$$\text{Area} = \int_{-\sqrt{3}/2}^{\sqrt{3}/2} \int_{1-\sqrt{1-y^2}}^{\sqrt{1-y^2}} dx \, dy.$$

(b) In polar coordinates, the circle centered at the origin has equation $r = 1$. See Figure 16.24. The other circle, $(x - 1)^2 + y^2 = 1$, can be written as

$$x^2 - 2x + 1 + y^2 = 1$$
$$x^2 + y^2 = 2x,$$

so its equation in polar coordinates is

$$r^2 = 2r\cos\theta,$$

and, since $r \neq 0$,

$$r = 2\cos\theta.$$

At the top point of intersection of the two circles, $x = 1/2$, $y = \sqrt{3}/2$, so $\tan\theta = \sqrt{3}$, giving $\theta = \pi/3$.

Figure 16.24 shows that if we integrate with respect to r first, we have to write the integral as the sum of two integrals. Thus, we integrate with respect to θ first. To do this, we rewrite

$$r = 2\cos\theta \qquad \text{as} \qquad \theta = \arccos\left(\frac{r}{2}\right).$$

This gives the top half of the circle; the bottom half is given by

$$\theta = -\arccos\left(\frac{r}{2}\right).$$

Thus the area is given by

$$\text{Area} = \int_0^1 \int_{-\arccos(r/2)}^{\arccos(r/2)} r\,d\theta\,dr.$$

Solutions for Section 16.5

Exercises

1.

$$\int_W f\,dV = \int_{-1}^1 \int_{\pi/4}^{3\pi/4} \int_0^4 (r^2 + z^2)\,r\,dr\,d\theta\,dz$$

$$= \int_{-1}^1 \int_{\pi/4}^{3\pi/4} (64 + 8z^2)\,d\theta\,dz$$

$$= \int_{-1}^1 \frac{\pi}{2}(64 + 8z^2)\,dz$$

$$= 64\pi + \frac{8}{3}\pi = \frac{200}{3}\pi$$

5. Using Cartesian coordinates, we get:

$$\int_0^3 \int_0^1 \int_0^5 f\,dz\,dy\,dx$$

9. Using spherical coordinates, we get:

$$\int_0^{2\pi} \int_0^{\pi/6} \int_0^3 f \cdot \rho^2 \sin\phi\,d\rho\,d\phi\,d\theta$$

13. **(a)** The region of integration is the region between the cone $z = r$, the xy-plane and the cylinder $r = 3$. In spherical coordinates, $r = 3$ becomes $\rho \sin \phi = 3$, so $\rho = 3/\sin \phi$. The cone is $\phi = \pi/4$ and the xy-plane is $\phi = \pi/2$. See Figure 16.25. Thus, the integral becomes

$$\int_0^{2\pi} \int_{\pi/4}^{\pi/2} \int_0^{3/\sin \phi} \rho^2 \sin \phi \, d\rho d\phi d\theta.$$

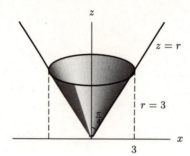

Figure 16.25: Region of integration is between the cone and the xy-plane

(b) The original integral is easier to evaluate, so

$$\int_0^{2\pi} \int_0^3 \int_0^r r \, dz dr d\theta = \int_0^{2\pi} \int_0^3 zr \Big|_{z=0}^{z=r} dr d\theta = \int_0^{2\pi} \int_0^3 r^2 \, dr d\theta = 2\pi \cdot \frac{r^3}{3} \Big|_0^3 = 18\pi.$$

Problems

17. **(a)** The angle ϕ takes on values in the range $0 \leq \phi \leq \pi$. Thus, $\sin \phi$ is nonnegative everywhere in W_1, and so its integral is positive.

   (b) The function ϕ is symmetric across the xy plane, such that for any point (x, y, z) in W_1, with $z \neq 0$, the point $(x, y, -z)$ has a $\cos \phi$ value with the same magnitude but opposite sign of the $\cos \phi$ value for (x, y, z). Furthermore, if $z = 0$, then (x, y, z) has a $\cos \phi$ value of 0. Thus, with $\cos \phi$ positive on the top half of the sphere and negative on the bottom half, the integral will cancel out and be equal to zero.

21. **(a)** In cylindrical coordinates, the cone is $z = r$ and the sphere is $r^2 + z^2 = 4$. The surfaces intersect where $z^2 + z^2 = 2z^2 = 4$. So $z = \sqrt{2}$ and $r = \sqrt{2}$.

$$\text{Volume} = \int_0^{2\pi} \int_0^{\sqrt{2}} \int_r^{\sqrt{4-r^2}} r \, dz dr d\theta.$$

   (b) In spherical coordinates, the cone is $\phi = \pi/4$ and the sphere is $\rho = 2$.

$$\text{Volume} = \int_0^{2\pi} \int_0^{\pi/4} \int_0^2 \rho^2 \sin \phi \, d\rho d\phi d\theta.$$

25. **(a)** In Cartesian coordinates, the bottom half of the sphere $x^2 + y^2 + z^2 = 1$ is given by $z = -\sqrt{1 - x^2 - y^2}$. Thus

$$\int_W dV = \int_0^1 \int_0^{\sqrt{1-x^2}} \int_{-\sqrt{1-x^2-y^2}}^0 dz dy dx.$$

   (b) In cylindrical coordinates, the sphere is $r^2 + z^2 = 1$ and the bottom half is given by $z = -\sqrt{1 - r^2}$. Thus

$$\int_W dV = \int_0^{\pi/2} \int_0^1 \int_{-\sqrt{1-r^2}}^0 r \, dz dr d\theta.$$

   (c) In spherical coordinates, the sphere is $\rho = 1$. Thus,

$$\int_W dV = \int_0^{\pi/2} \int_{\pi/2}^{\pi} \int_0^1 \rho^2 \sin \phi \, d\rho d\phi d\theta.$$

29. Using spherical coordinates:

$$M = \int_0^\pi \int_0^{2\pi} \int_0^3 (3 - \rho)\rho^2 \sin\phi \, d\rho \, d\theta \, d\phi$$

$$= \int_0^\pi \int_0^{2\pi} \left[\rho^3 - \frac{\rho^4}{4} \right]_0^3 \sin\phi \, d\theta \, d\phi$$

$$= \frac{27}{4} \int_0^\pi \int_0^{2\pi} \sin\phi \, d\theta \, d\phi$$

$$= \frac{27}{4} \cdot 2\pi \cdot (-\cos\phi) \Big|_0^\pi = \frac{27}{2}\pi \cdot [-(-1) + 1] = 27\pi.$$

33. (a) We use the axes shown in Figure 16.26. Then the sphere is given by $r^2 + z^2 = 25$, so

$$\text{Volume} = \int_0^{2\pi} \int_1^5 \int_{-\sqrt{25-r^2}}^{\sqrt{25-r^2}} r \, dz \, dr \, d\theta.$$

(b) Evaluating gives

$$\text{Volume} = 2\pi \int_1^5 rz \Big|_{z=-\sqrt{25-r^2}}^{z=\sqrt{25-r^2}} dr = 2\pi \int_1^5 2r\sqrt{25-r^2} \, dr$$

$$= 2\pi \left(-\frac{2}{3} \right) (25 - r^2)^{3/2} \Big|_1^5$$

$$= \frac{4\pi}{3} (24)^{3/2} = 64\sqrt{6}\pi = 492.5 \text{ mm}^3.$$

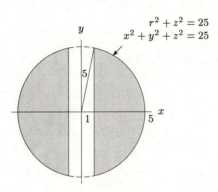

$$r^2 + z^2 = 25$$
$$x^2 + y^2 + z^2 = 25$$

Figure 16.26

37. We first need to find the mass of the solid, using cylindrical coordinates:

$$m = \int_0^{2\pi} \int_0^1 \int_0^{\sqrt{z/a}} r \, dr \, dz \, d\theta$$

$$= \int_0^{2\pi} \int_0^1 \frac{z}{2a} \, dz \, d\theta$$

$$= \int_0^{2\pi} \frac{1}{4a} \, d\theta = \frac{\pi}{2a}$$

It makes sense that the mass would vary inversely with a, since increasing a makes the paraboloid skinnier. Now for the z-coordinate of the center of mass, again using cylindrical coordinates:

$$\bar{z} = \frac{2a}{\pi} \int_0^{2\pi} \int_0^1 \int_0^{\sqrt{z/a}} zr \, dr \, dz \, d\theta$$

$$= \frac{2a}{\pi} \int_0^{2\pi} \int_0^1 \frac{z^2}{2a} \, dz \, d\theta$$

$$= \frac{2a}{\pi} \int_0^{2\pi} \frac{1}{6a} \, d\theta = \frac{2}{3}$$

41. Assume the base of the cylinder sits on the xy-plane with center at the origin. Because the cylinder is symmetric about the z-axis, the force in the horizontal x or y direction is 0. Thus we need only compute the vertical z component of the force. We are going to use cylindrical coordinates; since the force is $G \cdot \text{mass}/(\text{distance})^2$, a piece of the cylinder of volume dV located at (r, θ, z) exerts on the unit mass a force with magnitude $G(\delta \, dV)/(r^2 + z^2)$. See Figure 16.27.

$$\begin{array}{c} \text{Vertical component} \\ \text{of force} \end{array} = \frac{G(\delta \, dV)}{r^2 + z^2} \cdot \cos\phi = \frac{G\delta \, dV}{r^2 + z^2} \cdot \frac{z}{\sqrt{r^2 + z^2}} = \frac{G\delta z \, dV}{(r^2 + z^2)^{3/2}}.$$

Adding up all the contributions of all the dV's, we obtain

$$\text{Vertical force} = \int_0^H \int_0^{2\pi} \int_0^R \frac{G\delta z r}{(r^2 + z^2)^{3/2}} \, dr \, d\theta \, dz$$

$$= \int_0^H \int_0^{2\pi} (G\delta z) \left(-\frac{1}{\sqrt{r^2 + z^2}} \right) \Big|_0^R \, d\theta \, dz$$

$$= \int_0^H \int_0^{2\pi} (G\delta z) \cdot \left(-\frac{1}{\sqrt{R^2 + z^2}} + \frac{1}{z} \right) \, d\theta \, dz$$

$$= \int_0^H 2\pi G\delta \left(1 - \frac{z}{\sqrt{R^2 + z^2}} \right) \, dz$$

$$= 2\pi G\delta (z - \sqrt{R^2 + z^2}) \Big|_0^H$$

$$= 2\pi G\delta (H - \sqrt{R^2 + H^2} + R) = 2\pi G\delta (H + R - \sqrt{R^2 + H^2})$$

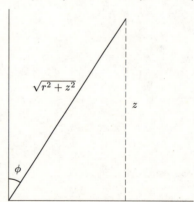

Figure 16.27

45. Using spherical coordinates,

$$\text{Stored energy} = \frac{1}{2} \int_a^b \int_0^\pi \int_0^{2\pi} \epsilon E^2 \rho^2 \sin\phi \, d\theta \, d\phi \, d\rho = \frac{q^2}{32\pi^2\epsilon} \int_a^b \int_0^\pi \int_0^{2\pi} \frac{1}{\rho^2} \sin\phi \, d\theta \, d\phi \, d\rho$$

$$= \frac{q^2}{8\pi\epsilon} \int_a^b \frac{1}{\rho^2} \, d\rho = \frac{q^2}{8\pi\epsilon} \left(\frac{1}{a} - \frac{1}{b} \right).$$

Solutions for Section 16.6

Exercises

1. No, p is not a joint density function. Since $p(x, y) = 0$ outside the region R, the volume under the graph of p is the same as the volume under the graph of p over the region R, which is 2 not 1.

5. Yes, p is a joint density function. In the region R we have $1 \geq x^2 + y^2$, so $p(x, y) = (2/\pi)(1 - x^2 - y^2) \geq 0$ for all x and y in R, and $p(x, y) = 0$ for all other (x, y). To check that p is a joint density function, we check that the total volume under the graph of p over the region R is 1. Using polar coordinates, we get:

$$\int_R p(x, y)dA = \frac{2}{\pi} \int_0^{2\pi} \int_0^1 (1 - r^2)r \, dr \, d\theta = \frac{2}{\pi} \int_0^{2\pi} \left(\frac{r^2}{2} - \frac{r^4}{4} \right) \bigg|_0^1 d\theta = \frac{2}{\pi} \int_0^{2\pi} \frac{1}{4} d\theta = 1.$$

9. Since $x + y \leq 3$ for all points (x, y) in the region R, the fraction of the population satisfying $x + y \leq 3$ is 1.

13. The fraction of the population is given by the double integral:

$$\int_0^{1/2} \int_0^1 xy \, dx \, dy = \int_0^{1/2} \frac{x^2 y}{2} \bigg|_0^1 dy = \int_0^{1/2} \frac{y}{2} dy = \frac{y^2}{4} \bigg|_0^{1/2} = \frac{1}{16}.$$

Problems

17. (a) We know that $\int_{-\infty}^{\infty} \int_{-\infty}^{\infty} f(x, y)dydx = 1$ for a joint density function. So,

$$1 = \int_{-\infty}^{\infty} \int_{-\infty}^{\infty} f(x, y)dydx = \int_0^1 \int_x^1 kxy \, dy \, dx$$
$$= \frac{1}{8}k$$

hence $k = 8$.

(b) The region where $x < y < \sqrt{x}$ is sketched in Figure 16.28

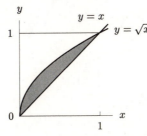

Figure 16.28

So the probability that (x, y) satisfies $x < y < \sqrt{x}$ is given by:

$$\int_0^1 \int_x^{\sqrt{x}} 8xy \, dy \, dx = \int_0^1 4x(y^2) \big|_x^{\sqrt{x}} dx$$
$$= \int_0^1 4x(x - x^2)dx$$
$$= 4 \left(\frac{1}{3}x^3 - \frac{1}{4}x^4 \right) \bigg|_0^1$$
$$= 4 \left(\frac{1}{3} - \frac{1}{4} \right)$$
$$= \frac{1}{3}$$

This tells us that in choosing points from the region defined by $0 \leq x \leq y \leq 1$, that $1/3$ of the time we would pick a point from the region defined by $x < y < \sqrt{x}$. These regions are shown in Figure 16.28.

21. (a) Since the exponential function is always positive and λ is positive, $p(t) \geq 0$ for all t, and

$$\int_0^\infty p(t)dt = \lim_{b \to \infty} -e^{-\lambda t}\Big|_0^b = \lim_{b \to \infty} -e^{-bt} + 1 = 1.$$

(b) The density function for the probability that the first substance decays at time t and the second decays at time s is

$$p(t,s) = \lambda e^{-\lambda t}\mu e^{-\mu s} = \lambda \mu e^{-\lambda t - \mu s},$$

for $s \geq 0$ and $t \geq 0$, and is zero otherwise.

(c) We want the probability that the decay time t of the first substance is less than or equal to the decay time s of the second, so we want to integrate the density function over the region $0 \leq t \leq s$. Thus, we compute

$$\int_0^\infty \int_t^\infty \lambda \mu e^{-\lambda t} e^{-\mu s} \, ds \, dt = \int_0^\infty \lambda e^{-\lambda t}(-e^{-\mu s})\Big|_t^\infty \, dt$$

$$= \int_0^\infty \lambda e^{-\lambda t} e^{-\mu t} \, dt$$

$$= \int_0^\infty \lambda e^{(-\lambda + \mu)t} \, dt$$

$$= \frac{-\lambda}{\lambda + \mu} e^{-(\lambda + \mu)t}\Big|_0^\infty = \frac{\lambda}{\lambda + \mu}.$$

So for example, if $\lambda = 1$ and $\mu = 4$, then the probability that the first substance decays first is $1/5$.

Solutions for Section 16.7

Exercises

1. We have

$$\frac{\partial(x,y)}{\partial(s,t)} = \begin{vmatrix} x_s & x_t \\ y_s & y_t \end{vmatrix} = \begin{vmatrix} 5 & 2 \\ 3 & 1 \end{vmatrix} = -1.$$

Therefore,

$$\left| \frac{\partial(x,y)}{\partial(s,t)} \right| = 1.$$

5. We have

$$\frac{\partial(x,y,z)}{\partial(s,t,u)} = \begin{vmatrix} x_s & x_t & x_u \\ y_s & y_t & y_u \\ z_s & z_t & z_u \end{vmatrix} = \begin{vmatrix} 3 & 1 & 2 \\ 1 & 5 & -1 \\ 2 & -1 & 1 \end{vmatrix}.$$

This 3×3 determinant is computed the same way as for the cross product, with the entries $3, 1, 2$ in the first row playing the same role as $\vec{i}, \vec{j}, \vec{k}$. We get

$$\frac{\partial(x,y,z)}{\partial(s,t,u)} = ((5)(1) - (-1)(-1))3 + ((-1)(2) - (1)(1))1 + ((1)(-1) - (2)(5))2 = -13.$$

Problems

9. Given

$$\begin{cases} x = \rho \sin \phi \cos \theta \\ y = \rho \sin \phi \sin \theta \\ z = \rho \cos \phi, \end{cases}$$

$$\frac{\partial(x,y,z)}{\partial(\rho,\phi,\theta)} = \begin{vmatrix} \frac{\partial x}{\partial \rho} & \frac{\partial x}{\partial \phi} & \frac{\partial x}{\partial \theta} \\ \frac{\partial y}{\partial \rho} & \frac{\partial y}{\partial \phi} & \frac{\partial y}{\partial \theta} \\ \frac{\partial z}{\partial \rho} & \frac{\partial z}{\partial \phi} & \frac{\partial z}{\partial \theta} \end{vmatrix} = \begin{vmatrix} \sin\phi\cos\theta & \rho\cos\phi\cos\theta & -\rho\sin\phi\sin\theta \\ \sin\phi\sin\theta & \rho\cos\phi\sin\theta & \rho\sin\phi\cos\theta \\ \cos\phi & -\rho\sin\phi & 0 \end{vmatrix}$$

$$= \cos\phi \begin{vmatrix} \rho\cos\phi\cos\theta & -\rho\sin\phi\sin\theta \\ \rho\cos\phi\sin\theta & \rho\sin\phi\cos\theta \end{vmatrix} + \rho\sin\phi \begin{vmatrix} \sin\phi\cos\theta & -\rho\sin\phi\sin\theta \\ \sin\phi\sin\theta & \rho\sin\phi\cos\theta \end{vmatrix}$$

$$= \cos\phi(\rho^2\cos^2\theta\cos\phi\sin\phi + \rho^2\sin^2\theta\cos\phi\sin\phi)$$

$$\quad + \rho\sin\phi(\rho\sin^2\phi\cos^2\theta + \rho\sin^2\phi\sin^2\theta)$$

$$= \rho^2\cos^2\phi\sin\phi + \rho^2\sin^3\phi$$

$$= \rho^2\sin\phi.$$

13. Given

$$\begin{cases} s = xy \\ t = xy^2, \end{cases}$$

we have

$$\frac{\partial(s,t)}{\partial(x,y)} = \begin{vmatrix} \frac{\partial s}{\partial x} & \frac{\partial s}{\partial y} \\ \frac{\partial t}{\partial x} & \frac{\partial t}{\partial y} \end{vmatrix} = \begin{vmatrix} y & x \\ y^2 & 2xy \end{vmatrix} = xy^2 = t.$$

Since

$$\frac{\partial(s,t)}{\partial(x,y)} \cdot \frac{\partial(x,y)}{\partial(s,t)} = 1,$$

$$\frac{\partial(x,y)}{\partial(s,t)} = t \qquad \text{so} \qquad \left|\frac{\partial(x,y)}{\partial(s,t)}\right| = \frac{1}{t}$$

So

$$\int_R xy^2\,dA = \int_T t\left|\frac{\partial(x,y)}{\partial(s,t)}\right|\,ds\,dt = \int_T t\left(\frac{1}{t}\right)\,ds\,dt = \int_T ds\,dt,$$

where T is the region bounded by $s = 1, s = 4, t = 1, t = 4$.
Then

$$\int_R xy^2\,dA = \int_1^4 ds \int_1^4 dt = 9.$$

Solutions for Chapter 16 Review

Exercises

1.

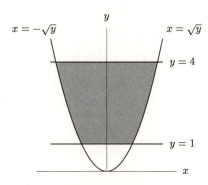

Figure 16.29

5. We use Cartesian coordinates, oriented so that the cube is in the first quadrant. See Figure 16.30. Then, if f is an arbitrary function, the integral is

$$\int_0^2 \int_0^3 \int_0^5 f \, dx \, dy \, dz.$$

Other answers are possible. In particular, the order of integration can be changed.

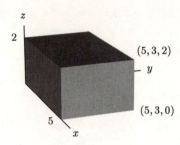

Figure 16.30

9.

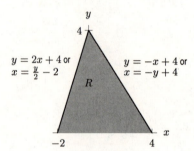

Figure 16.31

Integrating with respect to x first we get

$$\int_0^4 \int_{\frac{y}{2}-2}^{-y+4} f(x, y) \, dx \, dy$$

Integrating with respect to y first we get

$$\int_{-2}^0 \int_0^{2x+4} f(x, y) \, dy \, dx + \int_0^4 \int_0^{-x+4} f(x, y) \, dy \, dx.$$

13.

$$\int_0^1 \int_0^y (\sin^3 x)(\cos x)(\cos y) \, dx \, dy = \int_0^1 (\cos y) \left[\frac{\sin^4 x}{4} \right] \Bigg|_0^y dy$$

$$= \frac{1}{4} \int_0^1 (\sin^4 y)(\cos y) \, dy$$

$$= \frac{\sin^5 y}{20} \Bigg|_0^1$$

$$= \frac{\sin^5 1}{20}.$$

17. From Figure 16.32, we have the following iterated integrals:

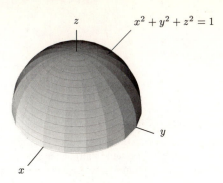

$$x^2 + y^2 + z^2 = 1$$

Figure 16.32

(a) $\displaystyle\int_R f\,dV = \int_{-1}^{1} \int_{-\sqrt{1-x^2}}^{\sqrt{1-x^2}} \int_{0}^{\sqrt{1-x^2-y^2}} f(x,y,z)\,dzdydx$

(b) $\displaystyle\int_R f\,dV = \int_{-1}^{1} \int_{-\sqrt{1-y^2}}^{\sqrt{1-y^2}} \int_{0}^{\sqrt{1-x^2-y^2}} f(x,y,z)\,dzdxdy$

(c) $\displaystyle\int_R f\,dV = \int_{-1}^{1} \int_{0}^{\sqrt{1-y^2}} \int_{-\sqrt{1-y^2-z^2}}^{\sqrt{1-y^2-z^2}} f(x,y,z)\,dxdzdy$

(d) $\displaystyle\int_R f\,dV = \int_{-1}^{1} \int_{0}^{\sqrt{1-x^2}} \int_{-\sqrt{1-x^2-z^2}}^{\sqrt{1-x^2-z^2}} f(x,y,z)\,dydzdx$

(e) $\displaystyle\int_R f\,dV = \int_{0}^{1} \int_{-\sqrt{1-z^2}}^{\sqrt{1-z^2}} \int_{-\sqrt{1-x^2-z^2}}^{\sqrt{1-x^2-z^2}} f(x,y,z)\,dydxdz$

(f) $\displaystyle\int_R f\,dV = \int_{0}^{1} \int_{-\sqrt{1-z^2}}^{\sqrt{1-z^2}} \int_{-\sqrt{1-y^2-z^2}}^{\sqrt{1-y^2-z^2}} f(x,y,z)\,dxdydz$

21. W is a cylindrical shell, so cylindrical coordinates should be used. See Figure 16.33.

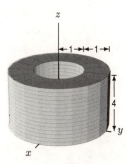

Figure 16.33

$$\int_W \frac{z}{(x^2+y^2)^{3/2}}\,dV = \int_0^4 \int_0^{2\pi} \int_1^2 \frac{z}{r^3}\,rdr\,d\theta\,dz$$

$$= \int_0^4 \int_0^{2\pi} \int_1^2 \frac{z}{r^2}\,dr\,d\theta\,dz$$

$$= \int_0^4 \int_0^{2\pi} \left(-\frac{z}{r}\right)\Big|_1^2\,d\theta\,dz$$

$$= \int_0^4 \int_0^{2\pi} \frac{z}{2} \, d\theta \, dz$$

$$= \int_0^4 \frac{z}{2} \cdot 2\pi \, dz = \frac{1}{2}\pi \cdot z^2 \bigg|_0^4 = 8\pi$$

Problems

25. Can't tell, since y^3 is both positive and negative for $x < 0$.

29. Zero. You can see this in several ways. One way is to observe that xy is positive on the part of the sphere above and below the first and third quadrants (where x and y are of the same sign) and negative (of equal absolute value) on the part of the sphere above and below the second and fourth quadrants (where x and y have opposite signs). These add up to zero in the integral of xy over all of W.

Another way to see that the integral is zero is to write the triple integral as an iterated integral, say integrating first with respect to x. For fixed y and z, the x-integral is over an interval symmetric about 0. The integral of x over such an interval is zero. If any of the inner integrals in an iterated integral is zero, then the triple integral is zero.

33. Negative. Since $z^2 - 1 \leq 0$ in the sphere, its integral is negative.

37. Let the lower left part of the forest be at $(0, 0)$. Then the other corners have coordinates as shown. The population density function is then given by

$$\rho(x, y) = 10 - 2y$$

The equations of the two diagonal lines are $x = -2y/5$ and $x = 6 + 2y/5$. So the total rabbit population in the forest is

$$\int_0^5 \int_{-\frac{2}{5}y}^{6+\frac{2}{5}y} (10 - 2y) \, dx \, dy = \int_0^5 (10 - 2y)(6 + \frac{4}{5}y) \, dy$$

$$= \int_0^5 (60 - 4y - \frac{8}{5}y^2) \, dy$$

$$= (60y - 2y^2 - \frac{8}{15}y^3) \bigg|_0^5$$

$$= 300 - 50 - \frac{8}{15} \cdot 125$$

$$= \frac{2750}{15} = \frac{550}{3} \approx 183$$

41. Since the hole resembles a cylinder, we will use cylindrical coordinates. Let the center of the sphere be at the origin, and let the center of the hole be the z-axis (see Figure 16.34).

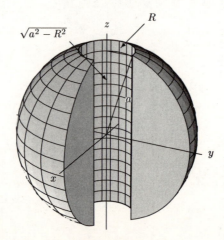

Figure 16.34

Then we will integrate from $z = -\sqrt{a^2 - R^2}$ to $z = \sqrt{a^2 - R^2}$, and each cross-section will be an annulus. So the volume is

$$\int_{-\sqrt{a^2-R^2}}^{\sqrt{a^2-R^2}} \int_0^{2\pi} \int_R^{\sqrt{a^2-z^2}} r\, dr\, d\theta\, dz = \int_{-\sqrt{a^2-R^2}}^{\sqrt{a^2-R^2}} \int_0^{2\pi} \frac{1}{2}(a^2 - z^2 - R^2)\, d\theta\, dz$$

$$= \pi \int_{-\sqrt{a^2-R^2}}^{\sqrt{a^2-R^2}} (a^2 - z^2 - R^2)\, dz$$

$$= \pi \left[(a^2 - R^2)(2\sqrt{a^2 - R^2}) - \frac{1}{3}(2(a^2 - R^2)^{\frac{3}{2}}) \right]$$

$$= \frac{4\pi}{3}(a^2 - R^2)^{\frac{3}{2}}$$

45. Let the ball be centered at the origin. Since a ball looks the same from all directions, we can choose the axis of rotation; in this case, let it be the z-axis. It is best to use spherical coordinates, so then

$$x^2 + y^2 = (\rho \sin\phi \cos\theta)^2 + (\rho \sin\phi \sin\theta)^2$$
$$= \rho^2 \sin^2\phi$$

Then $m/v = \text{Density} = 1$, so the moment of inertia is

$$I_z = \int_0^R \int_0^{2\pi} \int_0^\pi 1(\rho^2 \sin^2\phi)\rho^2 \sin\phi\, d\phi\, d\theta\, d\rho$$

$$= \int_0^R \int_0^{2\pi} \int_0^\pi \rho^4(\sin\phi)(1 - \cos^2\phi)\, d\phi\, d\theta\, d\rho$$

$$= \int_0^R \int_0^{2\pi} \rho^4 \left(-\cos\phi + \frac{1}{3}\cos^3\phi\right)\bigg|_0^\pi d\theta\, d\rho$$

$$= \int_0^R \int_0^{2\pi} \frac{4}{3}\rho^4\, d\theta\, d\rho$$

$$= \int_0^R \frac{8\pi}{3}\rho^4\, d\rho = \frac{8}{15}\pi R^5$$

CAS Challenge Problems

49. The region is the triangle to the right of the y-axis, below the line $y = 1$, and above the line $y = x$. Thus the integral can be written as $\int_0^1 \int_x^1 e^{y^2}\, dy dx$ or as $\int_0^1 \int_0^y e^{y^2}\, dx dy$. The second of these integrals can be evaluated easily by hand:

$$\int_0^1 \int_0^y e^{y^2}\, dx dy = \int_0^1 \left(e^{y^2}x \bigg|_{x=0}^{x=y}\right) dy = \int_0^1 y e^{y^2}\, dy$$

$$= \frac{1}{2}e^{y^2}\bigg|_0^1 = \frac{1}{2}(e - 1)$$

The other integral cannot be done by hand with the methods you have learned, but some computer algebra systems will compute it and give the same answer.

CHECK YOUR UNDERSTANDING

1. False. For example, if $f(x, y) < 0$ for all (x, y) in the region R, then $\int_R f\, dA$ is negative.

5. True. The double integral is the limit of the sum $\displaystyle\sum_{\Delta A \to 0} \rho(x,y)\Delta A$. Each of the terms $\rho(x,y)\Delta A$ is an approximation of the total population inside a small rectangle of area ΔA. Thus the limit of the sum of all of these numbers as $\Delta A \to \infty$ gives the total population of the region R.

9. False. There is no reason to expect this to be true, since the behavior of f on one half of R can be completely unrelated to the behavior of f on the other half. As a counterexample, suppose that f is defined so that $f(x,y) = 0$ for points (x,y) lying in S, and $f(x,y) = 1$ for points (x,y) lying in the part of R that is not in S. Then $\int_S f\,dA = 0$, since $f = 0$ on all of S. To evaluate $\int_R f\,dA$, note that $f = 1$ on the square S_1 which is $0 \le x \le 1, 1 \le y \le 2$. Then $\int_R f\,dA = \int_{S_1} f\,dA = \text{Area}(S_1) = 1$, since $f = 0$ on S.

13. True. For any point in the region of integration we have $1 \le x \le 2$, and so y is between the positive numbers 1 and 8.

17. False. The given limits describe only the upper half disk where $y \ge 0$. The correct limits are $\int_{-a}^{a}\int_{-\sqrt{a^2-x^2}}^{\sqrt{a^2-x^2}} f\,dy\,dx$.

21. False. The integral gives the total mass of the material contained in W.

25. True. Both sets of limits describe the solid region lying above the rectangle $-1 \le x \le 1, 0 \le y \le 1, z = 0$ and below the parabolic cylinder $z = 1 - x^2$.

29. False. As a counterexample, let W_1 be the solid cube $0 \le x \le 1, 0 \le y \le 1, 0 \le z \le 1$, and let W_2 be the solid cube $-\frac{1}{2} \le x \le 0, -\frac{1}{2} \le y \le 0, -\frac{1}{2} \le z \le 0$. Then volume$(W_1) = 1$ and volume$(W_2) = \frac{1}{8}$. Now if $f(x,y,z) = -1$, then $\int_{W_1} f\,dV = 1 \cdot -1$ which is less than $\int_{W_2} f\,dV = \frac{1}{8} \cdot -1$.

CHAPTER SEVENTEEN

Solutions for Section 17.1

Exercises

1. The direction vectors of the lines, $-\vec{i} + 4\vec{j} - 2\vec{k}$ and $2\vec{i} - 8\vec{j} + 4\vec{k}$, are multiplies of each other (the second is -2 times the first). Thus the lines are parallel. To see if they are the same line, we take the point corresponding to $t = 0$ on the first line, which has position vector $3\vec{i} + 3\vec{j} - \vec{k}$, and see if it is on the second line. So we solve

$$(1 + 2t)\vec{i} + (11 - 8t)\vec{j} + (4t - 5)\vec{k} = 3\vec{i} + 3\vec{j} - \vec{k}.$$

This has solution $t = 1$, so the two lines have a point in common and must be the same line, parameterized in two different ways.

5. The xz-plane is $y = 0$, so one possible answer is

$$x = 2\cos t, \quad y = 0, \quad z = 2\sin t.$$

9. The xy-plane is $z = 0$, so a possible answer is

$$x = t, \quad y = t^3, \quad z = 0.$$

13. Since its diameters lie along the x and y-axes and its center is the origin, the ellipse must lie in the xy-plane, hence at $z = 0$. The x-coordinate ranges between -3 and 3 and the y-coordinate between -2 and 2. One possible answer is

$$x = 3\cos t, \quad y = 2\sin t, \quad z = 0.$$

17. One possible parameterization is

$$x = 3 + t, \quad y = 2t, \quad z = -4 - t.$$

21. One possible parameterization is

$$x = 1, \quad y = 0, \quad z = t.$$

25. The line passes through $(3, 0, 0)$ and $(0, 0, -5)$. The displacement vector from the first of these points to the second is $\vec{v} = -3\vec{i} - 5\vec{k}$. The line through point $(3, 0, 0)$ and with direction vector $\vec{v} = -3\vec{i} - 5\vec{k}$ is given by parametric equations

$$x = 3 - 3t,$$
$$y = 0,$$
$$z = -5t.$$

Other parameterizations of the same line are also possible.

29. The vector from P_0 to P_1 is $\vec{v} = (5 + 1)\vec{i} + (2 + 3)\vec{j} = 6\vec{i} + 5\vec{j}$. Since $P_0 = -\vec{i} - 3\vec{j}$, the line is

$$\vec{r}(t) = -\vec{i} - 3\vec{j} + t(6\vec{i} + 5\vec{j}) \quad \text{for } 0 \le t \le 1.$$

In coordinate form, the equations are $x = -1 + 6t, y = -3 + 5t, 0 \le t \le 1$

33. The direction vectors for these two lines are $\vec{v}_1 = -2\vec{i} + \vec{j} + 2\vec{k}$ and $\vec{v}_2 = 6\vec{i} - 3\vec{j} - 6\vec{k}$. Since $\vec{v}_2 = -3\vec{v}_1$, the lines have the same direction. To see that the lines are not the same, check that a point on the first line (say $(3, 5, -4)$) is not on the second line: if $3 = 7 + 6t$, then $t = -2/3$, so $y = 1 - 3(-2/3) = 3 \ne 5$. So the lines are parallel and do not intersect.

Problems

37. We find the parameterization in terms of the displacement vector $\overrightarrow{OP} = 2\vec{i} + 5\vec{j}$ from the origin to the point P and the displacement vector $\overrightarrow{PQ} = 10\vec{i} + 4\vec{j}$ from P to Q.

$$\vec{r}(t) = \overrightarrow{OP} + (t/5)\overrightarrow{PQ} \text{ or } \vec{r}(t) = (2 + (t/5)10)\vec{i} + (5 + (t/5)4)\vec{j}$$

41. The graph is parameterized by $x = t$, $y = \sqrt{t}$. To obtain the segment, we restrict t to $1 \leq t \leq 16$. Thus one possible answer is

$$x = t, \quad y = \sqrt{t}, \quad 1 \leq t \leq 16.$$

45. It is a straight line through the point $(3, 5, 7)$ parallel to the vector $\vec{i} - \vec{j} + 2\vec{k}$. A linear parameterization of the same line is $x = 3 + t$, $y = 5 - t$, $z = 7 + 2t$.

49. Add the two equations to get $3x = 8$, or $x = \frac{8}{3}$. Then we have

$$-y + z = \frac{1}{3}.$$

So a possible parameterization is

$$x = \frac{8}{3}, \quad y = t, \quad z = \frac{1}{3} + t.$$

53. These equations parameterize a line. Since $(3 + t) + (2t) + 3(1 - t) = 6$, we have $x + y + 3z = 6$. Similarly, $x - y - z = (3 + t) - 2t - (1 - t) = 2$. That is, the curve lies entirely in the plane $x + y + 3z = 6$ and in the plane $x - y - z = 2$. Since the normals to the two planes, $\vec{n_1} = \vec{i} + \vec{j} + 3\vec{k}$ and $\vec{n_2} = \vec{i} - \vec{j} - \vec{k}$ are not parallel, the line is the intersection of two nonparallel planes, which is a straight line in 3-dimensional space.

57. The three shadows appear as a circle, a cosine wave and a sine wave, respectively.

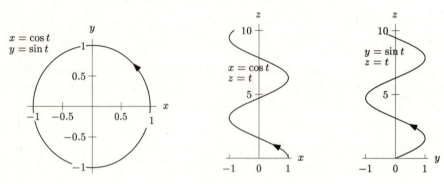

Figure 17.1

61. (a) Parametric equations are

$$x = 1 + 2t, \quad y = 5 + 3t, \quad z = 2 - t.$$

(b) We want to minimize D, the square of the distance of a point to the origin, where

$$D = (x - 0)^2 + (y - 0)^2 + (z - 0)^2 = (1 + 2t)^2 + (5 + 3t)^2 + (2 - t)^2.$$

Differentiating to find the critical points gives

$$\frac{dD}{dt} = 2(1 + 2t)2 + 2(5 + 3t)3 + 2(2 - t)(-1) = 0$$

$$2 + 4t + 15 + 9t - 2 + t = 0$$

$$t = \frac{-15}{14}.$$

Thus

$$x = 1 + 2\left(\frac{-15}{14}\right) = \frac{-8}{7}$$

$$y = 5 + 3\left(\frac{-15}{14}\right) = \frac{25}{14}$$

$$z = 2 - \left(\frac{-15}{14}\right) = \frac{43}{14}.$$

Since the distance of the point on the line from the origin increases without bound as the magnitude of x, y, z increase, the only critical point of D must be a global minimum. Therefore, the point $(-8/7, 25/14, 43/14)$ is the point on the line closest to the origin.

Solutions for Section 17.2

Exercises

1. To find $\vec{v}(t)$ we first find $dx/dt = 6t$ and $dy/dt = 3t^2$. Therefore, the velocity vector is $\vec{v} = 6t\vec{i} + 3t^2\vec{j}$. The speed of the particle is given by the magnitude of the vector,

$$\|\vec{v}\| = \sqrt{\left(\frac{dx}{dt}\right)^2 + \left(\frac{dy}{dt}\right)^2} = \sqrt{(6t)^2 + (3t^2)^2} = 3|t| \cdot \sqrt{4 + t^2}.$$

The particle stops when $\vec{v} = \vec{0}$, so when $6t = 3t^2 = 0$. Therefore, the particle stops when $t = 0$.

5. The velocity vector $\vec{v}$ is given by:

$$\vec{v} = \frac{d}{dt}(3\cos t)\vec{i} + \frac{d}{dt}(4\sin t)\vec{j} = -3\sin t\vec{i} + 4\cos t\vec{j}.$$

The acceleration vector $\vec{a}$ is given by:

$$\vec{a} = \frac{d\vec{v}}{dt} = \frac{d}{dt}(-3\sin t)\vec{i} + \frac{d}{dt}(4\cos t)\vec{j} = -3\cos t\vec{i} - 4\sin t\vec{j}.$$

9. At $t = 3$, the particle is at the point $(4, 5)$ since $x = 3^3 - 8 \cdot 3 + 1 = 4$ and $y = 3^2 - 4 = 5$. The velocity vector is

$$\vec{v} = \frac{dx}{dt}\vec{i} + \frac{dy}{dt}\vec{j} = (3t^2 - 8)\vec{i} + (2t)\vec{j}.$$

When $t = 3$, the velocity vector is $\vec{v} = 19\vec{i} + 6\vec{j}$. The speed of the particle at $t = 3$ is given by

$$\text{Speed} = \|\vec{v}\| = \sqrt{19^2 + 6^2} = \sqrt{397} = 19.92.$$

13. In vector form the parameterization is

$$\vec{r} = 2\vec{i} + 3\vec{j} + 5\vec{k} + t^2(\vec{i} - 2\vec{j} - \vec{k}).$$

Thus the motion is along the straight line through $(2, 3, 5)$ in the direction of $\vec{i} - 2\vec{j} - \vec{k}$. The velocity vector $\vec{v}$ is

$$\vec{v} = \frac{dx}{dt}\vec{i} + \frac{dy}{dt}\vec{j} + \frac{dz}{dt}\vec{k} = 2t(\vec{i} - 2\vec{j} - \vec{k})$$

The acceleration vector $\vec{a}$ is

$$\vec{a} = \frac{d^2x}{dt^2}\vec{i} + \frac{d^2y}{dt^2}\vec{j} + \frac{d^2z}{dt^2}\vec{k} = 2(\vec{i} - 2\vec{j} - \vec{k}).$$

The speed is

$$\|\vec{v}\| = 2|t|\|\vec{i} - 2\vec{j} - \vec{k}\| = 2\sqrt{6}|t|.$$

The acceleration vector is constant and points in the direction of $\vec{i} - 2\vec{j} - \vec{k}$. When $t < 0$ the absolute value $|t|$ is decreasing, hence the speed is decreasing. Also, when $t < 0$ the velocity vector $2t(\vec{i} - 2\vec{j} - \vec{k})$ points in the direction opposite to $\vec{i} - 2\vec{j} - \vec{k}$. When $t > 0$ the absolute value $|t|$ is increasing and hence the speed is increasing. Also, when $t > 0$ the velocity vector points in the same direction as $\vec{i} - 2\vec{j} - \vec{k}$.

17. We have

$$\text{Length} = \int_0^{2\pi} \sqrt{(-3\sin 3t)^2 + (5\cos 5t)^2}\, dt.$$

We cannot find this integral symbolically, but numerical methods show Length ≈ 24.6.

Problems

21. At $t = 2$, the position and velocity vectors are

$$\vec{r}(2) = (2-1)^2\vec{i} + 2\vec{j} + (2 \cdot 2^3 - 3 \cdot 2^2)\vec{k} = \vec{i} + 2\vec{j} + 4\vec{k},$$
$$\vec{v}(2) = 2 \cdot (2-1)\vec{i} + (6 \cdot 2^2 - 6 \cdot 2)\vec{k} = 2\vec{i} + 12\vec{k}.$$

So we want the line going through the point $(1, 2, 4)$ at the time $t = 2$, in the direction $2\vec{i} + 12\vec{k}$:

$$x = 1 + 2(t-2), \quad y = 2 \quad z = 4 + 12(t-2).$$

25. At time t the particle is $s = t - 7$ seconds from P, so the displacement vector from the point P to the particle is $\vec{d} = s\vec{v}$. To find the position vector of the particle at time t, we add this to the position vector $\vec{r}_0 = 5\vec{i} + 4\vec{j} + 3\vec{k}$ for the point P. Thus a vector equation for the motion is:

$$\begin{aligned} \vec{r} &= \vec{r}_0 + s\vec{v} \\ &= (5\vec{i} + 4\vec{j} + 3\vec{k}) + (t-7)(3\vec{i} + \vec{j} + 2\vec{k}), \end{aligned}$$

or equivalently,

$$x = 5 + 3(t-7), \quad y = 4 + 1(t-7), \quad z = 3 + 2(t-7).$$

Notice that these equations are linear. They describe motion on a straight line through the point $(5, 4, 3)$ that is parallel to the velocity vector $\vec{v} = 3\vec{i} + \vec{j} + 2\vec{k}$.

29. Parametric equations for a line in 2-space are

$$x = x_0 + at$$
$$y = y_0 + bt$$

where (x_0, y_0) is a point on the line and $\vec{v} = a\vec{i} + b\vec{j}$ is the direction of motion. Notice that the slope of the line is equal to $\Delta y/\Delta x = b/a$, so in this case we have

$$\frac{b}{a} = \text{Slope} = -2,$$
$$b = -2a.$$

In addition, the speed is 3, so we have

$$\|\vec{v}\| = 3$$
$$\sqrt{a^2 + b^2} = 3$$
$$a^2 + b^2 = 9.$$

Substituting $b = -2a$ gives

$$a^2 + (-2a)^2 = 9$$
$$5a^2 = 9$$
$$a = \frac{3}{\sqrt{5}}, -\frac{3}{\sqrt{5}}.$$

If we use $a = 3/\sqrt{5}$, then $b = -2a = -6/\sqrt{5}$. The point (x_0, y_0) can be any point on the line: we use $(0, 5)$. The parametric equations are

$$x = \frac{3}{\sqrt{5}}t, \quad y = 5 - \frac{6}{\sqrt{5}}t.$$

Alternatively, we can use $a = -3/\sqrt{5}$ giving $b = 6/\sqrt{5}$. An alternative answer, which represents the particle moving in the opposite direction is

$$x = -\frac{3}{\sqrt{5}}t, \quad y = 5 + \frac{6}{\sqrt{5}}t.$$

33. (a) No. The height of the particle is given by $2t$; the vertical velocity is the derivative $d(2t)/dt = 2$. Because this is a positive constant, the vertical component of the velocity vector is upward at a constant speed of 2.

(b) When $2t = 10$, so $t = 5$.

(c) The velocity vector is given by

$$\vec{v}(t) = \frac{d\vec{r}}{dt} = \frac{dx}{dt}\vec{i} + \frac{dy}{dt}\vec{j} + \frac{dz}{dt}\vec{k}$$
$$= -(\sin t)\vec{i} + (\cos t)\vec{j} + 2\vec{k}.$$

From (b), the particle is at 10 units above the ground when $t = 5$, so at $t = 5$,

$$\vec{v}(5) = 0.959\vec{i} + 0.284\vec{j} + 2\vec{k}.$$

Therefore, $\vec{v}(5) = -\sin(5)\vec{i} + \cos(5)\vec{j} + 2\vec{k}$.

(d) At this point, $t = 5$, the particle is located at

$$\vec{r}(5) = (\cos(5), \sin(5), 10) = (0.284, -0.959, 10).$$

The tangent vector to the helix at this point is given by the velocity vector found in part (c), that is, $\vec{v}(5) = 0.959\vec{i} + 0.284\vec{j} + 2\vec{k}$. So, the equation of the tangent line is

$$\vec{r}(t) = 0.284\vec{i} - 0.959\vec{j} + 10\vec{k} + (t-5)(0.959\vec{i} + 0.284\vec{j} + 2\vec{k}).$$

37. At time t object B is at the point with position vector $\vec{r}_B(t) = \vec{r}_A(2t)$, which is exactly where object A is at time $2t$. Thus B visits the same points as A, but does so at different times; A gets there later. While B covers the same path as A, it moves twice as fast. To see this, note for example that between $t = 1$ and $t = 3$, object B moves along the path from $\vec{r}_B(1) = \vec{r}_A(2)$ to $\vec{r}_B(3) = \vec{r}_A(6)$ which is traversed by object A during the time interval from $t = 2$ to $t = 6$. It takes A twice as long to cover the same ground.

In the case where $\vec{r}_A(t) = t\vec{i} + t^2\vec{j}$, both objects move on the parabola $y = x^2$. Both A and B are at the origin at time $t = 0$, but B arrives at the point $(2, 4)$ at time $t = 1$, whereas A does does not get there until $t = 2$.

41. The acceleration vector points from the object to the center of the orbit, and the velocity vector points from the object tangent to the circle in the direction of motion. From Figure 17.2 we see that the movement is counterclockwise.

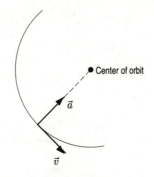

Figure 17.2

Solutions for Section 17.3

Exercises

1. Notice that for a repulsive force, the vectors point outward, away from the particle at the origin, for an attractive force, the vectors point toward the particle. So we can match up the vector field with the description as follows:

(a) IV

(b) III

(c) I

(d) II

5.

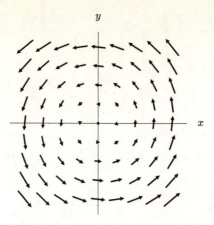

Figure 17.3: $\vec{F}(x, y) = -y\vec{i} + x\vec{j}$

9.

(I)

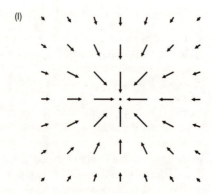

Figure 17.4: $\vec{F}(\vec{r}) = -\vec{r}/\|\vec{r}\|^3$

13. $\vec{V} = x\vec{i} + y\vec{j} = \vec{r}$

Problems

17. (a) The gradient is perpendicular to the level curves. See Figures 17.6 and 17.5. A function always increases in the direction of its gradient; this is why the values on the level curves of f and g increase as we approach the origin.

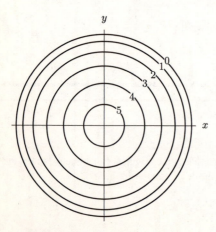

Figure 17.5: Level curves $z = f(x, y)$

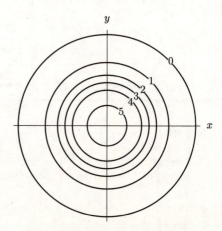

Figure 17.6: Level curves $z = g(x, y)$

(b) f climbs faster at outside, slower at center; g climbs slower at outside, faster at center:

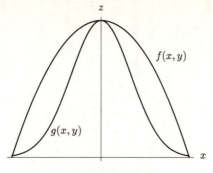

Figure 17.7

This can be understood if we notice that the magnitude of the gradient of f decreases as one approaches the origin whereas the magnitude of the gradient of g increases (at least for a while - what happens very close to the origin depends on the behavior of grad g in the region. One possibility for g is shown in Figure 17.7; the graph of g could also have a sharp peak at 0 or even blow up.)

21. (a) Since the velocity of the water is the sum of the velocities of the individual fields, then the total field should be

$$\vec{v} = \vec{v}_{\text{stream}} + \vec{v}_{\text{fountain}}.$$

It is reasonable to represent $\vec{v}_{\text{stream}}$ by the vector field $\vec{v}_{\text{stream}} = A\vec{i}$, since $A\vec{i}$ is a constant vector field flowing in the i-direction (provided $A > 0$). It is reasonable to represent $\vec{v}_{\text{fountain}}$ by

$$\vec{v}_{\text{fountain}} = K\vec{r}_r/r^2 = K(x^2 + y^2)^{-1}(x\vec{i} + y\vec{j}),$$

since this is a vector field flowing radially outward (provided $K > 0$), with decreasing velocity as r gets larger. We would expect the velocity to decrease as the water from the fountain spreads out. Adding the two vector fields together, we get

$$\vec{v} = A\vec{i} + K(x^2 + y^2)^{-1}(x\vec{i} + y\vec{j}), \quad A > 0, K > 0.$$

(b) The constants A and K signify the strength of the individual components of the field. A is the strength of the flow of the stream alone (in fact it is the speed of the stream), and K is the strength of the fountain acting alone.

(c)

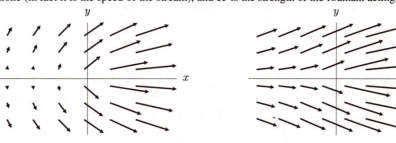

Figure 17.8: $A = 1, K = 1$ **Figure 17.9**: $A = 2, K = 1$

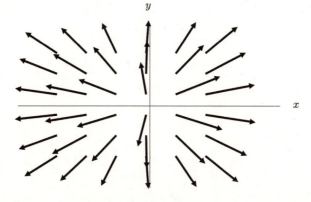

Figure 17.10: $A = 0.2, K = 2$

25. (a) The vector field $\vec{L} = 0\vec{F} + \vec{G} = -y\vec{i} + x\vec{j}$ is shown in Figure 17.11.

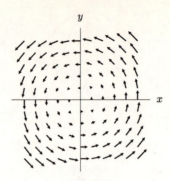

Figure 17.11

(b) The vector field $\vec{L} = a\vec{F} + \vec{G} = (ax - y)\vec{i} + (ay + x)\vec{j}$ where $a > 0$ is shown in Figure 17.12.

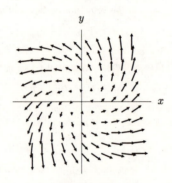

Figure 17.12

(c) The vector field $\vec{L} = a\vec{F} + \vec{G} = (ax - y)\vec{i} + (ay + x)\vec{j}$ where $a < 0$ is shown in Figure 17.13.

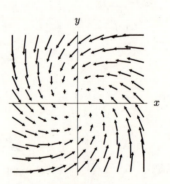

Figure 17.13

Solutions for Section 17.4

Exercises

1. Since $x'(t) = 3$ and $y'(t) = 0$, we have $x = 3t + x_0$ and $y = y_0$. Thus, the solution curves are $y = $ constant.

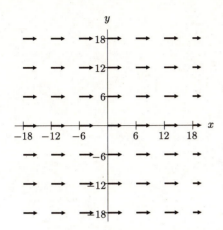

Figure 17.14: The field $\vec{v} = 3\vec{i}$

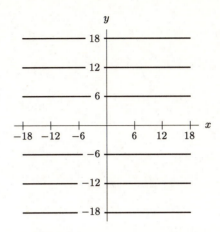

Figure 17.15: The flow $y =$ constant

5.

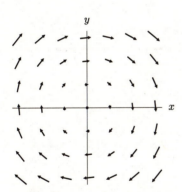

Figure 17.16: $\vec{v}(t) = y\vec{i} - x\vec{j}$

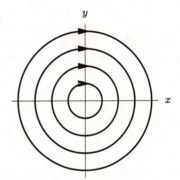

Figure 17.17: The flow $x = a \sin t$,
$y = a \cos t$

As

$$\vec{v}(t) = \frac{dx}{dt}\vec{i} + \frac{dy}{dt}\vec{j},$$

the system of differential equations is

$$\begin{cases} \frac{dx}{dt} = y \\ \frac{dy}{dt} = -x. \end{cases}$$

Since

$$\frac{dx(t)}{dt} = \frac{d}{dt}[a\sin t] = a\cos t = y(t)$$

and

$$\frac{dy(t)}{dt} = \frac{d}{dt}[a\cos t] = -a\sin t = -x(t),$$

the given flow satisfies the system. By eliminating the parameter t in $x(t)$ and $y(t)$, the solution curves obtained are $x^2 + y^2 = a^2$.

9. The directions of the flow lines are as shown.

   (a) III
   (b) I
   (c) II
   (d) V
   (e) VI
   (f) IV

(I) 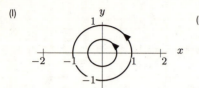   (II) 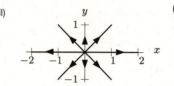   (III)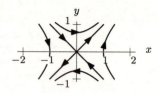

(IV) 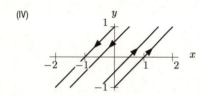   (V) 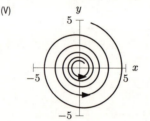   (VI)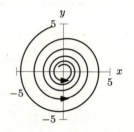

Solutions for Section 17.5

Exercises

1. A horizontal disk of radius 5 in the plane $z = 7$.

5. Since $z = r = \sqrt{x^2 + y^2}$, we have a cone around the z-axis. Since $0 \le r \le 5$, we have $0 \le z \le 5$, so the cone has height and maximum radius of 5.

9. The top half of the sphere ($z \ge 0$).

Problems

13. The cross sections of the cylinder perpendicular to the z-axis are circles which are vertical translates of the circle $x^2 + y^2 = a^2$, which is given parametrically by $x = a \cos \theta$, $y = a \sin \theta$. The vector $a \cos \theta \vec{i} + a \sin \theta \vec{j}$ traces out the circle, at any height. We get to a point on the surface by adding that vector to the vector $z\vec{k}$. Hence, the parameters are θ, with $0 \le \theta \le 2\pi$, and z, with $0 \le z \le h$. The parametric equations for the cylinder are

$$x\vec{i} + y\vec{j} + z\vec{k} = a \cos \theta \vec{i} + a \sin \theta \vec{j} + z\vec{k},$$

which can be written as

$$x = a \cos \theta, \quad y = a \sin \theta, \quad z = z.$$

17. (a) We want to find s and t so that

$$2 + s = 4$$
$$3 + s + t = 8$$
$$4t = 12$$

Since $s = 2$ and $t = 3$ satisfy these equations, the point $(4, 8, 12)$ lies on this plane.

(b) Are there values of s and t corresponding to the point $(1, 2, 3)$? If so, then

$$1 = 2 + s$$
$$2 = 3 + s + t$$
$$3 = 4t$$

From the first equation we must have $s = -1$ and from the third we must have $t = 3/4$. But these values of s and t do not satisfy the second equation. Therefore, no value of s and t corresponds to the point $(1, 2, 3)$, and so $(1, 2, 3)$ is not on the plane.

21. A vertical half-circle, going from the north to south poles.

25. The parameterization for a sphere of radius a using spherical coordinates is

$$x = a \sin \phi \cos \theta, \quad y = a \sin \phi \sin \theta, \quad z = a \cos \phi.$$

Think of the ellipsoid as a sphere whose radius is different along each axis and you get the parameterization:
$$\begin{cases} x = a \sin \phi \cos \theta, & 0 \leq \phi \leq \pi, \\ y = b \sin \phi \sin \theta, & 0 \leq \theta \leq 2\pi, \\ z = c \cos \phi. \end{cases}$$
To check this parameterization, substitute into the equation for the ellipsoid:

$$\frac{x^2}{a^2} + \frac{y^2}{b^2} + \frac{z^2}{c^2} = \frac{a^2 \sin^2 \phi \cos^2 \theta}{a^2} + \frac{b^2 \sin^2 \phi \sin^2 \theta}{b^2} + \frac{c^2 \cos^2 \phi}{c^2}$$
$$= \sin^2 \phi (\cos^2 \theta + \sin^2 \theta) + \cos^2 \phi = 1.$$

29. (a) The cone of height h, maximum radius a, vertex at the origin and opening upward is shown in Figure 17.18.

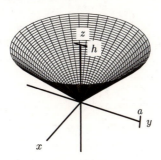

Figure 17.18

By similar triangles, we have

$$\frac{r}{z} = \frac{a}{h},$$

so

$$z = \frac{hr}{a}.$$

Therefore, one parameterization is

$$x = r \cos \theta, \quad 0 \leq r \leq a,$$
$$y = r \sin \theta, \quad 0 \leq \theta < 2\pi,$$
$$z = \frac{hr}{a}.$$

(b) Since $r = az/h$, we can write the parameterization in part (a) as

$$x = \frac{az}{h} \cos \theta, \quad 0 \leq z \leq h,$$
$$y = \frac{az}{h} \sin \theta, \quad 0 \leq \theta < 2\pi,$$
$$z = z.$$

33. (a) From the first two equations we get:

$$s = \frac{x+y}{2}, \qquad t = \frac{x-y}{2}.$$

Hence the equation of our surface is:

$$z = \left(\frac{x+y}{2}\right)^2 + \left(\frac{x-y}{2}\right)^2 = \frac{x^2}{2} + \frac{y^2}{2},$$

which is the equation of a paraboloid.

The conditions: $0 \le s \le 1, 0 \le t \le 1$ are equivalent to: $0 \le x+y \le 2, 0 \le x-y \le 2$. So our surface is defined by:

$$z = \frac{x^2}{2} + \frac{y^2}{2}, \qquad 0 \le x+y \le 2 \quad 0 \le x-y \le 2$$

(b) The surface is shown in Figure 17.19.

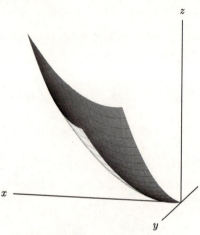

Figure 17.19: The surface $x = s+t$,
$y = s-t, z = s^2 + t^2$ for $0 \le s \le 1$,
$0 \le t \le 1$

37. The plane in which the circle lies is parameterized by

$$\vec{r}(p,q) = x_0\vec{i} + y_0\vec{j} + z_0\vec{k} + p\vec{u} + q\vec{v}.$$

Because $\vec{u}$ and $\vec{v}$ are perpendicular unit vectors, the parameters p and q establish a rectangular coordinate system on this plane exactly analogous to the usual xy-coordinate system, with $(p,q) = (0,0)$ corresponding to the point (x_0, y_0, z_0). Thus the circle we want to describe, which is the circle of radius a centered at $(p, q) = (0, 0)$, can be parameterized by

$$p = a\cos t, \qquad q = a\sin t.$$

Substituting into the equation of the plane gives the desired parameterization of the circle in 3-space,

$$\vec{r}(t) = x_0\vec{i} + y_0\vec{j} + z_0\vec{k} + a\cos t\vec{u} + a\sin t\vec{v},$$

where $0 \le t \le 2\pi$.

Solutions for Chapter 17 Review

Exercises

1. $x = t, y = 5$.

5. The vector $(\vec{i} + 2\vec{j} + 5\vec{k}) - (2\vec{i} - \vec{j} + 4\vec{k}) = -\vec{i} + 3\vec{j} + \vec{k}$ is parallel to the line, so a possible parameterization is

$$x = 2 - t, \quad y = -1 + 3t, \quad z = 4 + t.$$

9. Since the circle has radius 3, the equation must be of the form $x = 3\cos t, y = 5, z = 3\sin t$. But since the circle is being viewed from farther out on the y-axis, the circle we have now would be seen going clockwise. To correct this, we add a negative to the third component, giving us the equation $x = 3\cos t, y = 5, z = -3\sin t$.

13. The vector field points in a clockwise direction around the origin. Since

$$\left\| \left(\frac{y}{\sqrt{x^2 + y^2}} \right) \vec{i} - \left(\frac{x}{\sqrt{x^2 + y^2}} \right) \vec{j} \right\| = \frac{\sqrt{x^2 + y^2}}{\sqrt{x^2 + y^2}} = 1$$

the length of the vectors is constant everywhere.

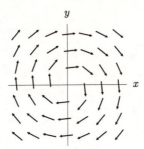

Figure 17.20

Problems

17. (a)   (I) has radius 1 and traces out a complete circle, so $I = C_4$.

    (II) has radius 2 and traces out the top half of a circle, so $II = C_1$

    (III) has radius 1 and traces out a quarter circle, so $III = C_2$.

    (IV) has radius 2 and traces out the bottom half of a circle, so $IV = C_6$.

(b) C_3 has radius $1/2$ and traces out a half circle below the x-axis, so

$$\vec{r} = 0.5 \cos t \vec{i} - 0.5 \sin t \vec{j}.$$

C_5 has radius 2 and traces out a quarter circle below the x-axis starting at the point $(-2, 0)$. Thus we have

$$\vec{r} = -2 \cos(t/2)\vec{i} - 2 \sin(t/2)\vec{j}.$$

21. (a) Separate the ant's path into three parts: from $(0, 0)$ to $(1, 0)$ along the x-axis; from $(1, 0)$ to $(0, 1)$ via the circle; and from $(0, 1)$ to $(0, 0)$ along the y-axis. (See Figure 17.21.) The lengths of the paths are 1, $\frac{2\pi}{4} = \frac{\pi}{2}$, and 1 respectively. Thus, the time it takes for the ant to travel the three paths are (using the formula $t = \frac{d}{v}$) $\frac{1}{2}$, $\frac{1}{3}$, and $\frac{1}{2}$ seconds.

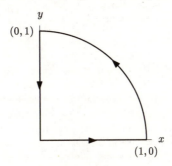

Figure 17.21

From $t = 0$ to $t = \frac{1}{2}$, the ant is heading toward $(1, 0)$ so its coordinate is $(2t, 0)$. From $t = \frac{1}{2}$ to $t = \frac{1}{2} + \frac{1}{3} = \frac{5}{6}$, the ant is veering to the left and heading toward $(0, 1)$. At $t = \frac{1}{2}$, it is at $(1, 0)$ and at $t = \frac{5}{6}$, it is at $(0, 1)$. Thus its position is $(\cos[\frac{3\pi}{2}(t - \frac{1}{2})], \sin[\frac{3\pi}{2}(t - \frac{1}{2})])$. Finally, from $t = \frac{5}{6}$ to $t = \frac{5}{6} + \frac{1}{2} = \frac{4}{3}$, the ant is headed home. Its coordinates are $(0, -2(t - \frac{4}{3}))$.

In summary, the function expressing the ant's coordinates is

$$(x(t), y(t)) = \begin{cases} (2t, 0) & \text{when } 0 \le t \le \frac{1}{2} \\ \left(\cos(\frac{3\pi}{2}(t - \frac{1}{2})), \sin(\frac{3\pi}{2}(t - \frac{1}{2}))\right) & \text{when } \frac{1}{2} < t \le \frac{5}{6} \\ (0, -2(t - \frac{4}{3})) & \text{when } \frac{5}{6} \le t \le \frac{4}{3}. \end{cases}$$

(b) To do the reverse path, observe that we can reverse the ant's path by interchanging the x and y coordinates (flipping it with respect to the line $y = x$), so the function is

$$(x(t), y(t)) = \begin{cases} (0, 2t) & \text{when } 0 \le t \le \frac{1}{2} \\ \left(\sin(\frac{3\pi}{2}(t - \frac{1}{2})), \cos(\frac{3\pi}{2}(t - \frac{1}{2}))\right) & \text{when } \frac{1}{2} < t \le \frac{5}{6} \\ (-2(t - \frac{4}{3}), 0) & \text{when } \frac{5}{6} < t \le \frac{4}{3}. \end{cases}$$

25. (a) The equation of the sphere is $x^2 + y^2 + z^2 = 1$. Substituting the parameterization gives

$$(x(t))^2 + (y(t))^2 + (z(t))^2 = \cos^4 t + \sin^2 t \cos^2 t + \sin^2 t = (\cos^2 t)(\cos^2 t + \sin^2 t) + \sin^2 t = \cos^2 t + \sin^2 t = 1.$$

Therefore the curve lies on this sphere.

(b) If $t = \pi/4$, we have $x(\pi/4) = 1/2$, $y(\pi/4) = 1/2$, $z(\pi/4) = 1/\sqrt{2}$. The gradient vector to the sphere at this point is perpendicular to the sphere and the curve. Since grad $f = 2x\vec{i} + 2y\vec{j} + 2z\vec{k}$, we have

$$\text{Normal} = \text{grad } f\left(\frac{1}{2}, \frac{1}{2}, \frac{1}{\sqrt{2}}\right) = 2\left(\frac{1}{2}\right)\vec{i} + 2\left(\frac{1}{2}\right)\vec{j} + 2\left(\frac{1}{\sqrt{2}}\right)\vec{k} = \vec{i} + \vec{j} + \sqrt{2}\vec{k}.$$

(c) A tangent vector is $x'(t)\vec{i} + y'(t)\vec{j} + z'(t)\vec{k} = (-2\cos t \sin t)\vec{i} + (\cos^2 t - \sin^2 t)\vec{j} + \cos t\vec{k}$. At $t = \pi/4$, we have

$$\text{Tangent vector} = -\vec{i} + \frac{1}{\sqrt{2}}\vec{k}.$$

29. Let's place the coordinate system so that the origin is at the center of the circle of radius R on which the object is moving. If $\theta(t)$ is the polar coordinate angle of the object at time t, the position vector of the object is given by

$$\vec{r}(t) = R\cos(\theta(t))\vec{i} + R\sin(\theta(t))\vec{j}.$$

(a) The velocity vector is given by the derivative

$$\vec{r}' = -R\theta' \sin\theta\vec{i} + R\theta' \cos\theta\vec{j} = R\theta'(-\sin\theta\vec{i} + \cos\theta\vec{j})$$

The velocity $\vec{r}'$ is perpendicular to the radius vector $\vec{r}$ from the center of the circle to the object because

$$\vec{r}' \cdot \vec{r} = R\theta'\left(-\sin\theta\vec{i} + \cos\theta\vec{j}\right) \cdot \left(R\cos\theta\vec{i} + R\sin\theta\vec{j}\right)$$
$$= R^2\theta'(-\sin\theta\cos\theta + \cos\theta\sin\theta) = 0.$$

So the velocity vector, $\vec{r}'$, is tangent to the circle.

(b) The acceleration vector is given by the second derivative

$$\vec{r}'' = R\theta''(-\sin\theta\vec{i} + \cos\theta\vec{j}) - R(\theta')^2(\cos\theta\vec{i} + \sin\theta\vec{j})$$

The acceleration vector $\vec{r}''$ is expressed as the sum of two vector components. Since $\vec{r} = R\cos\theta\vec{i} + R\sin\theta\vec{j}$, the component $-R(\theta')^2(\cos\theta\vec{i} + \sin\theta\vec{j})$ is in the direction opposite to $\vec{r}$, that is, toward the center of the circle. The component $R\theta''(-\sin\theta\vec{i} + \cos\theta\vec{j})$ is perpendicular to $\vec{r}$ since the dot product is $(-\sin\theta\vec{i} + \cos\theta\vec{j}) \cdot (\cos\theta\vec{i} + \sin\theta\vec{j}) = 0$. This means that $R\theta''\left(-\sin\theta\vec{i} + \cos\theta\vec{j}\right)$ is in the direction tangent to the circle. The sum $\vec{r}''$ is not in general directed toward the center of the circle. The acceleration is only directed toward the center of the circle in the case where $\theta''(t) = 0$, as it is for example in the case $\theta(t) = \omega t$ of uniform circular motion.

33. This corresponds to area C in Figure 17.22.

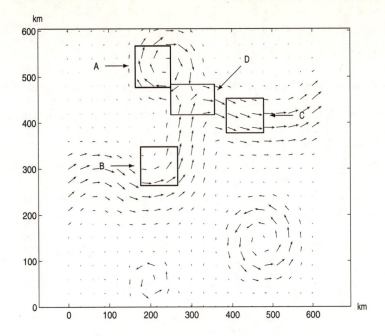

Figure 17.22

37. The sphere $x^2 + y^2 + z^2 = 1$ is shown in Figure 17.23.

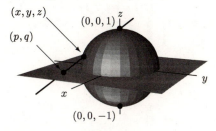

Figure 17.23

    (a) The origin corresponds to the south pole.
    (b) The circle $x^2 + y^2 = 1$ corresponds to the equator.
    (c) We get all the points of the sphere by this parameterization except the north pole itself.
    (d) $x^2 + y^2 > 1$ corresponds to the upper hemisphere.
    (e) $x^2 + y^2 < 1$ corresponds to the lower hemisphere.

CAS Challenge Problems

  41. Answers may differ depending on the method and CAS used.

    (a) Using a CAS to solve for x and y in terms of z and letting $z = t$, we get $x = \frac{20}{13} - \frac{6t}{13}, y = \frac{-1}{13} - \frac{t}{13}, z = t$.

    (b) Using a CAS to solve for y and z in terms of x and letting $x = t$, we get $x = t, y = \frac{1}{6}(-2 - 2t + 3t^2), z = \frac{1}{6}(20 - 10t - 3t^2)$.

    (c) Using a CAS to solve for x and z in terms of y, we get two solutions

$$x = \sqrt{2 - t^2}, \quad y = t, \quad z = 5 + 5t - 3\sqrt{2 - t^2}$$

and

$$x = -\sqrt{2 - t^2}, \quad y = t, \quad z = 5 + 5t + 3\sqrt{2 - t^2}$$

Each of these is a parameterization of one half of the intersection curve.

CHECK YOUR UNDERSTANDING

1. False. The y coordinate is zero when $t = 0$, but when $t = 0$ we have $x = 2$ so the curve never passes through $(0, 0)$.

5. False. When $t = 0$, we have $(x, y) = (0, -1)$. When $t = \pi/2$, we have $(x, y) = (-1, 0)$. Thus the circle is being traced out clockwise.

9. True. To find an intersection point, we look for values of s and t that make the coordinates in the first line the same as the coordinates in the second. Setting $x = t$ and $x = 2s$ equal, we see that $t = 2s$. Setting $y = 2 + t$ equal to $y = 1 - s$, we see that $t = -1 - s$. Solving both $t = 2s$ and $t = -1 - s$ yields $t = -\frac{2}{3}$, $s = -\frac{1}{3}$. These values of s and t will give equal x and y coordinates on both lines. We need to check if the z coordinates are equal also. In the first line, setting $t = -\frac{2}{3}$ gives $z = \frac{7}{3}$. In the second line, setting $s = -\frac{1}{3}$ gives $z = -\frac{1}{3}$. As these are not the same, the lines do not intersect.

13. False. The velocity vector is $\vec{v}(t) = \vec{r}'(t) = 2t\vec{i} - \vec{j}$. Then $\vec{v}(-1) = -2\vec{i} - \vec{j}$ and $\vec{v}(1) = 2\vec{i} - \vec{j}$, which are not equal.

17. False. As a counterexample, consider the curve $\vec{r}(t) = t^2\vec{i} + t^2\vec{j}$ for $0 \leq t \leq 1$. In this case, when t is replaced by $-t$, the parameterization is the same, and is not reversed.

21. True, since the vectors $x\vec{j}$ are parallel to the y-axis.

25. True. Any flow line which stays in the first quadrant has $x, y \to \infty$.

29. True. If (x, y) were a point where the y-coordinate along a flow line reached a relative maximum, then the tangent vector to the flow line, namely $\vec{F}(x, y)$, there would have to be horizontal (or $\vec{0}$), that is its $\vec{j}$ component would have to be 0. But the $\vec{j}$ component of $\vec{F}$ is always 2.

33. False. There is only one parameter, s. The equations parameterize a line.

37. True. If the surface is parameterized by $\vec{r}(s, t)$ and the point has parameters (s_0, t_0) then the parameter curves $\vec{r}(s_0, t)$ and $\vec{r}(s, t_0)$.

41. False. Suppose $\vec{r}(t) = t\vec{i} + t\vec{j}$. Then $\vec{r}'(t) = \vec{i} + \vec{j}$ and

$$\vec{r}'(t) \cdot \vec{r}(t) = (t\vec{i} + t\vec{j}) \cdot (\vec{i} + \vec{j}) = 2t.$$

So $\vec{r}'(t) \cdot \vec{r}(t) \neq 0$ for $t \neq 0$.

CHAPTER EIGHTEEN

Solutions for Section 18.1

Exercises

1. Positive, because the vectors are longer on the portion of the path that goes in the same direction as the vector field.

5. Since $\vec{F}$ is perpendicular to the curve at every point along it,

$$\int_C \vec{F} \cdot d\vec{r} = 0.$$

9. At every point on the path, $\vec{F}$ is parallel to $\Delta\vec{r}$. Suppose r is the distance from the point (x, y) to the origin, so $\|\vec{r}\| = r$. Then $\vec{F} \cdot \Delta\vec{r} = \|\vec{F}\|\|\Delta\vec{r}\| = r\Delta r$. At the start of the path, $r = \sqrt{2^2 + 2^2} = 2\sqrt{2}$ and at the end $r = 6\sqrt{2}$. Thus,

$$\int_C \vec{F} \cdot d\vec{r} = \int_{2\sqrt{2}}^{6\sqrt{2}} r\,dr = \left.\frac{r^2}{2}\right|_{2\sqrt{2}}^{6\sqrt{2}} = 32.$$

Problems

13. Since it appears that C_1 is everywhere perpendicular to the vector field, all of the dot products in the line integral are zero, hence $\int_{C_1} \vec{F} \cdot d\vec{r} \approx 0$. Along the path C_2 the dot products of $\vec{F}$ with $\Delta\vec{r_i}$ are all positive, so their sum is positive and we have $\int_{C_1} \vec{F} \cdot d\vec{r} < \int_{C_2} \vec{F} \cdot d\vec{r}$. For C_3 the vectors $\Delta\vec{r_i}$ are in the opposite direction to the vectors of $\vec{F}$, so the dot products $\vec{F} \cdot \Delta\vec{r_i}$ are all negative; so, $\int_{C_3} \vec{F} \cdot d\vec{r} < 0$. Thus, we have

$$\int_{C_3} \vec{F} \cdot d\vec{r} < \int_{C_1} \vec{F} \cdot d\vec{r} < \int_{C_2} \vec{F} \cdot d\vec{r}$$

17. Since it does not depend on y, this vector field is constant along vertical lines, $x = $ constant. Now let us consider two points P and Q on C_1 which lie on the same vertical line. Because C_1 is symmetric with respect to the x-axis, the tangent vectors at P and Q will be symmetric with respect to the vertical axis so their sum is a vertical vector. But $\vec{F}$ has only horizontal component and thus $\vec{F} \cdot (\Delta\vec{r}(P) + \Delta\vec{r}(Q)) = 0$. As $\vec{F}$ is constant along vertical lines (so $\vec{F}(P) = \vec{F}(Q)$), we obtain

$$\vec{F}(P) \cdot \Delta\vec{r}(P) + \vec{F}(Q) \cdot \Delta\vec{r}(Q) = 0.$$

Summing these products and making $\|\Delta\vec{r}\| \to 0$ gives us

$$\int_{C_1} \vec{F} \cdot d\vec{r} = 0.$$

The same thing happens on C_3, so $\int_{C_3} \vec{F} \cdot d\vec{r} = 0$.

Now let P be on C_2 and Q on C_4 lying on the same vertical line. The respective tangent vectors are symmetric with respect to the vertical axis hence they add up to a vertical vector and a similar argument as before gives

$$\vec{F}(P) \cdot \Delta\vec{r}(P) + \vec{F}(Q) \cdot \Delta\vec{r}(Q) = 0$$

and

$$\int_{C_2} \vec{F} \cdot d\vec{r} + \int_{C_4} \vec{F} \cdot d\vec{r} = 0$$

and so

$$\int_C \vec{F} \cdot d\vec{r} = 0.$$

See Figure 18.1.

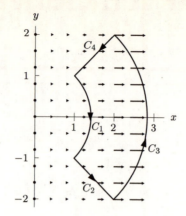

Figure 18.1

21. Suppose $\int_C \vec{F} \cdot d\vec{r} = 0$ for every closed curve C. Pick any two fixed points P_1, P_2 and curves C_1, C_2 each going from P_1 to P_2. See Figure 18.2. Define $-C_2$ to be the same curve as C_2 except in the opposite direction. Therefore, the curve formed by traversing C_1, followed by C_2 in the opposite direction, written as $C_1 - C_2$, is a closed curve, so by our assumption, $\int_{C_1-C_2} \vec{F} \cdot d\vec{r} = 0$. However, we can write

$$\int_{C_1-C_2} \vec{F} \cdot d\vec{r} = \int_{C_1} \vec{F} \cdot d\vec{r} - \int_{C_2} \vec{F} \cdot d\vec{r}$$

since C_2 and $-C_2$ are the same except for direction. Therefore,

$$\int_{C_1} \vec{F} \cdot d\vec{r} - \int_{C_2} \vec{F} \cdot d\vec{r} = 0,$$

so

$$\int_{C_1} \vec{F} \cdot d\vec{r} = \int_{C_2} \vec{F} \cdot d\vec{r}.$$

Since C_1 and C_2 are any two curves with the endpoints P_1, P_2, this gives the desired result – namely, that fixing endpoints and direction uniquely determines the value of $\int_C \vec{F} \cdot d\vec{r}$. In other words, the value of the integral $\int_C \vec{F} \cdot d\vec{r}$ does not depend on the path taken. We say the line integral is *path-independent*.

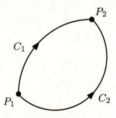

Figure 18.2

25. Let $r = \|\vec{r}\|$. Since $\Delta \vec{r}$ points outward, in the opposite direction to $\vec{F}$, we expect a negative answer. We take the upper limit to be $r = \infty$, so the integral is improper.

$$\int_C \vec{F} \cdot d\vec{r} = \int_C -\frac{GMm\vec{r}}{r^3} \cdot d\vec{r} = \int_{8000}^{\infty} -\frac{GMm}{r^2} dr$$

$$= \lim_{b \to \infty} \int_{8000}^{b} -\frac{GMm}{r^2} dr = \lim_{b \to \infty} \left.\frac{GMm}{r}\right|_{8000}^{b} = \lim_{b \to \infty} GMm\left(\frac{1}{b} - \frac{1}{8000}\right)$$

$$= -\frac{GMm}{8000}$$

29. (a) Suppose P is b units from the origin. Then P can be reached by a path, C, consisting of two pieces, C_1 and C_2, one lying on the sphere of radius a and one going straight along a line radiating from the origin (see Figure 18.3). We have $\vec{E} \cdot \Delta\vec{r} = 0$ on C_1, and, writing $r = \|\vec{r}\|$, we have $\vec{E} \cdot \Delta\vec{r} = \|\vec{E}\|\Delta r$ on C_2, so

$$\phi(P) = -\int_C \vec{E} \cdot d\vec{r} = -\int_{C_1} \vec{E} \cdot d\vec{r} - \int_{C_2} \vec{E} \cdot d\vec{r}$$

$$= 0 - \int_a^b \|\vec{E}\| \, dr = 0 - \int_a^b \frac{Q}{4\pi\epsilon} \frac{1}{r^2} \, dr$$

$$= \frac{Q}{4\pi\epsilon} \frac{1}{r} \Big|_a^b = \frac{Q}{4\pi\epsilon} \frac{1}{b} - \frac{Q}{4\pi\epsilon} \frac{1}{a}.$$

Let P be the point with position vector $\vec{r}$. Then

$$\phi(\vec{r}) = -\frac{Q}{4\pi\epsilon} \frac{1}{a} + \frac{Q}{4\pi\epsilon} \frac{1}{\|\vec{r}\|}.$$

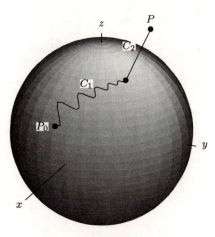

Figure 18.3

(b) If we let $a \to \infty$ in the formula for ϕ, the first term goes to zero and we get the simpler expression

$$\phi(\vec{r}) = \frac{Q}{4\pi\epsilon} \frac{1}{\|\vec{r}\|}.$$

Solutions for Section 18.2

Exercises

1. The curve C is parameterized by $(x, y) = (t, t)$ for $0 \le t \le 3$. Thus,

$$\int_C \vec{F} \cdot d\vec{r} = \int_0^3 (t\vec{i} + t\vec{j}) \cdot (\vec{i} + \vec{j}) \, dt = \int_0^3 2t \, dt = t^2 \Big|_0^3 = 9.$$

5. Use $x(t) = t$, $y(t) = t^2$, so $x'(t) = 1$, $y'(t) = 2t$, with $0 \le t \le 2$. Then

$$\int \vec{F} \cdot d\vec{r} = \int_0^2 (-t^2 \sin t\vec{i} + \cos t\vec{j}) \cdot (\vec{i} + 2t\vec{j}) \, dt$$

$$= \int_0^2 (-t^2 \sin t + 2t \cos t) \, dt = t^2 \cos t \Big|_0^2 = 4 \cos 2.$$

9. The path can be broken into three line segments: C_1, from $(1, 0)$ to $(-1, 0)$, and C_2, from $(-1, 0)$ to $(0, 1)$, and C_3, from $(0, 1)$ to $(1, 0)$. (See Figure 18.4.)

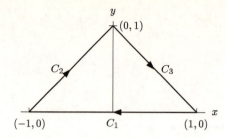

Figure 18.4

Along C_1 we have $y = 0$ so the vector field $xy\vec{i} + (x - y)\vec{j}$ is perpendicular to C_1; Thus, the line integral along C_1 is 0.

C_2 can be parameterized by $(-1 + t, t)$, for $0 \leq t \leq 1$ so the integral is

$$\int_{C_2} \vec{F} \cdot d\vec{r} = \int_0^1 \vec{F}(-1 + t, t) \cdot (\vec{i} + \vec{j}) \, dt$$

$$= \int_0^1 [t(-1 + t)\vec{i} + (-1)\vec{j}] \cdot (\vec{i} + \vec{j}) \, dt$$

$$= \int_0^1 (-t + t^2 - 1) \, dt$$

$$= (-t^2/2 + t^3/3 - t)\big|_0^1$$

$$= -1/2 + 1/3 - 1 - (0 + 0 + 0) = -7/6$$

C_3 can be parameterized by $(t, 1 - t)$, for $0 \leq t \leq 1$ so the integral is

$$\int_{C_3} \vec{F} \cdot d\vec{r} = \int_0^1 \vec{F}(t, 1 - t) \cdot (\vec{i} - \vec{j}) \, dt$$

$$= \int_0^1 (t(1 - t)\vec{i} + (2t - 1)\vec{j}) \cdot (\vec{i} - \vec{j}) \, dt$$

$$= \int_0^1 (-t^2 - t + 1) \, dt$$

$$= (-t^3/3 - t^2/2 + t)\big|_0^1$$

$$= -1/3 - 1/2 + 1 - (0 + 0 + 0) = 1/6$$

So the total line integral is

$$\int_C \vec{F} \cdot d\vec{r} = \int_{C_1} \vec{F} \cdot d\vec{r} + \int_{C_2} \vec{F} \cdot d\vec{r} + \int_{C_3} \vec{F} \cdot d\vec{r} = 0 + \left(-\frac{7}{6}\right) + \frac{1}{6} = -1$$

Problems

13. (a) Since $\vec{r}(t) = t\vec{i} + t^2\vec{j}$, we have $\vec{r}'(t) = \vec{i} + 2t\vec{j}$. Thus,

$$\int_C \vec{F} \cdot d\vec{r} = \int_0^1 \vec{F}(t, t^2) \cdot (\vec{i} + 2t\vec{j}) \, dt$$

$$= \int_0^1 [(3t - t^2)\vec{i} + t\vec{j}] \cdot (\vec{i} + 2t\vec{j}) \, dt$$

$$= \int_0^1 (3t + t^2)\, dt$$

$$= \left(\frac{3t^2}{2} + \frac{t^3}{3}\right)\Big|_0^1 = \frac{3}{2} + \frac{1}{3} - (0 + 0) = \frac{11}{6}$$

(b) Since $\vec{r}(t) = t^2\vec{i} + t\vec{j}$, we have $\vec{r}'(t) = 2t\vec{i} + \vec{j}$. Thus,

$$\int_C \vec{F} \cdot d\vec{r} = \int_0^1 \vec{F}(t^2, t) \cdot (2t\vec{i} + \vec{j})\, dt$$

$$= \int_0^1 [(3t^2 - t)\vec{i} + t^2\vec{j}] \cdot (2t\vec{i} + \vec{j})\, dt$$

$$= \int_0^1 (6t^3 - t^2)\, dt$$

$$= \left(\frac{3t^4}{2} - \frac{t^3}{3}\right)\Big|_0^1$$

$$= \frac{3}{2} - \frac{1}{3} - (0 - 0) = \frac{7}{6}$$

17. The line integral is defined by chopping the curve C into little pieces, C_i, and forming the sum

$$\sum_{C_i} \vec{F} \cdot \Delta\vec{r}.$$

When the pieces are small, $\Delta\vec{r}$ is approximately tangent to C_i, and its magnitude is approximately equal to the length of the little piece of curve C_i. This means that $\vec{F}$ and $\Delta\vec{r}$ are almost parallel, the dot product is approximately equal to the product of their magnitudes, i.e.,

$$\vec{F} \cdot \Delta\vec{r} \approx m \cdot (\text{Length of } C_i).$$

When we sum all the dot products, we get

$$\sum_{C_i} \vec{F} \cdot \Delta\vec{r} \approx \sum_{C_i} m \cdot (\text{Length of } C_i)$$

$$= m \cdot \sum_{C_i} (\text{Length of } C_i)$$

$$= m \cdot (\text{Length of } C)$$

21. The integral corresponding to $A(t) = (t, t)$ is

$$\int_0^1 3t\, dt.$$

The integral corresponding to $D(t) = (e^t - 1, e^t - 1)$ is

$$3\int_0^{\ln 2} (e^{2t} - e^t)\, dt.$$

The substitution $s = e^t - 1$ has $ds = e^t\, dt$. Also $s = 0$ when $t = 0$ and $s = 1$ when $t = \ln 2$. Thus, substituting into the integral corresponding to $D(t)$ and using the fact that $e^{2t} = e^t \cdot e^t$ gives

$$3\int_0^{\ln 2} (e^{2t} - e^t)\, dt = 3\int_0^{\ln 2} (e^t - 1)e^t\, dt = \int_0^1 3s\, ds.$$

The integral on the right-hand side is the same as the integral corresponding to $A(t)$. Therefore we have

$$3\int_0^{\ln 2} (e^{2t} - e^t)\, dt = \int_0^1 3s\, ds = \int_0^1 3t\, dt.$$

Alternatively, the substitution $t = e^w - 1$ converts the integral corresponding to $A(t)$ into the integral corresponding to $B(t)$.

Solutions for Section 18.3

Exercises

1. The vector field $\vec{F}$ points radially outward, and so is everywhere perpendicular to A; thus, $\int_A \vec{F} \cdot d\vec{r} = 0$.

Along the first half of B, the terms $\vec{F} \cdot \Delta \vec{r}$ are negative; along the second half the terms $\vec{F} \cdot \Delta \vec{r}$ are positive. By symmetry the positive and negative contributions cancel out, giving a Riemann sum and a line integral of 0.

The line integral is also 0 along C, by cancellation. Here the values of $\vec{F}$ along the x-axis have the same magnitude as those along the y-axis. On the first half of C the path is traversed in the opposite direction to $\vec{F}$; on the second half of C the path is traversed in the same direction as $\vec{F}$. So the two halves cancel.

5. No. Suppose there were a function f such that $\text{grad } f = \vec{F}$. Then we would have

$$\frac{\partial f}{\partial x} = \frac{-z}{\sqrt{x^2 + z^2}}.$$

Hence we would have

$$\frac{\partial^2 f}{\partial y \partial x} = \frac{\partial}{\partial y}\left(\frac{-z}{\sqrt{x^2 + z^2}}\right) = 0.$$

In addition, since $\text{grad } f = \vec{F}$, we have that

$$\frac{\partial f}{\partial y} = \frac{y}{\sqrt{x^2 + z^2}}.$$

Thus we also know that

$$\frac{\partial^2 f}{\partial x \partial y} = \frac{\partial}{\partial x}\left(\frac{y}{\sqrt{x^2 + y^2}}\right) = -xy(x^2 + z^2)^{-3/2}.$$

Notice that

$$\frac{\partial^2 f}{\partial y \partial x} \neq \frac{\partial^2 f}{\partial x \partial y}.$$

Since we expect $\frac{\partial^2 f}{\partial y \partial x} = \frac{\partial^2 f}{\partial x \partial y}$, we have got a contradiction. The only way out of this contradiction is to conclude there is no function f with $\text{grad } f = \vec{F}$. Thus $\vec{F}$ is not a gradient vector field.

9. Path independent

13. Since $\vec{F} = 2x\vec{i} - 4y\vec{j} + (2z - 3)\vec{k} = \text{grad}(x^2 - 2y^2 + z^2 - 3z)$, the Fundamental Theorem of Line Integrals gives

$$\int_C \vec{F} \cdot d\vec{r} = (x^2 - 2y^2 + z^2 - 3z)\Big|_{(1,1,1)}^{(2,3,-1)} = (4 - 2 \cdot 3^2 + (-1)^2 + 3) - (1^2 - 2 \cdot 1^2 + 1^2 - 3) = -7.$$

Problems

17. Since $\vec{F} = \text{grad}\left(\dfrac{x^2 + y^2}{2}\right)$, the line integral can be calculated using the Fundamental Theorem of Line Integrals:

$$\int_c \vec{F} \cdot d\vec{r} = \frac{x^2 + y^2}{2}\Big|_{(0,0)}^{(3/\sqrt{2}, 3/\sqrt{2})} = \frac{9}{2}.$$

21. Since $\vec{F} = \text{grad } f$ is a gradient vector field, the Fundamental Theorem of Line Integrals give us

$$\int_C \vec{F} \cdot d\vec{r} = f(\text{end}) - f(\text{start}) = \left(x^2 + 2y^3 + 3z^4\right)\Big|_{(4,0,0)}^{(0,0,5)} = 3 \cdot (5)^4 - 4^2 = 1859.$$

25. **(a)** Work done by the force is the line integral, so

$$\text{Work done against force} = -\int_C \vec{F} \cdot d\vec{r} = -\int_C (-mg\vec{k}) \cdot d\vec{r}.$$

Since $\vec{r} = (\cos t)\vec{i} + (\sin t)\vec{j} + t\vec{k}$, we have $\vec{r}\,' = -(\sin t)\vec{i} + (\cos t)\vec{j} + \vec{k}$,

$$\text{Work done against force} = \int_0^{2\pi} mg\vec{k} \cdot (-\sin t\vec{i} + \cos t\vec{j} + \vec{k})dt$$

$$= \int_0^{2\pi} mg\, dt = 2\pi\, mg.$$

(b) We know from physical principles that the force is conservative. (Because the work done depends only on the vertical distance moved, not on the path taken.) Alternatively, we see that

$$\vec{F} = -mg\vec{k} = \text{grad}(-mgz),$$

so $\vec{F}$ is a gradient field and therefore path independent, or conservative.

29. **(a)** The vector field ∇f is perpendicular to the level curves, in direction of increasing f. The length of ∇f is the rate of change of f in that direction. See Figure 18.5

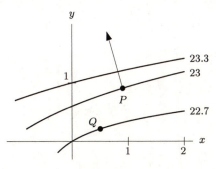

Figure 18.5

(b) Longer.
(c) Using the Fundamental Theorem of Calculus for Line Integrals, we have

$$\int_C \nabla f \cdot d\vec{r} = f(Q) - f(P) = 22.7 - 23 = -0.3.$$

Solutions for Section 18.4

Exercises

1. We know that

$$\frac{\partial f}{\partial x} = 2xy \quad \text{and} \quad \frac{\partial f}{\partial y} = x^2 + 8y^3$$

Now think of y as a constant in the equation for $\partial f/\partial x$ and integrate, giving

$$f(x, y) = x^2y + C(y).$$

Since the constant of integration may depend on y, it is written $C(y)$. Differentiating this expression for $f(x, y)$ with respect to y and using the fact that $\partial f/\partial y = x^2 + 8y^3$, we get

$$\frac{\partial f}{\partial y} = x^2 + C'(y) = x^2 + 8y^3.$$

Therefore

$$C'(y) = 8y^3 \quad \text{so} \quad C(y) = 2y^4 + K.$$

for some constant K. Thus,

$$f(x, y) = x^2 y + 2y^4 + K.$$

5. The domain of the vector field $\vec{F} = (2xy^3 + y)\vec{i} + (3x^2y^2 + x)\vec{j}$ is the whole xy-plane. We apply the curl test:

$$\frac{\partial F_1}{\partial y} = 6xy^2 + 1 = \frac{\partial F_2}{\partial x}$$

so $\vec{F}$ is the gradient of a function f. In order to compute f we first integrate

$$\frac{\partial f}{\partial x} = 2xy^3 + y$$

with respect to x thinking of y as a constant. We get

$$f(x, y) = x^2 y^3 + xy + C(y)$$

Differentiating with respect to y and using the fact that $\partial f / \partial y = 3x^2y^2 + x$ gives

$$\frac{\partial f}{\partial y} = 3x^2y^2 + x + C'(y) = 3x^2y^2 + x$$

Thus $C'(y) = 0$ so C is constant and

$$f(x, y) = x^2 y^3 + xy + C.$$

9. By Green's Theorem, with R representing the interior of the circle,

$$\int_C \vec{F} \cdot d\vec{r} = \int_R \left(\frac{\partial}{\partial x}(-x) - \frac{\partial}{\partial y}(y) \right) dA = -2 \int_R dA$$

$$= -2 \cdot \text{Area of circle} = -2\pi(1^2) = -2\pi.$$

Problems

13. Since the level curves must be perpendicular to the gradient vectors, if there were a contour diagram fitting this gradient field, it would have to look like Figure 18.6. However, this diagram could not be the contour diagram because the origin is on all contours. This means that $f(0, 0)$ would have to take on more than one value, which is impossible. At a point P other than the origin, we have the same problem. The values on the contours increase as you go counterclockwise around, since the gradient vector points in the direction of greatest increase of a function. But, starting at P, and going all the way around the origin, you would eventually get back to P again, and with a larger value of f, which is impossible.

An additional problem arises from the fact that the vectors in the original vector field are longer as you go away from the origin. This means that if there were a potential function f then $\| \text{grad } f \|$ would increase as you went away from the origin. This would mean that the level curves of f would get closer together as you go outward which does not happen in the contour diagram in Figure 18.6.

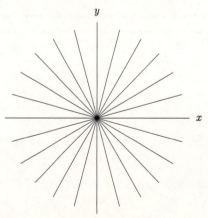

Figure 18.6

17. (a) C_1 is a line along the vertical axis; C_2 is a half circle from the positive y to the negative y-axis. See Figure 18.7.

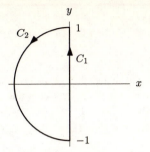

Figure 18.7

(b) Either use Green's Theorem or calculate directly. Using Green's Theorem, with R as the region inside C, we get

$$\int_{C_1+C_2} \vec{F} \cdot d\vec{r} = \int_R \left(\frac{\partial}{\partial x}(y) - \frac{\partial}{\partial y}(x + 3y) \right) dA$$

$$= \int_R -3\,dA = -3(\text{Area of region}) = -3\frac{\pi \cdot 1^2}{2} = -\frac{3\pi}{2}.$$

21. Using $\vec{F} = x\vec{j} = a\cos^3 t$ and $\vec{r}'(t) = -3a\cos^2 t \sin t\,\vec{i} + 3a\sin^2 t \cos t\,\vec{j}$, we have

$$A = \int_C \vec{F} \cdot d\vec{r} = \int_0^{2\pi} (a\cos^3 t)(3a\sin^2 t \cos t)\,dt$$

$$= 3a^2 \int_0^{2\pi} \cos^4 t \sin^2 t\,dt = 3a^2 \int_0^{2\pi} \cos^2 t(\sin t \cos t)^2\,dt = 3a^2 \int_0^{2\pi} \cos^2 t\,\frac{\sin^2 2t}{4}\,dt$$

$$= \frac{3a^2}{16} \int_0^{2\pi} (1 + \cos 2t)(1 - \cos 4t)\,dt$$

$$= \frac{3a^2}{16} \int_0^{2\pi} (1 + \cos 2t - \cos 4t - \cos 2t \cos 4t)\,dt$$

$$= \frac{3a^2}{16} \int_0^{2\pi} \left(1 + \cos 2t - \cos 4t - \frac{1}{2}\cos 6t - \frac{1}{2}\cos 2t\right) dt$$

$$= \frac{3a^2}{16}\left(t - \frac{1}{2}\sin 2t - \frac{1}{4}\sin 4t + \frac{1}{12}\sin 6t + \frac{1}{4}\sin 2t\right)\Big|_0^{2\pi} = \frac{3\pi a^2}{8}$$

For the last integral we use the trigonometric formula $\cos 2t \cos 4t = \frac{1}{2}(\cos 6t + \cos 2t)$. The hypocycloid is shown in Figure 18.8.

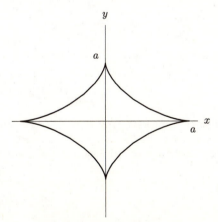

Figure 18.8: $x^{2/3} + y^{2/3} = a^{2/3}$

Solutions for Chapter 18 Review

Exercises

1. On the top half of the circle, the angle between the vector field and the curve is less than $90°$, so the line integral is positive. On the bottom half of the circle, the angle is more than $90°$, so the line integral is negative. However the magnitude of the vector field is larger on the top half of the curve, so the positive contribution to the line integral is larger than the negative. Thus the line integral $\int_C \vec{F} \cdot d\vec{r}$ is positive.

5. We parameterize the path C by (t, t, t), for $0 \leq t \leq 1$. Then

$$\int_C \vec{F} \cdot d\vec{r} = \int_0^1 \vec{F}(t, t, t) \cdot (\vec{i} + \vec{j} + \vec{k}) dt = \int_0^1 (t\vec{i} + 3t\vec{j} - t^2\vec{k}) \cdot (\vec{i} + \vec{j} + \vec{k}) dt$$

$$= \int_0^1 (4t - t^2) dt = \left(2t^2 - \frac{t^3}{3}\right)\Big|_0^1 = \frac{5}{3}.$$

9. Since $\vec{F} = \text{grad}\left(\frac{x^2}{2} + \frac{y^2}{2} + \frac{z^2}{2}\right)$, the Fundamental Theorem of Line Integrals gives

$$\int_C \vec{F} \cdot d\vec{r} = \left(\frac{x^2}{2} + \frac{y^2}{2} + \frac{z^2}{2}\right)\Big|_{(2,3,0)}^{(0,0,7)} = \frac{7^2}{2} - \left(\frac{2^2}{2} + \frac{3^2}{2}\right) = 18.$$

13. The domain is all 3-space. Since $F_1 = y$, $F_2 = x$,

$$\text{curl } y\vec{i} + x\vec{j} = \left(\frac{\partial F_3}{\partial y} - \frac{\partial F_2}{\partial z}\right)\vec{i} + \left(\frac{\partial F_1}{\partial z} - \frac{\partial F_3}{\partial x}\right)\vec{j} + \left(\frac{\partial F_2}{\partial x} - \frac{\partial F_1}{\partial y}\right)\vec{k} = \vec{0},$$

so $\vec{F}$ is path-independent

Problems

17. **(a)** Since $\vec{F} = x\vec{i} + y\vec{j} = \text{grad}\left(\frac{x^2 + y^2}{2}\right)$, we know that $\vec{F}$ is a gradient vector field. Thus, by the Fundamental Theorem of Line Integrals,

$$\int_{OA} \vec{F} \cdot d\vec{r} = \frac{x^2 + y^2}{2}\Big|_{(0,0)}^{(3,0)} = \frac{9}{2}.$$

(b) We know that $\vec{F}$ is path independent. If C is the closed curve consisting of the line in part (a) followed by the two-part curve in part (b), then

$$\int_C \vec{F} \cdot d\vec{r} = 0.$$

Thus, if ABO is the two-part curve of part (b) and OA is the line in part (a),

$$\int_{ABO} \vec{F} \cdot d\vec{r} = -\int_{OA} \vec{F} \cdot d\vec{r} = -\frac{9}{2}.$$

21. Yes, the line integral over C_1 is the negative of the line integral over C_2. One way to see this is to observe that the vector field $x\vec{i} + y\vec{j}$ is symmetric in the y-axis and that C_1 and C_2 are reflections in the y axis (except for orientation). See Figure 18.9. Since the orientation of C_2 is the reverse of the orientation of a mirror image of C_1, the two line integrals are opposite in sign.

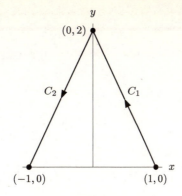

Figure 18.9

25. We'll assume that the rod is positioned along the z-axis, and look at the magnetic field $\vec{B}$ in the xy-plane. If C is a circle of radius r in the plane, centered at the origin, then we are told that the magnetic field is tangent to the circle and has constant magnitude $\|\vec{B}\|$. We divide the curve C into little pieces C_i and then we sum $\vec{B} \cdot \Delta \vec{r}$ computed on each piece C_i. But $\Delta \vec{r}$ points nearly in the same direction as $\vec{B}$, that is, tangent to C, and has magnitude nearly equal to the length of C_i. So the dot product is nearly equal to $\|\vec{B}\| \times$ length of C_i. When all of these dot products are summed and the limit is taken as $\|\Delta \vec{r}\| \to 0$, we get

$$\int_C \vec{B} \cdot d\vec{r} = \|\vec{B}\| \times \text{length of } C = \|\vec{B}\| \times 2\pi r$$

Now Ampère's Law also tells us that

$$\int_C \vec{B} \cdot d\vec{r} = kI$$

Setting these expressions for the line integral equal to each other and solving for $\|\vec{B}\|$ gives $kI = 2\pi r\|\vec{B}\|$, so

$$\|\vec{B}\| = \frac{kI}{2\pi r}.$$

CAS Challenge Problems

29. (a) We parameterize C_a by $\vec{r}(t) = a\cos t\vec{i} + a\sin t\vec{j}$. Then, using a CAS, we find

$$\int_{C_a} \vec{F}(\vec{r}(t)) \cdot \vec{r}'(t)\,dt = \int_0^{2\pi} a\cos t\left(2a\cos t - \frac{a^3\cos t^3}{3} + a^3\cos t\sin t^2\right)$$

$$-a\sin t\left(-(a\sin t) + \frac{2a^3\sin t^3}{3}\right)dt$$

$$= -\frac{\pi}{2}(-6a^2 + a^4)$$

The derivative of the expression on the right with respect to a is $-(2\pi)(-3a + a^3)$, which is zero at $a = 0, \pm\sqrt{3}$. Checking at $a = 0$ and as $a \to \infty$, we find the maximum is at $a = \sqrt{3}$.

(b) We have

$$\frac{\partial F_2}{\partial x} - \frac{\partial F_1}{\partial y} = (2 - x^2 + y^2) - (-1 + 2y^2) = 3 - x^2 - y^2.$$

So, by Green's theorem,

$$\int_{C_a} \vec{F} \cdot d\vec{r} = \int\int_{D_a} (3 - x^2 - y^2)\,dA,$$

where D_a is the disc of radius a centered at the origin. The integrand is positive for $x^2 + y^2 < 3$, so it is positive inside the disc of radius $\sqrt{3}$ and negative outside it. Thus the integral has its maximum value when $a = \sqrt{3}$.

CHECK YOUR UNDERSTANDING

1. A path-independent vector field must have zero circulation around all closed paths. Consider a vector field like $\vec{F}(x, y) = |x|\vec{j}$, shown in Figure 18.10.

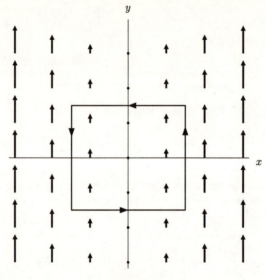

Figure 18.10

A rectangular path that is symmetric about the y-axis will have zero circulation: on the horizontal sides, the field is perpendicular, so the line integral is zero. The line integrals on the vertical sides are equal in magnitude and opposite in sign, so they cancel out, giving a line integral of zero. However, this field is not path-independent, because it is possible to find two paths with the same endpoints but different values of the line integral of $\vec{F}$. For example, consider the two points $(0, 0)$ and $(0, 1)$. The path C_1 in Figure 18.11 along the y axis gives zero for the line integral, because the field is 0 along the y axis, whereas a path like C_2 will have a nonzero line integral. Thus the line integral depends on the path between the points, so $\vec{F}$ is not path-independent.

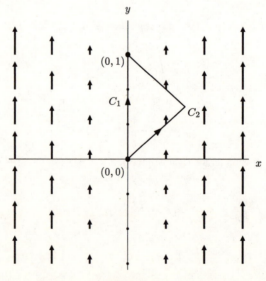

Figure 18.11

5. False. Because $\vec{F} \cdot \Delta\vec{r}$ is a scalar quantity, $\int_C \vec{F} \cdot d\vec{r}$ is also a scalar quantity.

9. True. The line integral is the limit of a sum of dot products, hence is a scalar.

13. False. All of the dot products $\vec{F}(\vec{r}_i) \cdot \Delta\vec{r}_i$ in this line integral are zero, since the vector field (the constant $\vec{i}$) points perpendicular to $\Delta\vec{r}_i$.

17. False. The line integrals of many vector fields (so called *path independent* or *conservative fields*) are zero around closed curves, but this is not true of all fields. For example, a vector field that is flowing in the same direction as the curve C all along the curve has a positive line integral. A specific example is given by $\vec{F} = -y\vec{i} + x\vec{j}$, where C is the unit circle centered at the origin, oriented counterclockwise.

21. False. As a counterexample, consider the unit circle C, centered at the origin, oriented counterclockwise and the vector field $\vec{F} = -y\vec{i} + x\vec{j}$. The vector field is always tangent to the circle, and in the same direction as C, so the line integral is positive.

25. True. The curves C_1 and C_2 are the same (they follow the graph of $y = x^2$ between $(0,0)$ and $(2,4)$), except that their orientations are opposite.

29. True. The construction at the end of Section 18.3 shows how to make a potential function from a path-independent vector field.

33. False. For example, take $\vec{F} = y\vec{i}$. By symmetry, the line integral of $\vec{F}$ over any circle centered at the origin is zero. But the curve consisting of the upper semicircle connecting $(-a, 0)$ to $(a, 0)$ has a positive line integral, while the line connecting these points along the x-axis has a zero line integral, so the field cannot be path-independent.

37. False. As a counterexample, consider $\vec{F} = x\vec{j}$ and $\vec{G} = y\vec{i}$. Then both of these are path-dependent (they each have nonzero curl), but the curl of $\vec{F} + \vec{G} = y\vec{i} + x\vec{j}$ is zero everywhere, so $\vec{F} + \vec{G}$ is path-independent.

41. False. As a counterexample, consider the vector field $\vec{F} = 2x\vec{i}$, which is path-independent, since it is the gradient of $f(x, y) = x^2$. Multiplying $\vec{F}$ by the function $h(x, y) = y$ gives the field $y\vec{F} = 2xy\vec{i}$. The curl of this vector field is $-2x \neq 0$, so $y\vec{F}$ is path-dependent.

CHAPTER NINETEEN

Solutions for Section 19.1

Exercises

1. **(a)** The flux is positive, since $\vec{F}$ points in direction of positive x-axis, the same direction as the normal vector.
   (b) The flux is negative, since below the xy-plane $\vec{F}$ points towards negative x-axis, which is opposite the orientation of the surface.
   (c) The flux is zero. Since $\vec{F}$ has only an x-component, there is no flow across the surface.
   (d) The flux is zero. Since $\vec{F}$ has only an x-component, there is no flow across the surface.
   (e) The flux is zero. Since $\vec{F}$ has only an x-component, there is no flow across the surface.

5. $\vec{v} \cdot \vec{A} = (2\vec{i} + 3\vec{j} + 5\vec{k}) \cdot 2\vec{i} = 4.$

9. The square has area 16, so its area vector is $16\vec{j}$. Since $\vec{F} = 5\vec{j}$ on the square,

$$\text{Flux} = 5\vec{j} \cdot 16\vec{j} = 80.$$

13. The disk has area 25π, so its area vector is $25\pi\vec{j}$. Thus

$$\text{Flux} = (2\vec{i} + 3\vec{j}) \cdot 25\pi\vec{j} = 75\pi.$$

17. Since the disk is in the xy-plane and oriented upward, $d\vec{A} = \vec{k}\, dxdy$ and

$$\int_{\text{Disk}} \vec{F} \cdot d\vec{A} = \int_{\text{Disk}} (x^2 + y^2)\vec{k} \cdot \vec{k}\, dxdy = \int_{\text{Disk}} (x^2 + y^2)\, dxdy.$$

Using polar coordinates

$$\int_{\text{Disk}} \vec{F} \cdot d\vec{A} = \int_0^{2\pi} \int_0^3 r^2 \cdot r\, drd\theta = 2\pi \left.\frac{r^4}{4}\right|_0^3 = \frac{81\pi}{2}.$$

Problems

21. On the curved sides of the cylinder, the $\vec{k}$ component of $\vec{F}$ does not contribute to the flux. Since the $\vec{i}$ and $\vec{j}$ components are constant, these components contribute 0 to the flux on the entire cylinder. Therefore the only nonzero contribution to the flux results from the $\vec{k}$ component through the top, where $z = 2$ and $d\vec{A} = \vec{k}\, dA$, and from the $\vec{k}$ component through the bottom, where $z = -2$ and $d\vec{A} = -\vec{k}\, dA$:

$$\text{Flux} = \int_{\text{Top}} \vec{F} \cdot d\vec{A} + \int_{\text{Bottom}} \vec{F} \cdot d\vec{A}$$

$$= \int_{\text{Top}} 2\vec{k} \cdot \vec{k}\, dA + \int_{\text{Bottom}} (-2\vec{k}) \cdot (-\vec{k}\, dA)$$

$$= 4 \int_{\text{Top}} dA = 4 \cdot \text{Area of top} = 4 \cdot \pi(3^2) = 36\pi.$$

25. **(a)** At the north pole, the area vector of the plate is upward (away from the center of the earth), and so is in the opposite direction to the magnetic field. Thus the magnetic flux is negative.
   (b) At the south pole, the area vector of the plate is again away from the center of the earth (because that is upward in the southern hemisphere), and so is in the same direction as the magnetic field. Thus, the magnetic flux is positive.
   (c) At the equator the magnetic field is parallel to the plate, so the flux is zero.

29. **(a)** Figure 19.1 shows the electric field $\vec{E}$. Note that $\vec{E}$ points radially outward from the z-axis.

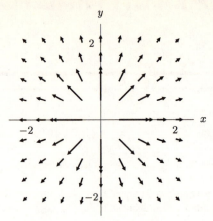

Figure 19.1: The electric field in the xy-plane due to a line of positive charge uniformly distributed along the z-axis: $\vec{E}(x, y, 0) = 2\lambda \dfrac{x\vec{i} + y\vec{j}}{x^2 + y^2}$

(b) On the cylinder $x^2 + y^2 = R^2$, the electric field $\vec{E}$ points in the same direction as the outward normal $\vec{n}$, and

$$\|\vec{E}\| = \frac{2\lambda}{R^2}\|x\vec{i} + y\vec{j}\| = \frac{2\lambda}{R}.$$

So

$$\int_S \vec{E} \cdot d\vec{A} = \int_S \vec{E} \cdot \vec{n}\, dA = \int_S \|\vec{E}\|\, dA = \int_S \frac{2\lambda}{R}\, dA$$
$$= \frac{2\lambda}{R}\int_S dA = \frac{2\lambda}{R} \cdot \text{Area of } S = \frac{2\lambda}{R} \cdot 2\pi R h = 4\pi\lambda h,$$

which is positive, as we expected.

33. **(a)** From Newton's law of cooling, we know that the temperature gradient will be proportional to the heat flow. If the constant of proportionality is k then we have the equation $\vec{F} = k\,\text{grad}\,T$. Since $\text{grad}\,T$ points in the direction of increasing T, but heat flows towards lower temperatures, the constant k must be negative.

(b) This form of Newton's law of cooling is saying that heat will be flowing in the direction in which temperature is decreasing most rapidly, in other words, in the direction exactly opposite to $\text{grad}\,T$. This agrees with our intuition which tells us that a difference in temperature causes heat to flow from the higher temperature to the lower temperature, and the rate at which it flows depends on the temperature gradient.

(c) The rate of heat loss from W is given by the flux of the heat flow vector field through the surface of the body. Thus,

$$\begin{array}{ccc} \text{Rate of heat} \\ \text{loss from } W \end{array} = \begin{array}{ccc} \text{Flux of } \vec{F} \\ \text{out of } S \end{array} = \int_S \vec{F} \cdot d\vec{A} = k\int_S (\text{grad}\,T) \cdot d\vec{A}$$

Solutions for Section 19.2

Exercises

1. Using $z = f(x, y) = x + y$, we have $d\vec{A} = (-\vec{i} - \vec{j} + \vec{k})\, dx\, dy$. As S is oriented upward, we have

$$\int_S \vec{F} \cdot d\vec{A} = \int_0^3 \int_0^2 ((x - y)\vec{i} + (x + y)\vec{j} + 3x\vec{k}) \cdot (-\vec{i} - \vec{j} + \vec{k})\, dx\, dy$$
$$= \int_0^3 \int_0^2 (-x + y - x - y + 3x)\, dx\, dy = \int_0^3 \int_0^2 x\, dx\, dy = 6.$$

5.

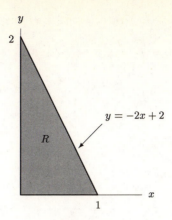

Figure 19.2

Writing the surface S as $z = f(x, y) = -2x - 4y + 1$, we have

$$d\vec{A} = (-f_x\vec{i} - f_y\vec{j} + \vec{k})dxdy.$$

With R as shown in Figure 19.2, we have

$$\int_S \vec{F} \cdot d\vec{A} = \int_R \vec{F}(x, y, f(x, y)) \cdot (-f_x\vec{i} - f_y\vec{j} + \vec{k}) dxdy$$

$$= \int_R (3x\vec{i} + y\vec{j} + (-2x - 4y + 1)\vec{k}) \cdot (2\vec{i} + 4\vec{j} + \vec{k}) dxdy$$

$$= \int_R (4x + 1) dxdy = \int_0^1 \int_0^{-2x+2} (4x + 1) dydx$$

$$= \int_0^1 (4x + 1)(-2x + 2) dx$$

$$= \int_0^1 (-8x^2 + 6x + 2) dx = (-\frac{8x^3}{3} + 3x^2 + 2x)\Big|_0^1 = \frac{7}{3}.$$

9. We have $0 \le z \le 6$ so $0 \le x^2 + y^2 \le 36$. Let R be the disk of radius 6 in the xy-plane centered at the origin. Because of the cone's point, the flux integral is improper; however, it does converge. We have

$$\int_S \vec{F} \cdot d\vec{A} = \int_R \vec{F}(x, y, f(x, y)) \cdot (-f_x\vec{i} - f_y\vec{j} + \vec{k}) dxdy$$

$$= \int_R (-x\sqrt{x^2 + y^2}\vec{i} - y\sqrt{x^2 + y^2}\vec{j} + (x^2 + y^2)\vec{k})$$

$$\cdot \left(-\frac{x}{\sqrt{x^2 + y^2}}\vec{i} - \frac{y}{\sqrt{x^2 + y^2}}\vec{j} + \vec{k}\right) dxdy$$

$$= \int_R 2(x^2 + y^2) dxdy$$

$$= 2\int_0^6 \int_0^{2\pi} r^3 d\theta dr$$

$$= 4\pi \int_0^6 r^3 dr = 1296\pi.$$

13. Since the radius of the cylinder is 1, using cylindrical coordinates we have

$$d\vec{A} = (\cos\theta\vec{i} + \sin\theta\vec{j})d\theta dz.$$

Thus,

$$\int_S \vec{F} \cdot d\vec{A} = \int_0^6 \int_0^{2\pi} (\cos\theta\vec{i} + \sin\theta\vec{j}) \cdot (\cos\theta\vec{i} + \sin\theta\vec{j}) \, d\theta \, dz$$

$$= \int_0^6 \int_0^{2\pi} 1 \, d\theta \, dz = 12\pi.$$

Problems

17. On the disk, $z = 0$ and $d\vec{A} = \vec{k} \, dx \, dy$, so

$$\int_S \vec{F} \cdot d\vec{A} = \int_{x^2+y^2 \le 1} (xze^{yz}\vec{i} + x\vec{j} + (5 + x^2 + y^2)\vec{k}) \cdot \vec{k} \, dx \, dy$$

$$= \int_{x^2+y^2 \le 1} (5 + x^2 + y^2) \, dx \, dy = \int_0^{2\pi} \int_0^1 (5 + r^2)r \, dr \, d\theta$$

$$= 2\pi \left(\frac{5r^2}{2} + \frac{r^4}{4} \right) \Big|_0^1 = \frac{11\pi}{2}.$$

Solutions for Section 19.3 ——————————————

Exercises

1. Since S is given by

$$\vec{r}(s,t) = (s+t)\vec{i} + (s-t)\vec{j} + (s^2 + t^2)\vec{k},$$

we have

$$\frac{\partial \vec{r}}{\partial s} = \vec{i} + \vec{j} + 2s\vec{k} \quad \text{and} \quad \frac{\partial \vec{r}}{\partial t} = \vec{i} - \vec{j} + 2t\vec{k},$$

and

$$\frac{\partial \vec{r}}{\partial s} \times \frac{\partial \vec{r}}{\partial t} = \begin{vmatrix} \vec{i} & \vec{j} & \vec{k} \\ 1 & 1 & 2s \\ 1 & -1 & 2t \end{vmatrix} = (2s + 2t)\vec{i} + (2s - 2t)\vec{j} - 2\vec{k}.$$

Since the $\vec{i}$ component of this vector is positive for $0 < s < 1, 0 < t < 1$, it points away from the z-axis, and so has the opposite orientation to the one specified. Thus, we use

$$d\vec{A} = -\frac{\partial \vec{r}}{\partial s} \times \frac{\partial \vec{r}}{\partial t} \, ds \, dt,$$

and so we have

$$\int_S \vec{F} \cdot d\vec{A} = -\int_0^1 \int_0^1 (s^2 + t^2)\vec{k} \cdot ((2s + 2t)\vec{i} + (2s - 2t)\vec{j} - 2\vec{k}) \, ds \, dt$$

$$= 2\int_0^1 \int_0^1 (s^2 + t^2) \, ds \, dt = 2\int_0^1 \left(\frac{s^3}{3} + st^2 \right) \Big|_{s=0}^{s=1} dt$$

$$= 2\int_0^1 (\frac{1}{3} + t^2) \, dt = 2(\frac{1}{3}t + \frac{t^3}{3}) \Big|_0^1 = 2(\frac{1}{3} + \frac{1}{3}) = \frac{4}{3}.$$

5. Using cylindrical coordinates, we see that the surface S is parameterized by

$$\vec{r}(r, \theta) = r \cos \theta \vec{i} + r \sin \theta \vec{j} + r \vec{k}.$$

We have

$$\frac{\partial \vec{r}}{\partial r} \times \frac{\partial \vec{r}}{\partial \theta} = \begin{vmatrix} \vec{i} & \vec{j} & \vec{k} \\ \cos \theta & \sin \theta & 1 \\ -r \sin \theta & r \cos \theta & 0 \end{vmatrix} = -r \cos \theta \vec{i} - r \sin \theta \vec{j} + r \vec{k}.$$

Since the vector $\partial \vec{r}/\partial r \times \partial \vec{r}/\partial \theta$ points upward, in the direction opposite to the specified orientation, we use $d\vec{A} = -(\partial \vec{r}/\partial r \times \partial \vec{r}/\partial \theta) \, dr \, d\theta$. Hence

$$\int_S \vec{F} \cdot d\vec{A} = -\int_0^{2\pi} \int_0^R (r^5 \cos^2 \theta \sin^2 \theta \vec{k}) \cdot (-r \cos \theta \vec{i} - r \sin \theta \vec{j} + r \vec{k}) \, dr \, d\theta$$

$$= -\int_0^{2\pi} \int_0^R r^6 \cos^2 \theta \sin^2 \theta \, dr \, d\theta$$

$$= -\frac{R^7}{7} \int_0^{2\pi} \sin^2 \theta \cos^2 \theta \, d\theta$$

$$= -\frac{R^7}{7} \int_0^{2\pi} \sin^2 \theta (1 - \sin^2 \theta) \, d\theta$$

$$= -\frac{R^7}{7} \int_0^{2\pi} (\sin^2 \theta - \sin^4 \theta) \, d\theta$$

$$= -(\frac{R^7}{7})(\frac{\pi}{4}) = \frac{-\pi}{28} R^7.$$

The cone is not differentiable at the point $(0, 0)$. However the flux integral, which is improper, converges.

Problems

9. If S is the part of the graph of $z = f(x, y)$ lying over a region R in the xy-plane, then S is parameterized by

$$\vec{r}(x, y) = x\vec{i} + y\vec{j} + f(x, y)\vec{k}, \qquad (x, y) \text{ in } R.$$

So

$$\frac{\partial \vec{r}}{\partial x} \times \frac{\partial \vec{r}}{\partial y} = (\vec{i} + f_x \vec{k}) \times (\vec{j} + f_y \vec{k}) = -f_x \vec{i} - f_y \vec{j} + \vec{k}.$$

Since the $\vec{k}$ component is positive, this points upward, so if S is oriented upward

$$d\vec{A} = (-f_x \vec{i} - f_y \vec{j} + \vec{k}) \, dx \, dy$$

and therefore we have the expression for the flux integral obtained on page 874:

$$\int_S \vec{F} \cdot d\vec{A} = \int_R \vec{F}(x, y, f(x, y)) \cdot (-f_x \vec{i} - f_y \vec{k} + \vec{k}) \, dx \, dy.$$

Solutions for Chapter 19 Review ———————————————

Exercises

1. Since $\vec{G}$ is constant, the net flux through the sphere is 0, so

$$\int_S \vec{G} \cdot d\vec{A} = 0.$$

5. Since the vector field is everywhere perpendicular to the surface of the sphere, and $||\vec{F}|| = \pi$ on the surface, we have

$$\int_S \vec{F} \cdot d\vec{A} = ||\vec{F}|| \cdot \text{Area of sphere} = \pi \cdot 4\pi(\pi)^2 = 4\pi^4.$$

9. Only the $\vec{k}$ component contributes to the flux. In the plane $z = 4$, we have $\vec{F} = 2\vec{i} + 3\vec{j} + 4\vec{k}$. On the square $d\vec{A} = \vec{k}\, dA$, so we have

$$\text{Flux} = \int \vec{F} \cdot d\vec{A} = 4\vec{k} \cdot (\vec{k}\ \text{Area of square}) = 4(5^2) = 100.$$

13. We have $d\vec{A} = \vec{k}\, dA$, so

$$\int_S \vec{F} \cdot d\vec{A} = \int_S (z\vec{i} + y\vec{j} + 2x\vec{k}) \cdot \vec{k}\, dA = \int_S 2x\, dA$$
$$= \int_0^3 \int_0^2 2x\, dx dy = 12.$$

17. There is no flux through the base or top of the cylinder because the vector field is parallel to these faces. For the curved surface, consider a small patch with area $\Delta \vec{A}$. The vector field is pointing radially outward from the z-axis and so is parallel to $\Delta \vec{A}$. Since $||\vec{F}|| = \sqrt{x^2 + y^2} = 2$ on the curved surface of the cylinder, we have $\vec{F} \cdot \Delta \vec{A} = ||\vec{F}|| ||\Delta \vec{A}|| = 2\Delta A$. Replacing ΔA with dA, we get

$$\int_S \vec{F} \cdot d\vec{A} = \int_{\substack{\text{Curved} \\ \text{surface}}} 2\, dA = 2(\text{Area of curved surface}) = 2(2\pi \cdot 2 \cdot 3) = 24\pi.$$

Problems

21. The vector field $\vec{D}$ has constant magnitude on S, equal to $Q/4\pi R^2$, and points radially outward, so

$$\int_S \vec{D} \cdot d\vec{A} = \frac{Q}{4\pi R^2} \cdot 4\pi R^2 = Q.$$

25. **(a)** Consider two opposite faces of the cube, S_1 and S_2. The corresponding area vectors are $\vec{A}_1 = 4\vec{i}$ and $\vec{A}_2 = -4\vec{i}$ (since the side of the cube has length 2). Since $\vec{E}$ is constant, we find the flux by taking the dot product, giving

$$\text{Flux through } S_1 = \vec{E} \cdot \vec{A}_1 = (a\vec{i} + b\vec{j} + c\vec{k}) \cdot 4\vec{i} = 4a.$$
$$\text{Flux through } S_2 = \vec{E} \cdot \vec{A}_2 = (a\vec{i} + b\vec{j} + c\vec{k}) \cdot (-4\vec{i}) = -4a.$$

Thus the fluxes through S_1 and S_2 cancel. Arguing similarly, we conclude that, for any pair of opposite faces, the sum of the fluxes of $\vec{E}$ through these faces is zero. Hence, by addition, $\int_S \vec{E} \cdot d\vec{A} = 0$.

(b) The basic idea is the same as in part (a), except that we now need to use Riemann sums. First divide S into two hemispheres H_1 and H_2 by the equator C located in a plane perpendicular to $\vec{E}$. For a tiny patch S_1 in the hemisphere H_1, consider the patch S_2 in the opposite hemisphere which is symmetric to S_1 with respect to the center O of the sphere. The area vectors $\Delta \vec{A}_1$ and $\Delta \vec{A}_2$ satisfy $\Delta \vec{A}_2 = -\Delta \vec{A}_1$, so if we consider S_1 and S_2 to be approximately flat, then $\vec{E} \cdot \Delta \vec{A}_1 = -\vec{E} \cdot \Delta \vec{A}_2$. By decomposing H_1 and H_2 into small patches as above and using Riemann sums, we get

$$\int_{H_1} \vec{E} \cdot d\vec{A} = -\int_{H_2} \vec{E} \cdot d\vec{A}, \quad \text{so} \int_S \vec{E} \cdot d\vec{A} = 0.$$

(c) The reasoning in part (b) can be used to prove that the flux of $\vec{E}$ through any surface with a center of symmetry is zero. For instance, in the case of the cylinder, cut it in half with a plane $z = 1$ and denote the two halves by H_1 and H_2. Just as before, take patches in H_1 and H_2 with $\Delta A_1 = -\Delta A_2$, so that $\vec{E} \cdot \Delta \vec{A}_1 = -\vec{E} \cdot \Delta \vec{A}_2$. Thus, we get

$$\int_{H_1} \vec{E} \cdot d\vec{A} = -\int_{H_2} \vec{E} \cdot d\vec{A},$$

which shows that

$$\int_S \vec{E} \cdot d\vec{A} = 0.$$

CAS Challenge Problems

29. (a) When $x > 0$, the vector $x\vec{i}$ points in the positive x-direction, and when $x < 0$ it points in the negative x-direction. Thus it always points from the inside of the ellipsoid to the outside, so we expect the flux integral to be positive. The upper half of the ellipsoid is the graph of $z = f(x, y) = \frac{1}{\sqrt{2}}(1 - x^2 - y^2)$, so the flux integral is

$$\int_S \vec{F} \cdot d\vec{A} = \int_{-1/2}^{1/2} \int_{-1/2}^{1/2} x\vec{i} \cdot (-f_x\vec{i} - f_y\vec{j} + \vec{k}) \, dx dy$$

$$= \int_{-1/2}^{1/2} \int_{-1/2}^{1/2} (-xf_x) \, dx dy = \int_{-1/2}^{1/2} \int_{-1/2}^{1/2} \frac{x^2}{\sqrt{1 - x^2 - y^2}} \, dx dy$$

$$= \frac{-\sqrt{2} + 11 \arcsin(\frac{1}{\sqrt{3}}) + 10 \arctan(\frac{1}{\sqrt{2}}) - 8 \arctan(\frac{5}{\sqrt{2}})}{12} = 0.0958.$$

Different CASs may give the answer in different forms. Note that we could have predicted the integral was positive without evaluating it, since the integrand is positive everywhere in the region of integration.

(b) For $x > -1$, the quantity $x + 1$ is positive, so the vector field $(x + 1)\vec{i}$ always points in the direction of the positive x-axis. It is pointing into the ellipsoid when $x < 0$ and out of it when $x > 0$. However, its magnitude is smaller when $-1/2 < x < 0$ than it is when $0 < x < 1/2$, so the net flux out of the ellipsoid should be positive. The flux integral is

$$\int_S \vec{F} \cdot d\vec{A} = \int_{-1/2}^{1/2} \int_{-1/2}^{1/2} (x + 1)\vec{i} \cdot (-f_x\vec{i} - f_y\vec{j} + \vec{k}) \, dx dy$$

$$= \int_{-1/2}^{1/2} \int_{-1/2}^{1/2} -(x + 1)f_x \, dx dy = \int_{-1/2}^{1/2} \int_{-1/2}^{1/2} \frac{x(1 + x)}{\sqrt{1 - x^2 - y^2}} \, dx dy$$

$$= \frac{\sqrt{2} - 11 \arcsin(\frac{1}{\sqrt{3}}) - 10 \arctan(\frac{1}{\sqrt{2}}) + 8 \arctan(\frac{5}{\sqrt{2}})}{12} = 0.0958$$

The answer is the same as in part (a). This makes sense because the difference between the integrals in parts (a) and (b) is the integral of $\int_{-1/2}^{1/2} \int_{-1/2}^{1/2} (x/\sqrt{1 - x^2 - y^2}) \, dx dy$, which is zero because the integrand is odd with respect to x.

(c) This integral should be positive for the same reason as in part (a). The vector field $y\vec{j}$ points in the positive y-direction when $y > 0$ and in the negative y-direction when $y < 0$, thus it always points out of the ellipsoid. Evaluating the integral we get

$$\int_S \vec{F} \cdot d\vec{A} = \int_{-1/2}^{1/2} \int_{-1/2}^{1/2} y\vec{j} \cdot (-f_x\vec{i} - f_y\vec{j} + \vec{k}) \, dx dy$$

$$= \int_{-1/2}^{1/2} \int_{-1/2}^{1/2} (-yf_y) \, dx dy = \int_{-1/2}^{1/2} \int_{-1/2}^{1/2} \frac{y^2}{\sqrt{1 - x^2 - y^2}} \, dx dy$$

$$= \frac{\sqrt{2} - 2 \arcsin(\frac{1}{\sqrt{3}}) - 19 \arctan(\frac{1}{\sqrt{2}}) + 8 \arctan(\frac{5}{\sqrt{2}})}{12} = 0.0958.$$

The symbolic answer appears different but has the same numerical value as in parts (a) and (b). In fact the answer is the same because the integral here is the same as in part (a) except that the roles of x and y have been exchanged. Different CASs may give different symbolic forms.

CHECK YOUR UNDERSTANDING

1. True. By definition, the flux integral is the limit of a sum of dot products, hence is a scalar.

5. True. The flow of this field is in the same direction as the orientation of the surface everywhere on the surface, so the flux is positive.

9. True. In the sum defining the flux integral for $\vec{F}$, we have terms like $\vec{F} \cdot \Delta\vec{A} = (2\vec{G}) \cdot \Delta\vec{A} = 2(\vec{G} \cdot \Delta\vec{A})$. So each term in the sum approximating the flux of $\vec{F}$ is twice the corresponding term in the sum approximating the flux of $\vec{G}$, making the sum for $\vec{F}$ twice that of the sum for $\vec{G}$. Thus the flux of $\vec{F}$ is twice the flux of $\vec{G}$.

13. False. Both surfaces are oriented upward, so $\vec{A}(x, y)$ and $\vec{B}(x, y)$ both point upward. But they could point in different directions, since the graph of $z = -f(x, y)$ is the graph of $z = f(x, y)$ turned upside down.

CHAPTER TWENTY

Solutions for Section 20.1

Exercises

1. Two vector fields that have positive divergence everywhere are as follows:

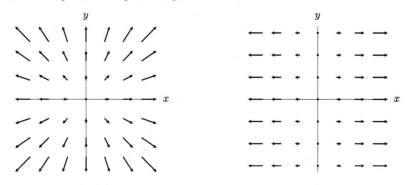

Figure 20.1   **Figure 20.2**

5. $\text{div } \vec{F} = \dfrac{\partial}{\partial x}(-y) + \dfrac{\partial}{\partial y}(x) = 0$

9. In coordinates, we have

$$\vec{F}(x, y, z) = \frac{(x - x_0)}{\sqrt{(x - x_0)^2 + (y - y_0)^2 + (z - z_0)^2}}\vec{i} + \frac{(y - y_0)}{\sqrt{(x - x_0)^2 + (y - y_0)^2 + (z - z_0)^2}}\vec{j}$$
$$+ \frac{(z - z_0)}{\sqrt{(x - x_0)^2 + (y - y_0)^2 + (z - z_0)^2}}\vec{k}.$$

So if $(x, y, z) \neq (x_0, y_0, z_0)$, then

$$\text{div } \vec{F} = \left(\frac{1}{\sqrt{(x - x_0)^2 + (y - y_0)^2 + (z - z_0)^2}} - \frac{(x - x_0)^2}{((x - x_0)^2 + (y - y_0)^2 + (z - z_0)^2)^{3/2}} \right)$$
$$+ \left(\frac{1}{\sqrt{(x - x_0)^2 + (y - y_0)^2 + (z - z_0)^2}} - \frac{(y - y_0)^2}{((x - x_0)^2 + (y - y_0)^2 + (z - z_0)^2)^{3/2}} \right)$$
$$+ \left(\frac{1}{\sqrt{(x - x_0)^2 + (y - y_0)^2 + (z - z_0)^2}} - \frac{(z - z_0)^2}{((x - x_0)^2 + (y - y_0)^2 + (z - z_0)^2)^{3/2}} \right)$$
$$= \left(\frac{(x - x_0)^2 + (y - y_0)^2 + (z - z_0)^2}{((x - x_0)^2 + (y - y_0)^2 + (z - z_0)^2)^{3/2}} - \frac{(x - x_0)^2}{((x - x_0)^2 + (y - y_0)^2 + (z - z_0)^2)^{3/2}} \right)$$
$$+ \left(\frac{(x - x_0)^2 + (y - y_0)^2 + (z - z_0)^2}{((x - x_0)^2 + (y - y_0)^2 + (z - z_0)^2)^{3/2}} - \frac{(y - y_0)^2}{((x - x_0)^2 + (y - y_0)^2 + (z - z_0)^2)^{3/2}} \right)$$
$$+ \left(\frac{(x - x_0)^2 + (y - y_0)^2 + (z - z_0)^2}{((x - x_0)^2 + (y - y_0)^2 + (z - z_0)^2)^{3/2}} - \frac{(z - z_0)^2}{((x - x_0)^2 + (y - y_0)^2 + (z - z_0)^2)^{3/2}} \right)$$
$$= \frac{3((x - x_0)^2 + (y - y_0)^2 + (z - z_0)^2) - ((x - x_0)^2 + (y - y_0)^2 + (z - z_0)^2)}{((x - x_0)^2 + (y - y_0)^2 + (z - z_0)^2)^{3/2}}$$
$$= \frac{2}{\sqrt{(x - x_0)^2 + (y - y_0)^2 + (z - z_0)^2}} = \frac{2}{\|\vec{r} - \vec{r}_0\|}.$$

Problems

13. Since div $F(1, 2, 3)$ is the flux density out of a small region surrounding the point $(1, 2, 3)$, we have

$$\text{div } \vec{F}\,(1, 2, 3) \approx \frac{\text{Flux out of small region around } (1, 2, 3)}{\text{Volume of region.}}$$

So

$$\text{Flux out of region} \approx (\text{div } \vec{F}\,(1, 2, 3)) \cdot \text{Volume of region}$$

$$= 5 \cdot \frac{4}{3}\pi(0.01)^3$$

$$= \frac{0.00002\pi}{3}.$$

17.

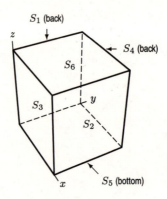

Figure 20.3

(a) The vector field is parallel to the x-axis and zero on the yz-plane. Thus the only contribution to the flux is from S_2. On S_2, $x = c$, the normal is outward. Since $\vec{F}$ is constant on S_2, the flux through face S_2 is

$$\int_{S_2} \vec{F} \cdot d\vec{A} = \vec{F} \cdot \vec{A}\, {}_{S_2}$$

$$= c\vec{i} \cdot c^2\vec{i}$$

$$= c^3.$$

Thus, total flux through box $= c^3$.

(b) Using the geometric definition of convergence

$$\text{div } \vec{F} = \lim_{c \to 0} \left(\frac{\text{Flux through box}}{\text{Volume of box}} \right)$$

$$= \lim_{c \to 0} \left(\frac{c^3}{c^3} \right)$$

$$= 1$$

(c)

$$\frac{\partial}{\partial x}(x) + \frac{\partial}{\partial y}(0) + \frac{\partial}{\partial z}(0) = 1 + 0 + 0 = 1.$$

21. Let $\vec{a} = a_1\vec{i} + a_2\vec{j} + a_3\vec{k}$ with a_1, a_2, and a_3 constant. Then $f\vec{a} = f(x,y,z)(a_1\vec{i} + a_2\vec{j} + a_3\vec{k}) = f(x,y,z)a_1\vec{i} + f(x,y,z)a_2\vec{j} + f(x,y,z)a_3\vec{k} = fa_1\vec{i} + fa_2\vec{j} + fa_3\vec{k}$. So

$$\text{div}(f\vec{a}) = \frac{\partial(fa_1)}{\partial x} + \frac{\partial(fa_2)}{\partial y} + \frac{\partial(fa_3)}{\partial z}$$

$$= a_1\frac{\partial f}{\partial x} + a_2\frac{\partial f}{\partial y} + a_3\frac{\partial f}{\partial z} \quad \text{since } a_1, a_2, a_3 \text{ are constants}$$

$$= (\frac{\partial f}{\partial x}\vec{i} + \frac{\partial f}{\partial y}\vec{j} + \frac{\partial f}{\partial z}\vec{k}) \cdot (a_1\vec{i} + a_2\vec{j} + a_3\vec{k})$$

$$= (\text{grad } f) \cdot \vec{a}.$$

25. Calculate $r_x = (1/2)(x^2 + y^2)^{-1/2}2x = x/r$ and $r_y = y/r$. We have

$$\text{div}\frac{h(r)}{r^2}(x\vec{i} + y\vec{j}) = \frac{\partial}{\partial x}\frac{xh(r)}{r^2} + \frac{\partial}{\partial y}\frac{yh(r)}{r^2}$$

$$= \frac{h(r)}{r^2} + x\frac{r^2h'(r)r_x - 2h(r)rr_x}{r^4} + \frac{h(r)}{r^2} + y\frac{r^2h'(r)r_y - 2h(r)rr_y}{r^4}$$

$$= 2\frac{h(r)}{r^2} + \left(\frac{h'(r)}{r^2} - \frac{2h(r)}{r^3}\right)(xr_x + yr_y)$$

$$= 2\frac{h(r)}{r^2} + \left(\frac{h'(r)}{r^2} - \frac{2h(r)}{r^3}\right)\left(\frac{x^2 + y^2}{r}\right)$$

$$= 2\frac{h(r)}{r^2} + \left(\frac{h'(r)}{r^2} - \frac{2h(r)}{r^3}\right)\left(\frac{r^2}{r}\right)$$

$$= \frac{h'(r)}{r}.$$

29. (a) Translating the vector field into rectangular coordinates gives, if $(x, y, z) \neq (0, 0, 0)$,

$$\vec{E}(x, y, z) = \frac{kx}{(x^2 + y^2 + z^2)^{3/2}}\vec{i} + \frac{ky}{(x^2 + y^2 + z^2)^{3/2}}\vec{j} + \frac{kz}{(x^2 + y^2 + z^2)^{3/2}}\vec{k}.$$

We now take the divergence of this to get

$$\text{div}\,\vec{E} = k\left(-3\frac{x^2 + y^2 + z^2}{(x^2 + y^2 + z^2)^{5/2}} + \frac{3}{(x^2 + y^2 + z^2)^{3/2}}\right)$$

$$= 0.$$

(b) Let S be the surface of a sphere centered at the origin. We have seen that for this field, the flux $\int \vec{E} \cdot d\vec{A}$ is the same for all such spheres, regardless of their radii. So let the constant c stand for $\int \vec{E} \cdot d\vec{A}$. Then

$$\text{div}\,\vec{E}(0, 0, 0) = \lim_{\text{vol}\to 0}\frac{\int \vec{E} \cdot d\vec{A}}{\text{Volume inside } S} = \lim_{\text{vol}\to 0}\frac{c}{\text{Volume}}.$$

(c) For a point charge, the charge density is not defined. The charge density is 0 everywhere else.

Solutions for Section 20.2

Exercises

1. No, because the surface S is not a closed surface.

5. First directly, since the vector field is totally in the $\vec{j}$ direction, there is no flux through the ends. On the side of the cylinder, a normal vector at (x, y, z) is $x\vec{i} + y\vec{j}$. This is in fact a unit normal, since $x^2 + y^2 = 1$ (the cylinder has radius 1). Also, using $x = \cos\theta, y = \sin\theta$, in this case, the element of area dA equals $1\,d\theta\,dz$. So

$$\text{Flux} = \int \vec{F} \cdot d\vec{A} = \int_0^2 \int_0^{2\pi} (y\vec{j}\,) \cdot (x\vec{i} + y\vec{j}\,)\,d\theta\,dz$$

$$= \int_0^2 \int_0^{2\pi} y^2\,d\theta\,dz = \int_0^2 \int_0^{2\pi} \sin^2\theta\,d\theta\,dz = \int_0^2 \pi\,dz = 2\pi.$$

Now we calculate the flux using the divergence theorem. The divergence of the field is given by the sum of the respective partials of the components, so the divergence is simply $\dfrac{\partial y}{\partial y} = 1$. Since the divergence is constant, we can simply calculate the volume of the cylinder and multiply by the divergence

$$\text{Flux} = 1\pi r^2 h = 2\pi.$$

Problems

9. Apply the Divergence Theorem to the solid cone, whose interior we call W. The surface of W consists of S and D. Thus

$$\int_S \vec{F} \cdot d\vec{A} + \int_D \vec{F} \cdot d\vec{A} = \int_W \text{div}\,\vec{F}\,dV.$$

But $\text{div}\,\vec{F} = 0$ everywhere, since $\vec{F}$ is constant. Thus

$$\int_D \vec{F} \cdot d\vec{A} = -\int_S \vec{F} \cdot d\vec{A} = -3.22.$$

13. (a) True. Flux that goes in one face goes out the other, since the vector field is constant and the surface is closed.
(b) True. The flux out of S_1 along the face shared with S_2 cancels with the flux out of S_2 over the same face. (The normals are in opposite directions.) The other five faces of S_1 and the other five faces of S_2 are each faces of S.

17. Any closed surface, S, oriented inward, will work. Then,

$$\int_{S(\text{inward})} \vec{F} \cdot d\vec{A} = -\int_{S(\text{outward})} \vec{F} \cdot d\vec{A},$$

so, by the Divergence Theorem, with W representing the region inside S,

$$\int_{S(\text{inward})} \vec{F} \cdot d\vec{A} = -\int_W \text{div}\,\vec{F}\,dV = -\int_W (x^2 + y^2 + 3)\,dV.$$

The integral on the right is positive because the integrand is positive everywhere. Therefore the flux through S oriented inward is negative.

21. (a) We cannot use the Divergence Theorem to calculate the flux through the sphere of radius 2 because $\vec{G}$ is not defined throughout the interior of the sphere. We calculate the flux directly. Since $\vec{G}$ is parallel to the area vector at the surface of the sphere, and since $\|\vec{G}\,\| = 4 \cdot 2^2 \cdot 2 = 32$ on the surface, we have

$$\text{Flux} = \int_S \vec{G} \cdot d\vec{A} = 32 \cdot \text{Area of surface} = 32 \cdot 4\pi 2^2 = 512\pi.$$

(b) The fact that the vectors of $\vec{G}$ get longer as we go away from the origin suggests that $\text{div}\,\vec{G} > 0$. This is confirmed by calculating the divergence using the formula

$$\vec{G} = 4r^2\vec{r} = 4(x^2 + y^2 + z^2)(x\vec{i} + y\vec{j} + z\vec{k}\,).$$

Since $\dfrac{\partial G_1}{\partial x} = 4(3x^2 + y^2 + z^2)$, and so on, $\text{div}\,\vec{G}$ is positive everywhere outside the unit sphere.

The sphere is completely contained within the box. Apply the Divergence Theorem to the region, W, between the sphere and the box. This region has surface area the sphere (oriented inward) and the box (oriented outward). The Divergence Theorem gives

$$\int_{\text{Box (outward)}} \vec{F} \cdot d\vec{A} + \int_{\text{Sphere (inward)}} \vec{F} \cdot d\vec{A} = \int_W \text{div}\, \vec{G}\, dV > 0.$$

So

$$\int_{\text{Box (outward)}} \vec{F} \cdot d\vec{A} - \int_{\text{Sphere (outward)}} \vec{F} \cdot d\vec{A} = \int_W \text{div}\, \vec{G}\, dV > 0.$$

Thus,

$$\int_{\text{Box (outward)}} \vec{F} \cdot d\vec{A} = \int_{\text{Sphere (outward)}} \vec{F} \cdot d\vec{A} + \int_W \text{div}\, \vec{G}\, dV.$$

So the flux through the box is larger than the flux through the sphere.

25. (a) Taking partial derivatives of $\vec{E}$ gives

$$\frac{\partial E_1}{\partial x} = \frac{\partial}{\partial x}[qx(x^2 + y^2 + z^2)^{-3/2}] = q[(x^2 + y^2 + z^2)^{-3/2} + x(-3/2)(2x)(x^2 + y^2 + z^2)^{-5/2}]$$

$$= q(y^2 + z^2 - 2x^2)(x^2 + y^2 + z^2)^{-5/2}.$$

Similarly,

$$\frac{\partial E_2}{\partial x} = q(x^2 + z^2 - 2y^2)(x^2 + y^2 + z^2)^{-5/2}$$

$$\frac{\partial E_3}{\partial x} = q(x^2 + y^2 - 2z^2)(x^2 + y^2 + z^2)^{-5/2}.$$

Summing, we obtain div $\vec{E} = 0$.

(b) Since on the surface of the sphere, the vector field $\vec{E}$ and the area vector $\Delta \vec{A}$ are parallel,

$$\vec{E} \cdot \Delta \vec{A} = \|\vec{E}\|\|\Delta \vec{A}\|.$$

Now, on the surface of a sphere of radius a,

$$\|\vec{E}\| = \frac{q\|\vec{r}\|}{\|\vec{r}\|^3} = \frac{q}{a^2}.$$

Thus,

$$\int_{S_a} \vec{E} \cdot d\vec{A} = \int \frac{q}{a^2}\|d\vec{A}\| = \frac{q}{a^2} \cdot \text{Surface area of sphere} = \frac{q}{a^2} \cdot 4\pi a^2 = 4\pi q.$$

(c) It is not possible to apply the Divergence Theorem in part (b) since $\vec{E}$ is not defined at the origin (which lies inside the region of space bounded by S_a), and the Divergence Theorem requires that the vector field be defined everywhere inside S.

(d) Let R be the solid region lying between a small sphere S_a, centered at the origin, and the surface S. Applying the Divergence Theorem and the result of part (a), we get:

$$0 = \int_R \text{div}\, \vec{E}\, dV = \int_{S_a} \vec{E} \cdot d\vec{A} + \int_S \vec{E} \cdot d\vec{A},$$

where S is oriented with the outward normal vector, and S_a with the inward normal vector (since this is "outward" with respect to the region R). Since

$$\int_{S_a,\,\text{inward}} \vec{E} \cdot d\vec{A} = -\int_{S_a,\,\text{outward}} \vec{E} \cdot d\vec{A},$$

the result of part (b) yields

$$\int_S \vec{E} \cdot d\vec{A} = 4\pi q.$$

[Note: It is legitimate to apply the Divergence Theorem to the region R since the vector field $\vec{E}$ is defined everywhere in R.]

Solutions for Section 20.3

Exercises

1. Using the definition in Cartesian coordinates, we have

$$\text{curl } \vec{F} = \begin{vmatrix} \vec{i} & \vec{j} & \vec{k} \\ \frac{\partial}{\partial x} & \frac{\partial}{\partial y} & \frac{\partial}{\partial z} \\ x^2 - y^2 & 2xy & 0 \end{vmatrix}$$

$$= \left(\frac{\partial}{\partial y}(0) - \frac{\partial}{\partial z}(2xy) \right) \vec{i} + \left(-\frac{\partial}{\partial x}(0) + \frac{\partial}{\partial z}(x^2 - y^2) \right) \vec{j} + \left(\frac{\partial}{\partial x}(2xy) - \frac{\partial}{\partial y}(x^2 - y^2) \right) \vec{k}$$

$$= 4y\vec{k} .$$

5. Using the definition of Cartesian coordinates,

$$\text{curl } \vec{F} = \begin{vmatrix} \vec{i} & \vec{j} & \vec{k} \\ \frac{\partial}{\partial x} & \frac{\partial}{\partial y} & \frac{\partial}{\partial z} \\ (-x + y) & (y + z) & (-z + x) \end{vmatrix}$$

$$= \left(\frac{\partial}{\partial y}(-z + x) - \frac{\partial}{\partial z}(y + z) \right) \vec{i} + \left(-\frac{\partial}{\partial x}(-z + x) + \frac{\partial}{\partial z}(-x + y) \right) \vec{j}$$

$$+ \left(\frac{\partial}{\partial x}(y + z) - \frac{\partial}{\partial y}(-x + y) \right) \vec{k}$$

$$= -\vec{i} - \vec{j} - \vec{k} .$$

9. This vector field shows no rotation, and the circulation around any closed curve appears to be zero, so we suspect a zero curl here.

Problems

13. The conjecture is that when the first component of $\vec{F}$ depends only on x, the second component depends only on y, and the third component depends only on z, that is, if

$$\vec{F} = F_1(x)\vec{i} + F_2(y)\vec{j} + F_3(z)\vec{k}$$

then

$$\text{curl } \vec{F} = \vec{0}$$

The reason for this is that if $\vec{F} = F_1(x)\vec{i} + F_2(y)\vec{j} + F_3(z)\vec{k}$, then

$$\text{curl } \vec{F} = \begin{vmatrix} \vec{i} & \vec{j} & \vec{k} \\ \frac{\partial}{\partial x} & \frac{\partial}{\partial y} & \frac{\partial}{\partial z} \\ F_1(x) & F_2(y) & F_3(z) \end{vmatrix}$$

$$= \left(\frac{\partial}{\partial y}F_3(z) - \frac{\partial}{\partial z}F_2(y) \right) \vec{i} + \left(-\frac{\partial}{\partial x}F_3(z) + \frac{\partial}{\partial z}F_1(x) \right) \vec{j} + \left(\frac{\partial}{\partial x}F_2(y) - \frac{\partial}{\partial y}F_1(x) \right) \vec{k}$$

$$= \vec{0} .$$

17. Investigate the velocity vector field of the atmosphere near the fire. If the curl of this vector field is non-zero, there is circulatory motion. Consequently, if the magnitude of the curl of this vector field is large near the fire, a fire storm has probably developed.

21. The vector curl $\vec{F}$ has its component in the x-direction given by

$$(\text{curl } \vec{F})_x \approx \frac{\text{Circulation around small circle around } x\text{-axis}}{\text{Area inside circle}}$$

$$= \frac{\text{Circulation around } C_2}{\text{Area inside } C_2} = \frac{0.5\pi}{\pi(0.1)^2} = 50.$$

Similar reasoning leads to

$$(\text{curl } \vec{F})_y \approx \frac{\text{Circulation around } C_3}{\text{Area inside } C_3} = \frac{3\pi}{\pi(0.1)^2} = 300,$$

$$(\text{curl } \vec{F})_z \approx \frac{\text{Circulation around } C_1}{\text{Area inside } C_1} = \frac{0.02\pi}{\pi(0.1)^2} = 2.$$

Thus,

$$\text{curl } \vec{F} \approx 50\vec{i} + 300\vec{j} + 2\vec{k}.$$

25. Let $\vec{c} = c_1\vec{i} + c_2\vec{j} + c_3\vec{k}$ and $\vec{F} = F_1\vec{i} + F_2\vec{j} + F_3\vec{k}$. We then show the desired result as follows:

$$\begin{aligned}
\text{div}(\vec{F} \times \vec{c}) &= \text{div}((F_1\vec{i} + F_2\vec{j} + F_3\vec{k}) \times (c_1\vec{i} + c_2\vec{j} + c_3\vec{k})) \\
&= \text{div}((F_2c_3 - F_3c_2)\vec{i} + (F_3c_1 - F_1c_3)\vec{j} + (F_1c_2 - F_2c_1)\vec{k}) \\
&= \frac{\partial}{\partial x}(F_2c_3 - F_3c_2) + \frac{\partial}{\partial y}(F_3c_1 - F_1c_3) + \frac{\partial}{\partial z}(F_1c_2 - F_2c_1) \\
&= c_3\frac{\partial F_2}{\partial x} - c_2\frac{\partial F_3}{\partial x} + c_1\frac{\partial F_3}{\partial y} - c_3\frac{\partial F_1}{\partial y} + c_2\frac{\partial F_1}{\partial z} - c_1\frac{\partial F_2}{\partial z} \\
&= c_1\left(\frac{\partial F_3}{\partial y} - \frac{\partial F_2}{\partial z}\right) + c_2\left(\frac{\partial F_1}{\partial z} - \frac{\partial F_3}{\partial x}\right) + c_3\left(\frac{\partial F_2}{\partial x} - \frac{\partial F_1}{\partial y}\right) \\
&= (c_1\vec{i} + c_2\vec{j} + c_3\vec{k}) \cdot \left(\left(\frac{\partial F_3}{\partial y} - \frac{\partial F_2}{\partial z}\right)\vec{i} + \left(\frac{\partial F_1}{\partial z} - \frac{\partial F_3}{\partial x}\right)\vec{j} + \left(\frac{\partial F_2}{\partial x} - \frac{\partial F_1}{\partial y}\right)\vec{k}\right) \\
&= \vec{c} \cdot \text{curl } \vec{F}.
\end{aligned}$$

29. (a) Figure 20.4 shows a cross-section of the vector field in xy-plane with $\omega = 1$, so $\vec{v} = -y\vec{i} + x\vec{j}$.

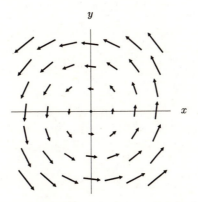

Figure 20.4: $\vec{v} = -y\vec{i} + x\vec{j}$

Figure 20.5 shows a cross-section of vector field in xy-plane with $\omega = -1$, so $\vec{v} = y\vec{i} - x\vec{j}$.

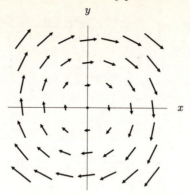

Figure 20.5: $\vec{v} = y\vec{i} - x\vec{j}$

(b) The distance from the center of the vortex is given by $r = \sqrt{x^2 + y^2}$. The velocity of the vortex at any point is $-\omega y\vec{i} + \omega x\vec{j}$, and the speed of the vortex at any point is the magnitude of the velocity, or $s = \|\vec{v}\| = \sqrt{(-\omega y)^2 + (\omega x)^2} = |\omega|\sqrt{x^2 + y^2} = |\omega|\,r$.

(c) The divergence of the velocity field is given by:

$$\operatorname{div}\vec{v} = \frac{\partial(-\omega y)}{\partial x} + \frac{\partial(\omega x)}{\partial y} = 0$$

The curl of the field is:

$$\operatorname{curl}\vec{v} = \operatorname{curl}(-\omega y\vec{i} + \omega x\vec{j}) = \left(\frac{\partial}{\partial x}(\omega x) - \frac{\partial}{\partial y}(-\omega y)\right)\vec{k} = 2\omega\vec{k}$$

(d) We know that $\vec{v}$ has constant magnitude $|\omega|R$ everywhere on the circle and is everywhere tangential to the circle. In addition, if $\omega > 0$, the vector field rotates counterclockwise; if $\omega < 0$, the vector field rotates clockwise. Thus if $\omega > 0$, $\vec{v}$ and $\Delta\vec{r}$ are parallel and in the same direction, so

$$\int_C \vec{v}\cdot d\vec{r} = |\vec{v}|\cdot(\text{Length of } C) = \omega R\cdot 2\pi R = 2\pi\omega R^2$$

If $\omega < 0$, then $|\omega| = -\omega$ and $\vec{v}$ and $\Delta\vec{r}$ are in opposite directions, so

$$\int_C \vec{v}\cdot d\vec{r} = -|\vec{v}|\cdot(\text{Length of } C) = -|\omega|R\cdot(2\pi R) = 2\pi\omega R^2.$$

33. Let $\vec{v} = a\vec{i} + b\vec{j} + c\vec{k}$ and try

$$\vec{F} = \vec{v}\times\vec{r} = (a\vec{i} + b\vec{j} + c\vec{k})\times(x\vec{i} + y\vec{j} + z\vec{k}) = (bz - cy)\vec{i} + (cx - az)\vec{j} + (ay - bx)\vec{k}.$$

Then

$$\operatorname{curl}\vec{F} = \begin{vmatrix} \vec{i} & \vec{j} & \vec{k} \\ \frac{\partial}{\partial x} & \frac{\partial}{\partial y} & \frac{\partial}{\partial z} \\ bz - cy & cx - az & ay - bx \end{vmatrix} = 2a\vec{i} + 2b\vec{j} + 2c\vec{k}.$$

Taking $a = 1$, $b = -\frac{3}{2}$, $c = 2$ gives $\operatorname{curl}\vec{F} = 2\vec{i} - 3\vec{j} + 4\vec{k}$, so the desired vector field is $\vec{F} = (-\frac{3}{2}z - 2y)\vec{i} + (2x - z)\vec{j} + (y + \frac{3}{2}x)\vec{k}$.

Solutions for Section 20.4

Exercises

1. No, because the curve C over which the integral is taken is not a closed curve, and so it is not the boundary of a surface.

5. The graph of $\vec{F} = \vec{r}/\|\vec{r}\|^3$ consists of vectors pointing radially outward. There is no swirl, so $\operatorname{curl}\vec{F} = \vec{0}$. From Stokes' Theorem,

$$\int_C \vec{F}\cdot d\vec{r} = \int_S \operatorname{curl}\vec{F}\cdot d\vec{A} = \int_S \vec{0}\cdot d\vec{A} = 0$$

Problems

9. (a) It appears that div $\vec{F} < 0$, and div $\vec{G} < 0$; div $\vec{G}$ is larger in magnitude (more negative) if the scales are the same.

(b) curl $\vec{F}$ and curl $\vec{G}$ both appear to be zero at the origin (and elsewhere).

(c) Yes, the cylinder with axis along the z-axis will have negative flux through it (ends parallel to xy-plane).

(d) Same as part(c).

(e) No, you cannot draw a closed curve around the origin such that $\vec{F}$ has a non-zero circulation around it because curl is zero. By Stokes' theorem, circulation equals the integral of the curl over the surface bounded by the curve.

(f) Same as part(e)

13. (a) We calculate

$$\text{curl}\,\vec{F} = \begin{vmatrix} \vec{i} & \vec{j} & \vec{k} \\ \frac{\partial}{\partial x} & \frac{\partial}{\partial y} & \frac{\partial}{\partial z} \\ y-z & x+z & xy \end{vmatrix} = (x-1)\vec{i} - (y+1)\vec{j} + (1-1)\vec{k} = (x-1)\vec{i} - (y+1)\vec{j}\,.$$

Since the circle is in the xy-plane and curl $\vec{F}$ has no $\vec{k}$ component, the line integral is zero.

(b) Even though the line integral around this closed curve is 0, we do not know that the line integral around every closed curve is zero, so we cannot conclude that $\vec{F}$ is path-independent (conservative). Since curl $\vec{F} \neq \vec{0}$, we know that $\vec{F}$ is not path-independent (conservative).

17. (a) Computing curl $\vec{G}$, we get

$$\text{curl}\,\vec{G} = \left(\frac{\partial(xz)}{\partial y} - \frac{\partial(3x)}{\partial z} \right)\vec{i} + \left(\frac{\partial(4yz^2)}{\partial z} - \frac{\partial(xz)}{\partial x} \right)\vec{j} + \left(\frac{\partial(3x)}{\partial x} - \frac{\partial(4yz^2)}{\partial y} \right)\vec{k}$$

$$= 0\vec{i} + (8yz - z)\vec{j} + (3 - 4z^2)\vec{k}$$

$$= \vec{F}$$

(b) Because $\vec{F}$ is a curl field, the flux of $\vec{F}$ through S is the same as its flux through any surface with the same boundary as S. Let us replace S by the circular disk S_1 of radius 5 in the xy-plane. On S_1, $z = 0$, and so the vector field reduces to $\vec{F} = 3\vec{k}$. Since the vector $3\vec{k}$ is normal to S_1 and in the direction of the orientation of S_1, we have

$$\int_S \vec{F} \cdot d\vec{A} = \int_{S_1} \vec{F} \cdot d\vec{A} = \int_{S_1} 3\vec{k} \cdot d\vec{A} = \|3\vec{k}\|(\text{area of } S_1) = 75\pi.$$

21. Notice that $\vec{F}$, div $\vec{F}$, and curl $\vec{F}$ are defined and continuous everywhere.

(a) By the Divergence Theorem, the fact that the flux through any closed surface is 0 tells us that everywhere

$$\text{div}\,\vec{F} = 0.$$

Since div $\vec{F} = a$, we know $a = 0$. We do not have any information about b, c, m.

(b) Since the circulation around any closed curve is 0, by Stokes' Theorem, we have

$$\text{curl}\,\vec{F} = \vec{0}.$$

Now

$$\text{curl}\,\vec{F} = \begin{vmatrix} \vec{i} & \vec{j} & \vec{k} \\ \frac{\partial}{\partial x} & \frac{\partial}{\partial y} & \frac{\partial}{\partial z} \\ ax+by+5z & x+cz & 3y+mx \end{vmatrix} = (3-c)\vec{i} - (m-5)\vec{j} + (1-b)\vec{k}\,,$$

so curl $\vec{F} = 0$ means $c = 3$, $m = 5$ and $b = 1$. We do not have any information about a.

Solutions for Section 20.5

Exercises

1. Since curl $\vec{F} = \vec{0}$ and $\vec{F}$ is defined everywhere, we know by the curl test that $\vec{F}$ is a gradient field. In fact, $\vec{F} = \text{grad}\,f$, where $f(x, y, z) = xyz + yz^2$, so f is a potential function for $\vec{F}$.

Problems

5. We must show $\operatorname{curl} \vec{A} = \vec{B}$.

$$\operatorname{curl} \vec{A} = \frac{\partial}{\partial y}\left(\frac{-I}{c}\ln(x^2 + y^2)\right)\vec{i} - \frac{\partial}{\partial x}\left(\frac{-I}{c}\ln(x^2 + y^2)\right)\vec{j}$$

$$= \frac{-I}{c}\left(\frac{2y}{x^2 + y^2}\right)\vec{i} + \frac{I}{c}\left(\frac{2x}{(x^2 + y^2)}\right)\vec{j}$$

$$= \frac{2I}{c}\left(\frac{-y\vec{i} + x\vec{j}}{x^2 + y^2}\right)$$

$$= \vec{B}.$$

9. (a) Yes. To show this, we use a version of the product rule for curl (Problem 26 on page 327):

$$\operatorname{curl}(\phi\vec{F}) = \phi\operatorname{curl}\vec{F} + (\operatorname{grad}\phi) \times \vec{F},$$

where ϕ is a scalar function and $\vec{F}$ is a vector field. So

$$\operatorname{curl}\left(q\frac{\vec{r}}{\|\vec{r}\|^3}\right) = \operatorname{curl}\left(\frac{q}{\|\vec{r}\|^3}\vec{r}\right) = \frac{q}{\|\vec{r}\|^3}\operatorname{curl}\vec{r} + \operatorname{grad}\left(\frac{q}{\|\vec{r}\|^3}\right) \times \vec{r}$$

$$= \vec{0} + q\operatorname{grad}\left(\frac{1}{\|\vec{r}\|^3}\right) \times \vec{r}$$

Since the level surfaces of $1/\|\vec{r}\|^3$ are spheres centered at the origin, $\operatorname{grad}(1/\|\vec{r}\|^3)$ is parallel to $\vec{r}$, so $\operatorname{grad}(1/\|\vec{r}\|^3) \times \vec{r} = \vec{0}$. Thus, $\operatorname{curl}\vec{E} = \vec{0}$.

(b) Yes. The domain of $\vec{E}$ is 3-space minus $(0, 0, 0)$. Any closed curve in this region is the boundary of a surface contained entirely in the region. (If the first surface you pick happens to contain $(0, 0, 0)$, change its shape slightly to avoid it.)

(c) Yes. Since $\vec{E}$ satisfies both conditions of the curl test, it must be a gradient field. In fact,

$$\vec{E} = \operatorname{grad}\left(-q\frac{1}{\|\vec{r}\|}\right).$$

13. (a) Since $\operatorname{curl}\operatorname{grad}\psi = 0$ for any function ψ, $\operatorname{curl}(\vec{A} + \operatorname{grad}\psi) = \operatorname{curl}\vec{A} + \operatorname{curl}\operatorname{grad}\psi = \operatorname{curl}\vec{A} = \vec{B}$.

(b) We have

$$\operatorname{div}(\vec{A} + \operatorname{grad}\psi) = \operatorname{div}\vec{A} + \operatorname{div}\operatorname{grad}\psi = \operatorname{div}\vec{A} + \nabla^2\psi.$$

Thus ψ should be chosen to satisfy the partial differential equation

$$\nabla^2\psi = -\operatorname{div}\vec{A}.$$

Solutions for Chapter 20 Review

Exercises

1. (a) We have

$$\operatorname{curl}\vec{F} = \begin{vmatrix} \vec{i} & \vec{j} & \vec{k} \\ \frac{\partial}{\partial x} & \frac{\partial}{\partial y} & \frac{\partial}{\partial z} \\ \cos x & e^y & x - y - z \end{vmatrix} = -\vec{i} - \vec{j}.$$

(b) If S is the disk on the plane within the circle C, Stokes' Theorem gives

$$\int_C \vec{F} \cdot d\vec{r} = \int_S \text{curl}\,\vec{F} \cdot d\vec{A}.$$

For Stokes' Theorem, the disk is oriented upward. Since the unit normal to the plane is $(\vec{i} + \vec{j} + \vec{k})/\sqrt{3}$ and the disk has radius 3, the area vector of the disk is

$$\vec{A} = \frac{\vec{i} + \vec{j} + \vec{k}}{\sqrt{3}}\pi(3^2) = 3\sqrt{3}\pi(\vec{i} + \vec{j} + \vec{k}).$$

Thus, using curl $\vec{F} = -\vec{i} - \vec{j}$, we have

$$\int_C \vec{F} \cdot d\vec{r} = \left(-\vec{i} - \vec{j}\right) \cdot 3\sqrt{3}\pi(\vec{i} + \vec{j} + \vec{k}) = -6\sqrt{3}\pi.$$

5. (a) We will compute separately the flux of the vector field $\vec{F} = x^3\vec{i} + 2y\vec{j} + 3\vec{k}$ through each of the six faces of the cube.

The face S_I where $x = 1$, which has normal vector $\vec{i}$. Only the $\vec{i}$ component $x^3\vec{i} = \vec{i}$ of $\vec{F}$ has flux through S_I.

$$\int_{S_I} \vec{F} \cdot d\vec{A} = \int_{S_I} \vec{i} \cdot d\vec{A} = \|\vec{i}\|(\text{area of } S_I) = 4.$$

The face S_{II} where $x = -1$, which has normal vector $-\vec{i}$. Only the $\vec{i}$ component $x^3\vec{i} = -\vec{i}$ of $\vec{F}$ has flux through S_{II}.

$$\int_{S_{II}} \vec{F} \cdot d\vec{A} = \int_{S_{II}} -\vec{i} \cdot d\vec{A} = \| -\vec{i}\|(\text{area of } S_{II}) = 4.$$

The face S_{III} where $y = 1$, which has normal vector $\vec{j}$. Only the $\vec{j}$ component $2y\vec{j} = 2\vec{j}$ of $\vec{F}$ has flux through S_{III}.

$$\int_{S_{III}} \vec{F} \cdot d\vec{A} = \int_{S_{III}} 2\vec{j} \cdot d\vec{A} = \|2\vec{j}\|(\text{area of } S_{III}) = 8.$$

The face S_{IV} where $y = -1$, which has normal vector $-\vec{j}$. Only the $\vec{j}$ component $2y\vec{j} = -2\vec{j}$ of $\vec{F}$ has flux through S_{IV}.

$$\int_{S_{IV}} \vec{F} \cdot d\vec{A} = \int_{S_{IV}} -2\vec{j} \cdot d\vec{A} = \| -2\vec{j}\|(\text{area of } S_{IV}) = 8.$$

The face S_V where $z = 1$, which has normal vector $\vec{k}$. Only the $\vec{k}$ component $3\vec{k}$ of $\vec{F}$ has flux through S_V.

$$\int_{S_V} \vec{F} \cdot d\vec{A} = \int_{S_V} 3\vec{k} \cdot d\vec{A} = \|3\vec{k}\|(\text{area of } S_V) = 12.$$

The face S_{VI} where $z = -1$, which has normal vector $-\vec{k}$. Only the $\vec{k}$ component $3\vec{k}$ of $\vec{F}$ has flux through S_{VI}.

$$\int_{S_{VI}} \vec{F} \cdot d\vec{A} = \int_{S_{VI}} 3\vec{k} \cdot d\vec{A} = -\|3\vec{k}\|(\text{area of } S_{VI}) = -12.$$

$$(\text{Total flux through } S) = 4 + 4 + 8 + 8 + 12 - 12 = 24.$$

(b) Since S is a closed surface the Divergence Theorem applies. Since $\text{div}\,\vec{F} = 3x^2 + 2$,

$$\int_S \vec{F} \cdot d\vec{A} = \int_{x=-1}^1 \int_{y=-1}^1 \int_{z=-1}^1 (3x^2 + 2)dzdydx = 24.$$

9. C_2, C_3, C_4, C_6, since line integrals around C_1 and C_5 are clearly nonzero. You can see directly that $\int_{C_2} \vec{F} \cdot d\vec{r}$ and $\int_{C_6} \vec{F} \cdot d\vec{r}$ are zero, because C_2 and C_6 are perpendicular to their fields at every point.

13. We have

$$\text{div } \vec{F} = \frac{\partial}{\partial x}(e^{y+z}) + \frac{\partial}{\partial y}(\sin(x+z)) + \frac{\partial}{\partial z}(x^2 + y^2) = 0$$

$$\text{curl } \vec{F} = \begin{vmatrix} \vec{i} & \vec{j} & \vec{k} \\ \frac{\partial}{\partial x} & \frac{\partial}{\partial y} & \frac{\partial}{\partial z} \\ e^{y+z} & \sin(x+z) & x^2+y^2 \end{vmatrix} = (2y - \cos(x+z))\vec{i} - \left(2x - e^{y+z}\right)\vec{j} + \left(\cos(x+z) - e^{y+z}\right)\vec{k}.$$

So $\vec{F}$ is solenoidal, but $\vec{F}$ is not irrotational.

17. Since $\text{div}\vec{F} = 3x^2 + 3y^2$, using cylindrical coordinates to calculate the triple integral gives

$$\int_S \vec{F} \cdot d\vec{A} = \int_{\substack{\text{Interior} \\ \text{of cylinder}}} (3x^2 + 3y^2)\, dV = 3\int_0^{2\pi}\int_0^5\int_0^2 r^2 \cdot r\, dr\, dz\, d\theta = 3 \cdot 2\pi \cdot 5\frac{r^4}{4}\Big|_0^2 = 120\pi.$$

21. If C is the rectangular path around the rectangle, traversed counterclockwise when viewed from above, Stokes' Theorem gives

$$\int_S \text{curl } \vec{F} \cdot d\vec{A} = \int_C \vec{F} \cdot d\vec{r}.$$

The $\vec{k}$ component of $\vec{F}$ does not contribute to the line integral, and the $\vec{j}$ component contributes to the line integral only along the segments of the curve parallel to the y-axis. Thus, if we break the line integral into four parts

$$\int_S \text{curl } \vec{F} \cdot d\vec{A} = \int_{(0,0)}^{(3,0)} \vec{F} \cdot d\vec{r} + \int_{(3,0)}^{(3,2)} \vec{F} \cdot d\vec{r} + \int_{(3,2)}^{(0,2)} \vec{F} \cdot d\vec{r} + \int_{(0,2)}^{(0,0)} \vec{F} \cdot d\vec{r},$$

we see that the first and third integrals are zero, and we can replace $\vec{F}$ by its $\vec{j}$ component in the other two

$$\int_S \text{curl } \vec{F} \cdot d\vec{A} = \int_{(3,0)}^{(3,2)} (x+7)\vec{j} \cdot d\vec{r} + \int_{(0,2)}^{(0,0)} (x+7)\vec{j} \cdot d\vec{r}.$$

Now $x = 3$ in the first integral and $x = 0$ in the second integral and the variable of integration is y in both, so

$$\int_S \text{curl}\vec{F} \cdot d\vec{A} = \int_0^2 10\, dy + \int_2^0 7\, dy = 20 - 14 = 6.$$

Problems

25. We use Stokes' Theorem. Since

$$\text{curl } \vec{F} = \begin{vmatrix} \vec{i} & \vec{j} & \vec{k} \\ \frac{\partial}{\partial x} & \frac{\partial}{\partial y} & \frac{\partial}{\partial z} \\ x+y & y+2z & z+3x \end{vmatrix} = -2\vec{i} - 3\vec{j} - \vec{k},$$

if S is the interior of the square, then

$$\int_C \vec{F} \cdot d\vec{r} = \int_S \text{curl } \vec{F} \cdot d\vec{A} = \int_S (-2\vec{i} - 3\vec{j} - \vec{k}) \cdot d\vec{A}.$$

Since the area vector of S is $49\vec{j}$, we have

$$\int_C \vec{F} \cdot d\vec{r} = \int_S (-2\vec{i} - 3\vec{j} - \vec{k}) \cdot d\vec{A} = -3\vec{j} \cdot 49\vec{j} = -147.$$

29. The flux of $\vec{E}$ through a small sphere of radius R around the point marked P is negative, because all the arrows are pointing into the sphere. The divergence at P is

$$\text{div } \vec{E}\,(P) = \lim_{\text{vol}\to 0} \left(\frac{\int_S \vec{E} \cdot d\vec{A}}{\text{Volume of sphere}} \right) = \lim_{R \to 0} \left(\frac{\text{Negative number}}{\frac{4}{3}\pi R^3} \right) \le 0.$$

By a similar argument, the divergence at Q must be positive or zero.

33. (a) Since $\vec{F}$ is radial, it is everywhere parallel to the area vector, $\Delta\vec{A}$. Also, $||\vec{F}|| = 1$ on the surface of the sphere $x^2 + y^2 + z^2 = 1$, so

$$\text{Flux through the sphere} = \int_S \vec{F} \cdot d\vec{A} = \lim_{||\Delta\vec{A}||\to 0} \sum \vec{F} \cdot \Delta\vec{A}$$

$$= \lim_{||\Delta\vec{A}||\to 0} \sum ||\vec{F}|| \, ||\Delta\vec{A}|| = \lim_{||\Delta\vec{A}||\to 0} \sum ||\Delta\vec{A}||$$

$$= \text{Surface area of sphere} = 4\pi \cdot 1^2 = 4\pi.$$

(b) In Cartesian coordinates,

$$\vec{F}(x,y,z) = \frac{x}{(x^2+y^2+z^2)^{3/2}}\vec{i} + \frac{y}{(x^2+y^2+z^2)^{3/2}}\vec{j} + \frac{z}{(x^2+y^2+z^2)^{3/2}}\vec{k}.$$

So,

$$\text{div}\,\vec{F}(x,y,z) = \left(\frac{1}{(x^2+y^2+z^2)^{3/2}} - \frac{3x^2}{(x^2+y^2+z^2)^{5/2}}\right)$$

$$+ \left(\frac{1}{(x^2+y^2+z^2)^{3/2}} - \frac{3y^2}{(x^2+y^2+z^2)^{5/2}}\right)$$

$$+ \left(\frac{1}{(x^2+y^2+z^2)^{3/2}} - \frac{3z^2}{(x^2+y^2+z^2)^{5/2}}\right)$$

$$= \left(\frac{x^2+y^2+z^2}{(x^2+y^2+z^2)^{5/2}} - \frac{3x^2}{(x^2+y^2+z^2)^{5/2}}\right)$$

$$+ \left(\frac{x^2+y^2+z^2}{(x^2+y^2+z^2)^{5/2}} - \frac{3y^2}{(x^2+y^2+z^2)^{5/2}}\right)$$

$$+ \left(\frac{x^2+y^2+z^2}{(x^2+y^2+z^2)^{5/2}} - \frac{3z^2}{(x^2+y^2+z^2)^{5/2}}\right)$$

$$= \frac{3(x^2+y^2+z^2) - 3(x^2+y^2+z^2)}{(x^2+y^2+z^2)^{5/2}}$$

$$= 0.$$

(c) We cannot apply the Divergence Theorem to the whole region within the box, because the vector field $\vec{F}$ is not defined at the origin. However, we can apply the Divergence Theorem to the region, W, between the sphere and the box. Since $\text{div}\,\vec{F} = 0$ there, the theorem tells us that

$$\underbrace{\int \vec{F} \cdot d\vec{A}}_{\substack{\text{Box} \\ \text{(outward)}}} + \underbrace{\int \vec{F} \cdot d\vec{A}}_{\substack{\text{Sphere} \\ \text{(inward)}}} = \int_W \text{div}\,\vec{F}\,dV = 0.$$

Therefore, the flux through the box and the sphere are equal if both are oriented outward:

$$\underbrace{\int \vec{F} \cdot d\vec{A}}_{\substack{\text{Box} \\ \text{(outward)}}} = -\underbrace{\int \vec{F} \cdot d\vec{A}}_{\substack{\text{Sphere} \\ \text{(inward)}}} = \underbrace{\int \vec{F} \cdot d\vec{A}}_{\substack{\text{Sphere} \\ \text{(outward)}}} = 4\pi.$$

37. We find $\text{div}\,\vec{F} = 1+1+1 = 3$. To use the Divergence Theorem, we need to have a closed surface, so we add a circular disk, D, oriented downward. With S representing the hemisphere and W the interior of the closed region, the Divergence Theorem gives

$$\int_{\text{Closed surface}} \vec{F} \cdot d\vec{A} = \int_S \vec{F} \cdot d\vec{A} + \int_D \vec{F} \cdot d\vec{A} = \int_W \text{div}\,\vec{F}\,dV = 3 \cdot \text{Volume of region}$$

$$= 3\left(\frac{1}{2} \cdot \frac{4}{3}\pi 5^3\right) = 250\pi.$$

Since D is in the plane $z = 0$ and is oriented downward, $d\vec{A} = -\vec{k}\,dA$ on D, giving

$$\text{Flux through disk } = \int_D \vec{F} \cdot d\vec{A} = \int_D \left((x + \cos y)\vec{i} + (y + \sin x)\vec{j} + 3\vec{k}\right) \cdot (-\vec{k}\,dA)$$

$$= -\int_D 3\,dA = -3\pi(5^2) = -75\pi.$$

Thus,

$$\int_S \vec{F} \cdot d\vec{A} = 250\pi - \int_D \vec{F} \cdot d\vec{A} = 250\pi - (-75\pi) = 325\pi.$$

41. (a) Since

$$\vec{F} = F_1\vec{i} + F_2\vec{j} + F_3\vec{k} = \frac{x\vec{i} + y\vec{j} + z\vec{k}}{(x^2 + y^2 + z^2)^{a/2}},$$

we have

$$\frac{\partial F_1}{\partial x} = \frac{1}{(x^2 + y^2 + z^2)^{a/2}} - \frac{a}{2} \cdot \frac{x(2x)}{(x^2 + y^2 + z^2)^{(a/2)+1}}$$

$$= \frac{x^2 + y^2 + z^2 - ax^2}{(x^2 + y^2 + z^2)^{(a/2)+1}}.$$

Similarly, calculating $\partial F_2/\partial y$ and $\partial F_3/\partial z$ and adding gives

$$\text{div}\,\vec{F} = \frac{\partial F_1}{\partial x} + \frac{\partial F_2}{\partial y} + \frac{\partial F_3}{\partial z} = \frac{3(x^2 + y^2 + z^2) - ax^2 - ay^2 - az^2}{(x^2 + y^2 + z^2)^{(a/2)+1}}$$

$$= \frac{3 - a}{(x^2 + y^2 + z^2)^{a/2}}.$$

(b) $\text{div}\,\vec{F} = 0$ if $a = 3$.

(c) The vector field $\vec{F}$ is radial, and shows no "swirl", so we expect $\text{curl}\,\vec{F} = \vec{0}$. See Figure 20.6.

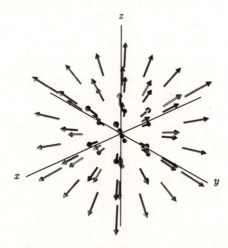

Figure 20.6

45. (a) Since $\vec{v} = \text{grad}\,\phi$ we have

$$\vec{v} = \left(1 + \frac{y^2 - x^2}{(x^2 + y^2)^2}\right)\vec{i} + \frac{-2xy}{(x^2 + y^2)^2}\vec{j}$$

(b) Differentiating the components of $\vec{v}$, we have

$$\text{div}\,\vec{v} = \frac{\partial}{\partial x}\left(1 + \frac{y^2 - x^2}{(x^2 + y^2)^2}\right) + \frac{\partial}{\partial y}\left(\frac{-2xy}{(x^2 + y^2)^2}\right) = \frac{2x(x^2 - 3y^2)}{(x^2 + y^2)^3} + \frac{2x(3y^2 - x^2)}{(x^2 + y^2)^3} = 0$$

(c) The vector $\vec{v}$ is tangent to the circle $x^2 + y^2 = 1$, if and only if the dot product of the field on the circle with any radius vector of that circle is zero. Let (x, y) be a point on the circle. We want to show: $\vec{v} \cdot \vec{r} = \vec{v}(x, y) \cdot (x\vec{i} + y\vec{j}) = 0$. We have:

$$\vec{v}(x, y) \cdot (x\vec{i} + y\vec{j}) = ((1 + \frac{y^2 - x^2}{(x^2 + y^2)^2})\vec{i} + \frac{-2xy}{(x^2 + y^2)^2}\vec{j}) \cdot (x\vec{i} + y\vec{j})$$

$$= x + x\frac{y^2 - x^2}{(x^2 + y^2)^2} - \frac{2xy^2}{(x^2 + y^2)^2}$$

$$= \frac{x(x^2 + y^2 - 1)}{x^2 + y^2},$$

but we know that for any point on the circle, $x^2 + y^2 = 1$, thus we have $\vec{v} \cdot \vec{r} = 0$. Therefore, the velocity field is tangent to the circle. Consequently, there is no flow through the circle and any water on the outside of the circle must flow around it.

(d)

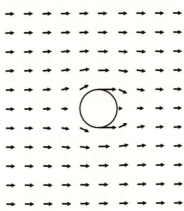

Figure 20.7

49. (a) The path along which we integrate is shown in Figure 20.8.

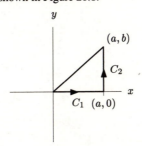

Figure 20.8

The path C_1 is given by

$$C_1 : \begin{cases} x = t & 0 \le t \le a \\ y = 0 \end{cases}$$

and path C_2 is given by

$$C_2 : \begin{cases} x = a \\ y = t & 0 \le t \le b \end{cases}.$$

We integrate along the path $C = C_1 + C_2$. Then,

$$\int_C \vec{E} \cdot d\vec{r} = \int_{C_1} \vec{E} \cdot d\vec{r} + \int_{C_2} \vec{E} \cdot d\vec{r}$$

$$= \int_0^a 5t^2\vec{j} \cdot \vec{i} \, dt + \int_0^b (10at\vec{i} + (5a^2 - 5t^2)\vec{j}) \cdot \vec{j} \, dt$$

$$= 5a^2b - \frac{5}{3}b^3$$

(b) The path along which we integrate is shown in Figure 20.9.

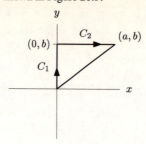

Figure 20.9

The path C_1 is given by

$$C_1 : \begin{cases} x = 0 \\ y = t \quad 0 \le t \le b \end{cases}$$

and Path C_2 is given by

$$C_2 : \begin{cases} x = t \quad 0 \le t \le a \\ y = b \end{cases}.$$

We integrate along the path $C = C_1 + C_2$. Then,

$$\int_C \vec{E} \cdot d\vec{r} = \int_{C_1} \vec{E} \cdot d\vec{r} + \int_{C_2} \vec{E} \cdot d\vec{r}$$

$$= \int_0^b -5t^2 \vec{j} \cdot \vec{j} \, dt + \int_0^a (10bt\vec{i} + (5t^2 - 5b^2)\vec{j}) \cdot \vec{i} \, dt$$

$$= -\frac{5}{3}b^3 + 5a^2 b$$

(c) Notice that the line integrals along both paths in part (a) and (b) are equal. Thus $\vec{E}$ could be path independent. This property is a property of electric fields.

(d) Using the calculations of the parts (a) and (b), with the point (a, b) replaced by (x, y), we see that the electric potential, ϕ, at (x, y) is given by

$$\phi = \frac{5}{3}y^3 - 5x^2 y.$$

Then, taking the gradient gives

$$\text{grad}\,\phi = \phi_x \vec{i} + \phi_y \vec{j}$$

$$= -10xy\vec{i} + (\frac{5}{3}(3y^2) - 5x^2)\vec{j}$$

$$= -10xy\vec{i} + (5y^2 - 5x^2)\vec{j}$$

$$= -\vec{E}.$$

Thus, we have confirmed that $\vec{E} = -\text{grad}\,\phi$.

CAS Challenge Problems

53. (a) Let W be the region enclosed by the sphere. We have div $\vec{F} = 2ax + bz + 2cy + p + q$ so by the Divergence Theorem $\int_S \vec{F} \cdot d\vec{A} = \int_W (2ax + bz + 2cy + p + q) \, dV$. Now $\int_W x \, dV = \int_W y \, dV = \int_W z \, dV = 0$, because W is symmetric about the origin and x, y, and z are odd functions. So $\int_W (2ax + bz + 2cy + p + q) \, dV = \int_B (p+q)dV = \frac{4(p+q)\pi R^3}{3}$.

(b) Using spherical coordinates, we calculate the flux integral directly as

$$\int_0^{2\pi} \int_0^\pi ((bR^2 \cos(\theta) \cos(\phi) \sin(\phi) + aR^2 \cos(\theta)^2 \sin(\phi)^2)\vec{i}$$

$$+ (pR \sin(\theta) \sin(\phi) + cR^2 \sin(\theta)^2 \sin(\phi)^2)\vec{j} + (qR \cos(\phi)$$

$$+ rR^3 \cos(\theta)^3 \sin(\phi)^3)\vec{k}) \cdot (\sin\phi \cos\theta \vec{i} + \sin\phi \sin\theta \vec{j} + \cos\phi \vec{k})R^2 \sin\phi \, d\phi d\theta = \frac{4(p+q)\pi R^3}{3}.$$

Rather than entering this integral directly into your CAS, it is better to define the vector field and parameterization separately and enter the formula for flux integral through a sphere.

CHECK YOUR UNDERSTANDING

1. True. By Stokes' Theorem, the circulation of $\vec{F}$ around C is the flux of curl $\vec{F}$ through the flat disc S in the xy-plane enclosed by the circle. An area element for S is $d\vec{A} = \pm\vec{k}\,dA$, where the sign depends on the orientation of the circle. Since curl $\vec{F}$ is perpendicular to the z-axis, curl $\vec{F} \cdot d\vec{A} = \pm(\text{curl } \vec{F} \cdot \vec{k})dA = 0$, so the flux of curl $\vec{F}$ through S is zero, hence the circulation of $\vec{F}$ around C is zero.

5. True. div $\vec{F}$ is a scalar whose value depends on the point at which it is calculated.

9. False. As a counterexample, consider $\vec{F} = 2x\vec{i} + 2y\vec{j} + 2z\vec{k}$. Then $\vec{F} = \text{grad}(x^2 + y^2 + z^2)$, and div $\vec{F} = 2 + 2 + 2 \neq 0$.

13. False. As a counterexample, note that $\vec{F} = \vec{i}$ and $\vec{G} = \vec{j}$ both have divergence zero, but are not the same vector fields.

17. True. The Divergence theorem says that $\int_W \text{div }\vec{F}\,dV = \int_S \vec{F} \cdot d\vec{A}$, where S is the outward oriented boundary of W. In this case, the boundary of W consists of the surfaces S_1 and S_2. To give this boundary surface a consistent outward orientation, we use a normal vector on S_1 that points towards the origin, and a normal on S_2 that points away from the origin. Thus $\int_W \text{div }\vec{F}\,dV = \int_{S_2} \vec{F} \cdot d\vec{A} + \int_{S_1} \vec{F} \cdot d\vec{A}$, with S_2 oriented outward and S_1 oriented inward. Reversing the orientation on S_1 so that both spheres are oriented outward yields $\int_W \text{div }\vec{F}\,dV = \int_{S_2} \vec{F} \cdot d\vec{A} - \int_{S_1} \vec{F} \cdot d\vec{A}$.

21. False. The left-hand side of the equation, $\text{div}(\text{grad } f)$, is a scalar function and the right hand side, $\text{grad}(\text{div } F)$, is a vector. There cannot be an equality between a scalar and a vector.

25. True. Writing $\vec{F} = F_1\vec{i} + F_2\vec{j} + F_3\vec{k}$ and $\vec{G} = G_1\vec{i} + G_2\vec{j} + G_3\vec{k}$, we have $\vec{F} + \vec{G} = (F_1 + G_1)\vec{i} + (F_2 + G_2)\vec{j} + (F_3 + G_3)\vec{k}$. Then the $\vec{i}$ component of $\text{curl}(\vec{F} + \vec{G})$ is

$$\frac{\partial(F_3 + G_3)}{\partial y} - \frac{\partial(F_2 + G_2)}{\partial z} = \frac{\partial F_3}{\partial y} - \frac{\partial F_2}{\partial z} + \frac{\partial G_3}{\partial y} - \frac{\partial G_2}{\partial z}$$

which is the $\vec{i}$ component of $\text{curl}\vec{F}$ plus the $\vec{i}$ component of $\text{curl}\vec{G}$. The $\vec{j}$ and $\vec{k}$ components work out in a similar manner.

29. False. For example, take $\vec{F} = z\vec{i} + x\vec{j}$. Then $\text{curl}\vec{F} = \vec{j} + \vec{k}$, which is not perpendicular to $\vec{F}$, since $(z\vec{i} + x\vec{j}) \cdot (\vec{j} + \vec{k}) = x \neq 0$.

33. False. The curl needs to be in the flux integral, not the line integral, for a correct statement of Stokes' theorem: $\int_C \vec{F} \cdot d\vec{r} = \int_S \text{curl }\vec{F} \cdot d\vec{A}$.

37. True. Let S be the rectangular region inside C, oriented by the right hand rule. By Stokes' theorem, $\int_C \vec{F} \cdot d\vec{r} = \int_S \text{curl}\vec{F} \cdot d\vec{A} = 0$.

41. False. The condition that $\int_S \text{curl}\vec{F} \cdot d\vec{A} = 0$ implies, by Stokes' theorem, that $\int_C \vec{F} \cdot d\vec{r} = 0$. However, $\vec{F}$ need not be a gradient field for this to occur. For example, let $\vec{F} = x\vec{k}$, and let S be the upper unit hemisphere $x^2 + y^2 + z^2 = 1, z \geq 0$ oriented upward. Then C is the circle $x^2 + y^2 = 1, z = 0$ oriented counterclockwise when viewed from above. The line integral $\int_C x\vec{k} \cdot d\vec{r} = 0$, since the field $\vec{F}$ is everywhere perpendicular to C. The curl of $\vec{F}$ is the constant field $-\vec{j}$, so $\vec{F}$ is not a gradient field. Yet we have $\int_S -\vec{j} \cdot d\vec{A} = 0$, since the constant field $-\vec{j}$ flows in, and then out of the hemisphere S.

APPENDIX

Solutions for Section A

1. The graph is

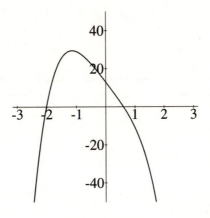

   (a) The range appears to be $y \leq 30$.
   (b) The function has two zeros.

5. The largest root is at about 2.5.

9. Using a graphing calculator, we see that when x is around 0.45, the graphs intersect.

13. (a) Only one real zero, at about $x = -1.15$.
    (b) Three real zeros: at $x = 1$, and at about $x = 1.41$ and $x = -1.41$.

17. (a) Since f is continuous, there must be one zero between $\theta = 1.4$ and $\theta = 1.6$, and another between $\theta = 1.6$ and $\theta = 1.8$. These are the only clear cases. We might also want to investigate the interval $0.6 \leq \theta \leq 0.8$ since $f(\theta)$ takes on values close to zero on at least part of this interval. Now, $\theta = 0.7$ is in this interval, and $f(0.7) = -0.01 < 0$, so f changes sign twice between $\theta = 0.6$ and $\theta = 0.8$ and hence has two zeros on this interval (assuming f is not *really* wiggly here, which it's not). There are a total of 4 zeros.
    (b) As an example, we find the zero of f between $\theta = 0.6$ and $\theta = 0.7$. $f(0.65)$ is positive; $f(0.66)$ is negative. So this zero is contained in $[0.65, 0.66]$. The other zeros are contained in the intervals $[0.72, 0.73]$, $[1.43, 1.44]$, and $[1.7, 1.71]$.
    (c) You've found all the zeros. A picture will confirm this; see Figure A.1.

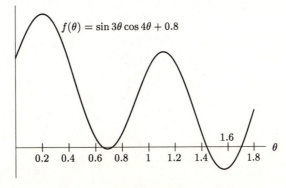

Figure A.1

21.

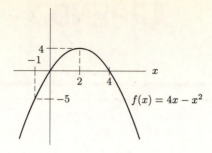

Bounded and $-5 \leq f(x) \leq 4$.

Solutions for Section B

1. $r = \sqrt{1^2 + 0^2} = 1, \quad \theta = 0.$

5. $r = \sqrt{(-3)^2 + (-3)^2} = 4.2.$
  $\tan \theta = (-3/-3) = 1.$ Since the point is in the third quadrant, $\theta = 5\pi/4$.

9. $(1,0)$

13. $\left(\frac{5\sqrt{3}}{2}, -\frac{5}{2}\right)$

17. The graph is a circle of radius 2 centered at the origin. See Figure B.2.

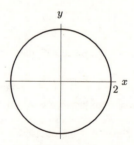

Figure B.2

21. The condition that $r \leq 2$ tells us that the region is inside a circle of radius 2 centered at the origin. The second condition $0 \leq \theta \leq \pi/2$ tells us that the points must be in the first quadrant. Thus, the region consists of the quarter of the circle in the first quadrant, as shown in Figure B.3.

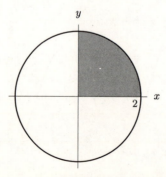

Figure B.3

25. The region consists of the portion of a circle of radius 1 centered at the origin that is between an angle of $\theta = 0$ and an angle of $\theta = \pi/4$, Therefore, the region is defined by $r \leq 1$ and $0 \leq \theta \leq \pi/4$.

29. Putting $\theta = \pi/3$ into $\tan \theta = y/x$ gives $\sqrt{3} = y/x$, or $y = \sqrt{3}x$. This is a line through the origin of slope $\sqrt{3}$. See Figure B.4.

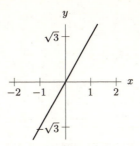

Figure B.4

33. For $r = \theta/10$, the radius r increases as the angle θ winds around the origin, so this is a spiral. See Figure B.5.

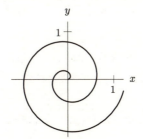

Figure B.5

Solutions for Section C

1. $2e^{i\pi/2}$

5. $0e^{i\theta}$, for any θ.

9. $-3 - 4i$

13. $\frac{1}{4} - \frac{9i}{8}$

17. $5^3(\cos \frac{3\pi}{2} + i \sin \frac{3\pi}{2}) = -125i$

21. One value of $\sqrt[3]{i}$ is $\sqrt[3]{e^{i\frac{\pi}{2}}} = (e^{i\frac{\pi}{2}})^{\frac{1}{3}} = e^{i\frac{\pi}{6}} = \cos \frac{\pi}{6} + i \sin \frac{\pi}{6} = \frac{\sqrt{3}}{2} + \frac{i}{2}$

25. One value of $(-4 + 4i)^{2/3}$ is $[\sqrt{32}e^{(i3\pi/4)}]^{(2/3)} = (\sqrt{32})^{2/3}e^{(i\pi/2)} = 2^{5/3} \cos \frac{\pi}{2} + i2^{5/3} \sin \frac{\pi}{2} = 2i\sqrt[3]{4}$

29. We have

$$i^{-1} = \frac{1}{i} = \frac{1}{i} \cdot \frac{i}{i} = -i,$$

$$i^{-2} = \frac{1}{i^2} = -1,$$

$$i^{-3} = \frac{1}{i^3} = \frac{1}{-i} \cdot \frac{i}{i} = i,$$

$$i^{-4} = \frac{1}{i^4} = 1.$$

The pattern is

$$i^n = \begin{cases} -i & n = -1, -5, -9, \cdots \\ -1 & n = -2, -6, -10, \cdots \\ i & n = -3, -7, -11, \cdots \\ 1 & n = -4, -8, -12, \cdots. \end{cases}$$

Since 36 is a multiple of 4, we know $i^{-36} = 1$.
Since $41 = 4 \cdot 10 + 1$, we know $i^{-41} = -i$.

33. To confirm that $z = \dfrac{a + bi}{c + di}$, we calculate the product

$$z(c + di) = \left(\frac{ac + bd}{c^2 + d^2} = \frac{bc - ad}{c^2 + d^2}i \right)(c + di)$$

$$= \frac{ac^2 + bcd - bcd + ad^2 + (bc^2 - acd + acd + bd^2)i}{c^2 + d^2}$$

$$= \frac{a(c^2 + d^2) + b(c^2 + d^2)i}{c^2 + d^2} = a + bi.$$

37. True, since $\sqrt{a}$ is real for all $a \geq 0$.

41. True. We can write any nonzero complex number z as $re^{i\beta}$, where r and β are real numbers with $r > 0$. Since $r > 0$, we can write $r = e^c$ for some real number c. Therefore, $z = re^{i\beta} = e^c e^{i\beta} = e^{c+i\beta} = e^w$ where $w = c + i\beta$ is a complex number.

45. Using Euler's formula, we have:

$$e^{i(2\theta)} = \cos 2\theta + i \sin 2\theta$$

On the other hand,

$$e^{i(2\theta)} = \left(e^{i\theta}\right)^2 = (\cos \theta + i \sin \theta)^2 = (\cos^2 \theta - \sin^2 \theta) + i(2 \cos \theta \sin \theta)$$

Equating real parts, we find

$$\cos 2\theta = \cos^2 \theta - \sin^2 \theta.$$

49. Replacing θ by $(x + y)$ in the formula for $\sin \theta$:

$$\sin(x + y) = \frac{1}{2i}\left(e^{i(x+y)} - e^{-i(x+y)}\right) = \frac{1}{2i}\left(e^{ix}e^{iy} - e^{-ix}e^{-iy}\right)$$

$$= \frac{1}{2i}\left((\cos x + i \sin x)(\cos y + i \sin y) - (\cos(-x) + i \sin(-x))(\cos(-y) + i \sin(-y))\right)$$

$$= \frac{1}{2i}\left((\cos x + i \sin x)(\cos y + i \sin y) - (\cos x - i \sin x)(\cos y - i \sin y)\right)$$

$$= \sin x \cos y + \cos x \sin y.$$

Solutions for Section D

1. **(a)** $f'(x) = 3x^2 + 6x + 3 = 3(x+1)^2$. Thus $f'(x) > 0$ everywhere except at $x = -1$, so it is increasing everywhere except perhaps at $x = -1$. The function is in fact increasing at $x = -1$ since $f(x) > f(-1)$ for $x > -1$, and $f(x) < f(-1)$ for $x < -1$.

   (b) The original equation can have at most one root, since it can only pass through the x-axis once if it never decreases. It must have one root, since $f(0) = -6$ and $f(1) = 1$.

   (c) The root is in the interval $[0, 1]$, since $f(0) < 0 < f(1)$.

   (d) Let $x_0 = 1$.

$$x_0 = 1$$
$$x_1 = 1 - \frac{f(1)}{f'(1)} = 1 - \frac{1}{12} = \frac{11}{12} \approx 0.917$$
$$x_2 = \frac{11}{12} - \frac{f\left(\frac{11}{12}\right)}{f'\left(\frac{11}{12}\right)} \approx 0.913$$
$$x_3 = 0.913 - \frac{f(0.913)}{f'(0.913)} \approx 0.913.$$

Since the digits repeat, they should be accurate. Thus $x \approx 0.913$.

5. Let $f(x) = \sin x - 1 + x$; we want to find all zeros of f, because $f(x) = 0$ implies $\sin x = 1 - x$.
   Graphing $\sin x$ and $1 - x$ in Figure D.6, we see that $f(x)$ has one solution at $x \approx \frac{1}{2}$.

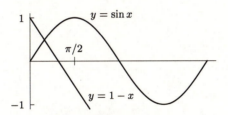

Figure D.6

Letting $x_0 = 0.5$, and using Newton's method, we have $f'(x) = \cos x + 1$, so that

$$x_1 = 0.5 - \frac{\sin(0.5) - 1 + 0.5}{\cos(0.5) + 1} \approx 0.511,$$

$$x_2 = 0.511 - \frac{\sin(0.511) - 1 + 0.511}{\cos(0.511) + 1} \approx 0.511.$$

Thus $\sin x = 1 - x$ has one solution at $x \approx 0.511$.

9. Let $f(x) = \ln x - \frac{1}{x}$, so $f'(x) = \frac{1}{x} + \frac{1}{x^2}$.
   Now use Newton's method with an initial guess of $x_0 = 2$.

$$x_1 = 2 - \frac{\ln 2 - \frac{1}{2}}{\frac{1}{2} + \frac{1}{4}} \approx 1.7425,$$
$$x_2 \approx 1.763,$$
$$x_3 \approx 1.763.$$

Thus $x \approx 1.763$ is a solution. Since $f'(x) > 0$ for positive x, f is increasing: it must be the only solution.